AF400848

Problem Solving in Chemical Reactor Design

**Author**

*Prof. Juan A. Conesa*
Universidad de Alicante
Alicante 03690
Spain

**Cover**: © Shaiith/Shutterstock



**Library of Congress Card No.:** applied for

**British Library Cataloguing-in-Publication Data:**
A catalogue record for this book is available from the British Library.

**Bibliographic information published by the Deutsche Nationalbibliothek**
The Deutsche Nationalbibliothek lists this publication in the Deutsche Nationalbibliografie; detailed bibliographic data are available on the Internet at <http://dnb.d-nb.de>.



**Print ISBN:** 978-3-527-35411-5
**ePDF ISBN:** 978-3-527-84802-7
**ePub ISBN:** 978-3-527-84801-0
**oBook ISBN:** 978-3-527-84803-4

**Typesetting**   Straive, Chennai, India

Printed and bound by
CPI Group (UK) Ltd, Croydon, CR04YY

C9783527354115_300824

# Problem Solving in Chemical Reactor Design

*Juan A. Conesa*

WILEY-VCH

*To my beloved wife, Marga, and children, Margarita, Juan, Pablo, and Miguel.*

# Contents

# Preface

This book has been written after more than two decades of teaching experience in chemical reactor engineering. During these years, I have been collecting problems posed to undergraduate and master's students, which are now compiled here along with their solutions. This is not a basic chemical reactor engineering (CRE) book; instead, it covers the design of nonideal, catalytic, multiphase, and biochemical reactors.

Following the publication of my book *Chemical Reactor Design: Mathematical Modeling and Applications* in 2019, this book addresses aspects not covered in basic textbooks dedicated to reactor design. The concepts introduced in my earlier book are further developed here through solved examples. This approach aims to help the reader gain a deeper understanding of reactor design procedures, especially in the context of nonideal and heterogeneous reactors.

This collection of solved problems in advanced chemical reactor engineering provides in-depth coverage of complex reactor design issues, such as the design of catalytic or biochemical reactors. It is not a basic book covering simple concepts.

The book is accompanied by electronic materials, including spreadsheets and Matlab® programs, which can be downloaded from the Wiley website at http://www.wiley-vch.de/ISBN9783527354115.

The tools used in this book are not complex, and the solutions are presented in a simplified manner. I primarily use spreadsheets and occasionally incorporate Matlab. Importantly, the programs are designed to be highly understandable.

Elche (Spain)  
May 2024

*Juan A. Conesa*

# Nomenclature

Suggested units are indicated for each variable.

$A_{m \times n}$ — matrix of convolution

$C$ — concentration (usually mol/m$^3$, or kg/m$^3$)

$C_{As}$ — concentration of A in the surface (mol/m$^3$)

$c_p$ — calorific capacity of the reacting flow (J/K $\cdot$ g)

$C_p$ — concentration of product "P" (mol/m$^3$)

$C_T$ — total concentration (mol/m$^3$)

$d_p$ — particle diameter (m)

$D$ — dilution rate (s$^{-1}$)

$D_A$ — diffusion coefficient of species A (m$^2$/s)

$D_{AB}$ — diffusion coefficient (diffusivity) of A in B (m$^2$/s)

$D_e$ — effective diffusion coefficient (m$^2$/s)

$E$ — (in the reaction rates context) $\rightarrow$ activation energy (J/mol)

$E$ — (in the RTD context) $\rightarrow$ residence time distribution function, RTD (—)

$F$ — integral form of the residence time distribution function (—)

$g_A$ — generation term of a mole balance (mol/s)

$h$ — heat transfer coefficient (J/K $\cdot$ s)

$H(s)$ — transfer function (—) (Laplace space)

$h(t)$ — time-dependent transfer function (—)

$k$ — kinetic constant (units depend on the kinetic law)

$k_0$ — pre-exponential factor of the kinetic constant (units depend on the reaction order)

$k_L$ — mass transfer coefficient in the liquid phase (mol/s $\cdot$ m$^2$/(mol/m$^3$) = m/s)

$K_M$ — Michaelis constant (g/l)

$K_S$ — Monod constant (g/l)

$L$ — characteristic length (m)

$L\{h(t)\}$ — Laplace transform of function $h(t)$

$M$ — tracer mass (g or mol)

$N$ — amount of substance (mol)

$n_A$ — molar flow of component "A" (mol/s or mol/s $\cdot$ m$^2$)

$N_{ad}$ — dimensionless adiabatic temperature increase (—)

$N_c$ — dimensionless cooling capacity (—)

$n_t$ — number of tanks in the tanks-in-series model (—)

$n_T$     total molar flow (mol/s)

$q$     heat flux (J/s)

$Q$     volumetric flowrate (m$^3$/s)

$q_C$     heat flux in cooling media (J/s)

$R$     radius (m)

$r_A'''$     reaction (or process) rate based on the volume of catalyst particles (mol/s $\cdot$ m$^3$)

$r_A''$     reaction (or process) rate based on the volume of reacting species (mol/s $\cdot$ m$^3$)

$r_A'$     reaction (or process) rate based on the weight of catalyst (mol/s $\cdot$ g)

$r_A$     reaction (or process) rate based on the external catalyst surface (mol/s $\cdot$ m$^2$)

$R_g$     universal gas constant (J/mol $\cdot$ K)

$s$     (in the Laplace transform context) $\rightarrow$ main variable in the Laplace space

$S$     section (m$^2$)

$T$     temperature (K)

$t$     time (s)

$\bar{t}$     average residence time ($= V/Q$) (s)

$T_C$     temperature of the cooling media (K)

$t_m$     first moment of the RTD; mean value of time (s)

$U$     global heat transfer coefficient (J/K $\cdot$ s)

$u$     linear velocity (m/s)

$V$     volume (m$^3$)

$V_d$     dead volume (m$^3$)

$V_p$     volume with plug flow regime (m$^3$)

$W$     weight of catalyst (g)

We     Weisz modulus (—)

$X(s)$     stimulus function (—) (Laplace space)

$x(t)$     time-dependent stimulus function (—)

$X_A$     molar conversion of reactant A (—)

$Y(s)$     response to stimulus function (—) (Laplace space)

$y(t)$     time-dependent response to stimulus function (—)

## Greek Symbols

$\tau$     tortuosity factor (—)

$\alpha$     fraction of volume of a subsystem (—)

$\beta$     fraction of flow passing through a subsystem (—)

$\delta$     dirac delta function (—)

$\mu$     growth rate per unit of cell (1/s)

$\varepsilon_s$     solid load (m$^3$ solid/m$^3$ reactor)

$\varepsilon_G$     gas fraction (m$^3$ gas/m$^3$ reactor)

$\varepsilon_L$     liquid fraction (m$^3$ liquid/m$^3$ reactor)

$\varepsilon_b$     bed porosity (m$^3$ void/m$^3$ reactor)

$\sigma^2$     variance (s$^2$)

$\sigma$     standard deviation (s)

$\eta$ — effectiveness (—)
$\Delta H_r$ — enthalpy of a reaction (kJ/mol)

## Subscripts

$i$ — actual position
$i+1$ — position of the following interval
$i-1$ — position of the preceding interval
$S$ — surface
$0$ — inlet conditions
F — fluid
C — catalyst
R — reactor

## Superscripts

$t$ — actual time
$t+1$ — time of the following interval
$t-1$ — time of the previous interval

# Part I

# Non-ideal Flow Characterization and Chemical Reaction

# 1

# Non-ideal Flow and Reactor Characterization

## Summary of Residence Time Distribution Properties and Most Important Models

### Residence Time Distribution

In a reactor, $C(t)$ is obtained by injecting pulse of tracer. From that:

$$E(t) = \frac{C(t)}{\int_0^\infty C(t) \cdot dt}$$

$$\int_0^\infty E(t)dt = 1$$

$$\int_0^t E(t)dt = F(t)$$

$$I(t) = 1 - F(t)$$

Total amount of tracer injected is:

$$M_0 = Q_0 \int_0^\infty C(t)dt$$

In a step tracer run:

$$E(t) = \frac{d}{dt}\left[\frac{C(t)}{C_0}\right]_{\text{step input}}$$

Mean residence time:

$$t_m = \frac{\int_0^\infty t \cdot C(t)dt}{\int_0^\infty C(t)dt} = \int_0^\infty t \cdot E(t)dt = \int_0^\infty t \cdot E(t)dt$$

Variance of the residence time distribution (RTD):

$$\sigma^2 = \int_0^\infty (t - t_m)^2 \cdot E(t)dt$$

The square root of the variance, $\sigma$, is called standard deviation.
Average residence time:

$$\bar{t} = \frac{V}{Q}$$

### RTD in Ideal Reactors

For the plug flow reactor (PFR):

$$E(t) = \delta(t - \bar{t})$$

For the continuous stirred tank reactor (CSTR):

$$E(t) = \frac{1}{\bar{t}} \cdot \exp(-t/\bar{t})$$

### Tanks-in-series (TIS) Model

$$E(t) = \frac{t^{n_t - 1}}{(n_t - 1)! \bar{t}_i^{n_t}} \cdot \exp\left(-\frac{t}{\bar{t}_i}\right)$$

$$\sigma^2 = \frac{t_m^2}{n_t}$$

$$n_t = \frac{t_m^2}{\sigma^2}$$

### Dispersion Model

$$\mathrm{Bo} = \frac{D_e}{u \cdot L}$$

### Bo < 0.01

$$E = \frac{1}{\bar{t} \cdot \sqrt{4 \cdot \pi \cdot \mathrm{Bo}}} \cdot \exp\left[-\frac{\left(1 - \frac{t}{\bar{t}}\right)^2}{4 \cdot \mathrm{Bo}}\right]$$

$$t_m = \bar{t} = \frac{V}{Q}$$

$$\sigma^2 = 2 \cdot \mathrm{Bo}$$

### Bo > 0.01, Closed–Closed Recipient

$E(t)$ is only integrable by numerical methods.

$$\bar{t} = t_m$$

$$\left(\frac{\sigma}{t_m}\right)^2 = 2 \cdot \mathrm{Bo} - 2 \cdot \mathrm{Bo}^2 \cdot \left[1 - \exp\left(-\frac{1}{\mathrm{Bo}}\right)\right]$$

### Bo > 0.01, Open–Open Recipient

$$E = \frac{1}{\sqrt{4 \cdot \pi \cdot \mathrm{Bo} \cdot t/\bar{t}}} \cdot \exp\left[-\frac{\left(1 - \frac{t}{\bar{t}}\right)^2}{4 \cdot \mathrm{Bo} \cdot t/\bar{t}}\right]$$

$$t_m = \bar{t} \cdot (1 + 2 \cdot \mathrm{Bo})$$

$$\left(\frac{\sigma}{t_m}\right)^2 = 2 \cdot \mathrm{Bo} + 8 \cdot \mathrm{Bo}^2$$

Designing $E_1$ to the $E(t)$ given by the first expression (valid for $\mathrm{Bo} < 0.01$) and $E_2$ to the one predicted by the open–open (o–o) assumption, the differences between these RTDs are small at low Bo, but at a high Bo number, $E_1$ is not valid (for more details, consult spreadsheet "dispersion model Bo fitting.xls"):

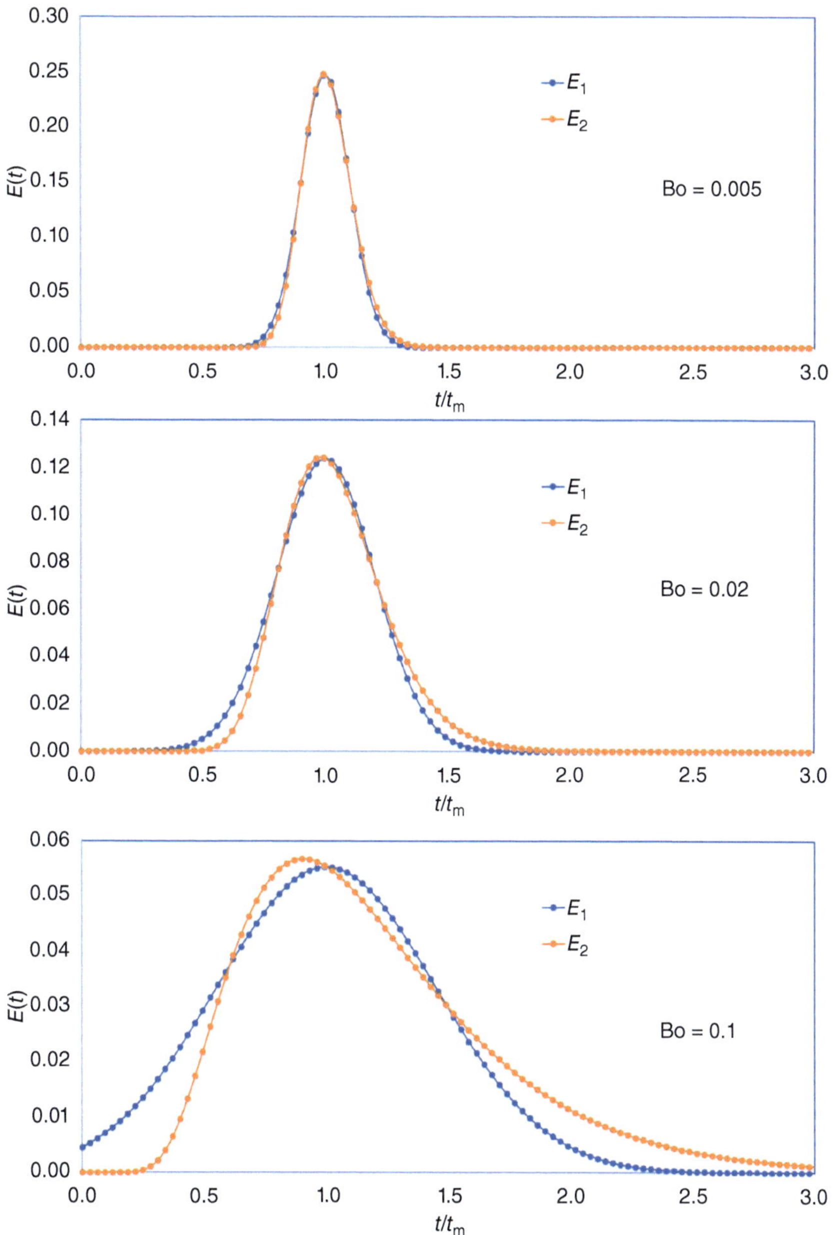

**Problem 1.1** A solution of potassium permanganate ($KMnO_4$) was rapidly injected into a water stream flowing through a circular tube at a linear velocity of 35.70 cm/s. A photoelectric cell located 2.74 m downstream from the injection point was utilized to monitor the local concentration of $KMnO_4$.

(a) By using the given effluent $KMnO_4$ concentrations, calculate the average residence time of the fluid as well as the variance, $E(t)$, $F(t)$, and $I(t)$.
(b) Determine the number of ideal tanks ($n_t$), the variance, the dispersion number, and the Peclet number.

| Time (s) | $KMnO_4$ concentration |
|---|---|
| 0.0 | 0.0 |
| 2.0 | 11.0 |
| 4.0 | 53.0 |
| 6.0 | 64.0 |
| 8.0 | 58.0 |
| 10.0 | 48.0 |
| 12.0 | 39.0 |
| 14.0 | 29.0 |
| 16.0 | 22.0 |
| 18.0 | 16.0 |
| 20.0 | 11.0 |
| 22.0 | 9.0 |
| 24.0 | 7.0 |
| 26.0 | 5.0 |
| 28.0 | 4.0 |
| 30.0 | 2.0 |
| 32.0 | 2.0 |
| 34.0 | 2.0 |
| 36.0 | 1.0 |
| 38.0 | 1.0 |
| 40.0 | 1.0 |
| 42.0 | 1.0 |

**Solution to Problem 1.1**

For details refer the Wiley website at http://www.wiley-vch.de/ISBN9783527354115

(a) To solve this problem, we should use a spreadsheet. First, data given in the statement are introduced, and then we should do the following calculations:

| $t$ (s) | $C(t)$ | $C(t)dt$ | $E(t)$ | $t \cdot E(t)$ | $t \cdot E(t) \cdot \Delta t$ | $(t - t_m)^2 \cdot E(t)$ | $(t - t_m)^2 \cdot E(t) \cdot \Delta t$ | $F(t)$ | $I(t)$ |
|---|---|---|---|---|---|---|---|---|---|
| 0 | 0 | 0 | 0.000 | 0.000 | 0.000 | 0.000 | 0.000 | 0.000 | 1.000 |
| 2 | 11 | 11 | 0.014 | 0.029 | 0.029 | 1.149 | 1.149 | 0.029 | 0.971 |
| 4 | 53 | 64 | 0.069 | 0.275 | 0.304 | 3.345 | 4.494 | 0.166 | 0.834 |
| 6 | 64 | 117 | 0.083 | 0.498 | 0.773 | 2.055 | 5.399 | 0.332 | 0.668 |
| 8 | 58 | 122 | 0.075 | 0.602 | 1.100 | 0.666 | 2.721 | 0.482 | 0.518 |
| 10 | 48 | 106 | 0.062 | 0.623 | 1.224 | 0.059 | 0.725 | 0.607 | 0.393 |
| 12 | 39 | 87 | 0.051 | 0.607 | 1.230 | 0.053 | 0.112 | 0.708 | 0.292 |
| 14 | 29 | 68 | 0.038 | 0.527 | 1.134 | 0.344 | 0.397 | 0.783 | 0.217 |
| 16 | 22 | 51 | 0.029 | 0.457 | 0.983 | 0.720 | 1.065 | 0.840 | 0.160 |
| 18 | 16 | 38 | 0.021 | 0.374 | 0.830 | 1.024 | 1.744 | 0.882 | 0.118 |
| 20 | 11 | 27 | 0.014 | 0.285 | 0.659 | 1.162 | 2.186 | 0.911 | 0.089 |
| 22 | 9 | 20 | 0.012 | 0.257 | 0.542 | 1.419 | 2.581 | 0.934 | 0.066 |
| 24 | 7 | 16 | 0.009 | 0.218 | 0.475 | 1.540 | 2.959 | 0.952 | 0.048 |
| 26 | 5 | 12 | 0.006 | 0.169 | 0.387 | 1.464 | 3.004 | 0.965 | 0.035 |
| 28 | 4 | 9 | 0.005 | 0.145 | 0.314 | 1.504 | 2.968 | 0.975 | 0.025 |
| 30 | 2 | 6 | 0.003 | 0.078 | 0.223 | 0.939 | 2.443 | 0.981 | 0.019 |
| 32 | 2 | 4 | 0.003 | 0.083 | 0.161 | 1.147 | 2.086 | 0.986 | 0.014 |
| 34 | 2 | 4 | 0.003 | 0.088 | 0.171 | 1.375 | 2.522 | 0.991 | 0.009 |
| 36 | 1 | 3 | 0.001 | 0.047 | 0.135 | 0.812 | 2.187 | 0.994 | 0.006 |
| 38 | 1 | 2 | 0.001 | 0.049 | 0.096 | 0.947 | 1.759 | 0.996 | 0.004 |
| 40 | 1 | 2 | 0.001 | 0.052 | 0.101 | 1.093 | 2.040 | 0.999 | 0.001 |
| 42 | 1 | 2 | 0.001 | 0.054 | 0.106 | 1.248 | 2.341 | | |
| $\Sigma(C(t)dt) = 771$ | | | | | $t_m = 10.98$ s | | $\sigma^2 = 46.88$ s$^2$ | | |

We can calculate time increment as the difference between time in two experimental data; in the third column, concentration and time increment are multiplied. For a better calculation, instead of calculating $C(t) \cdot \Delta t$ for each time directly, we can use the average value of $C(t)$ between two consecutive data. This is called Simpson's rule for evaluating areas graphically. The sum of the values in this column would give the area of the $C(t)$ curve, so in the fourth column, the following relationship is applied to calculate $E(t)$:

$$E(t) = \frac{C(t)}{\int_0^\infty C(t) \cdot dt} = \frac{C(t)}{771}$$

In the next columns, the average value of $E(t)$ between two consecutive data is multiplied by time, by time increment, and the sum would represent mean time, as we have:

$$t_m = \int_0^\infty t \cdot E(t)\mathrm{d}t = \sum_{\text{all points}} t \cdot E(t) \cdot \Delta t = 10.98 \text{ s}$$

In a similar way, variance is calculated:

$$\sigma^2 = \sum_{\text{all points}} (t - t_m)^2 \cdot E(t) \cdot \Delta t = 46.88 \text{ s}^2$$

Finally, we can calculate $F(t)$ and $I(t)$ according to their definitions:

$$F(t) = \sum_{\text{all points at time}<t} E(t) \cdot \Delta t$$

$$I(t) = 1 - F(t)$$

We can check the form of the graphs showing the distributions:

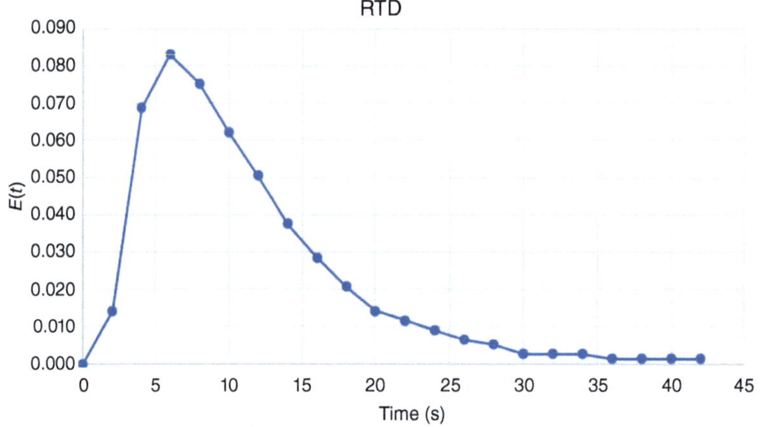


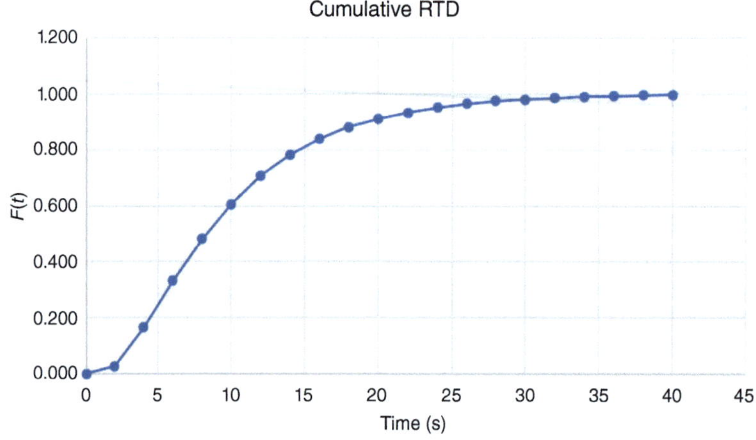

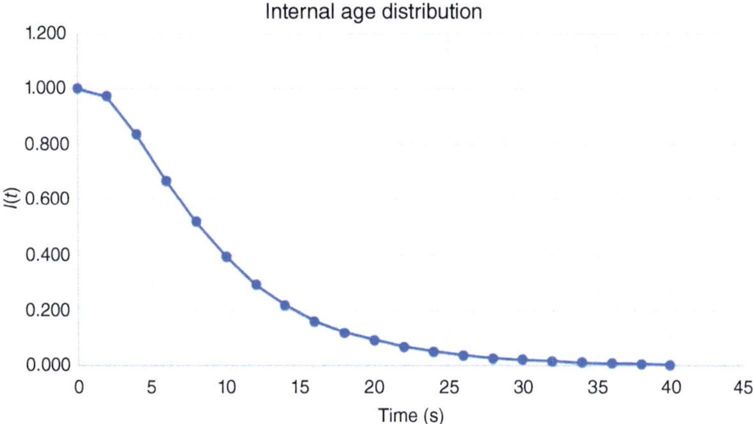


(b)  From the calculated parameters, it is easy to find the number of tanks in the TIS model:

$$\sigma^2 = \frac{t_m^2}{n_t}$$

$$n_t = \frac{t_m^2}{\sigma^2} = \frac{10.98^2}{46.88} = 2.57 \text{ tanks}$$

For the dispersion model to be applied, first we should assume $Bo < 0.01$, and then:

$$\sigma^2 = 2 \cdot Bo$$

We obtain $Bo = 11.24 \gg 0.01$, so this assumption is not valid.

Assuming now $Bo > 0.01$ and closed–closed recipient:

$$\left(\frac{\sigma}{t_m}\right)^2 = 2 \cdot Bo - 2 \cdot Bo^2 \cdot \left[1 - \exp\left(-\frac{1}{Bo}\right)\right]$$

From that:

$$\frac{46.88}{10.98^2} = 2 \cdot Bo - 2 \cdot Bo^2 \cdot \left[1 - \exp\left(-\frac{1}{Bo}\right)\right]$$

and we obtain $Bo = 0.2413$, and $Pe = 1/Bo = 4.14$.

On the other hand, if an o–o recipient is assumed:

$$\left(\frac{\sigma}{t_m}\right)^2 = 2 \cdot Bo + 8 \cdot Bo^2$$

Obtaining $Bo = 0.127$ and $Pe = 7.85$.

For checking which one of the conditions is fulfilled, data about average time in the recipient is needed in order to test if the equation $t_m = \bar{t} \cdot (1 + 2 \cdot Bo)$ is satisfied.

$$\bar{t} = \frac{\text{Length}}{\text{Linear velocity}} = \frac{274 \text{ cm}}{35.70 \frac{\text{cm}}{\text{s}}} = 7.67 \text{ s}$$

So, we have that:

$$10.98 = 7.67 \, (1 + 2\text{Bo})$$

Obtaining Bo = 0.215. This difference between $t_m$ and $\bar{t}$ indicates that probably the recipient is open.

**Problem 1.2**  An experiment to characterize a tubular reactor was conducted using a technique in which a tracer is continuously fed into the system. At a specific time, the tracer supply is halted. From this point onward, the exit signal is recorded, resembling what can be termed a "decreasing step input" (or negative step). In this experiment, the following data were obtained:

| $t$ (min) | 0 | 1 | 2 | 4 | 6 | 7 | 8 | 10 |
|---|---|---|---|---|---|---|---|---|
| $C$ (g/l) | 0.5 | 0.5 | 0.5 | 0.4 | 0.1 | 0 | 0 | 0 |

(a) Calculate the RTD of this reactor.
(b) What fraction of the fluid spends more than 3 minutes in the reactor? Make a plot of the procedure.
(c) What fraction of the fluid spends between 3 and 4 minutes in the reactor?

**Solution to Problem 1.2**

For details refer the Wiley website at http://www.wiley-vch.de/ISBN9783527354115

(a) Following this procedure, the exit signal will be related to the cumulative RTD of the system, $F(t)$, but the actual function obtained is the internal age function, $I(t)$. In the spreadsheet, we can do the following calculation:

| $t$ (min) | $C$ (g/l) | $C/C_{max}$ | $I(t)$ | $F(t)$ | $E(t) = \Delta F/\Delta t$ |
|---|---|---|---|---|---|
| 0 | 0.5 | 1 | 1 | 0 | 0.00 |
| 1 | 0.5 | 1 | 1 | 0 | 0.00 |
| 2 | 0.5 | 1 | 1 | 0 | 0.10 |
| 4 | 0.4 | 0.8 | 0.8 | 0.2 | 0.30 |
| 6 | 0.1 | 0.2 | 0.2 | 0.8 | 0.20 |
| 7 | 0 | 0 | 0 | 1 | 0.00 |
| 8 | 0 | 0 | 0 | 1 | 0.00 |
| 10 | 0 | 0 | 0 | 1 | |

In the table, $I(t)$ is derived directly from the data, and:

$$F(t) = 1 - I(t)$$

For the calculation of $E(t)$, a numerical derivative is done:

$$E(t) = \frac{F^{t+1} - F^t}{\Delta t}$$

Obtaining:

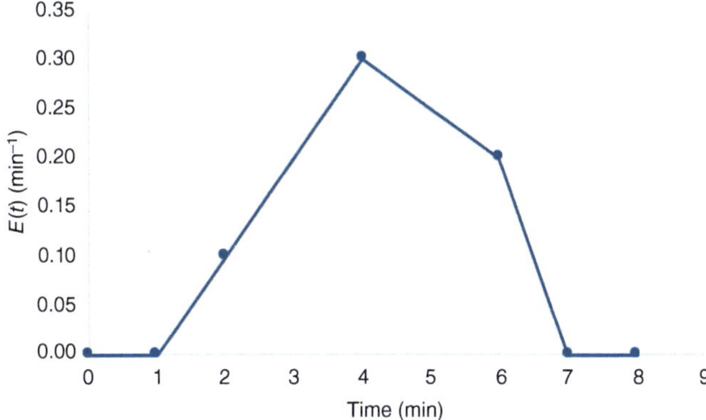

(b)  For calculating the fraction of fluid that spends a time in the reactor, we use the definition of $E(t)$:

$$\text{Fluid passing more than 3 minutes} = f_{>3} = \int_3^\infty E(t)dt$$

This integration can be done numerically, in increments:

$$f_{>3} = E(3) \cdot \Delta t + E(4) \cdot \Delta t + E(5) \cdot \Delta t + E(6) \cdot \Delta t + E(7) \cdot \Delta t$$

where $\Delta t$ is obviously 1 minute. Graphically, the approximation of the integral would be the following:

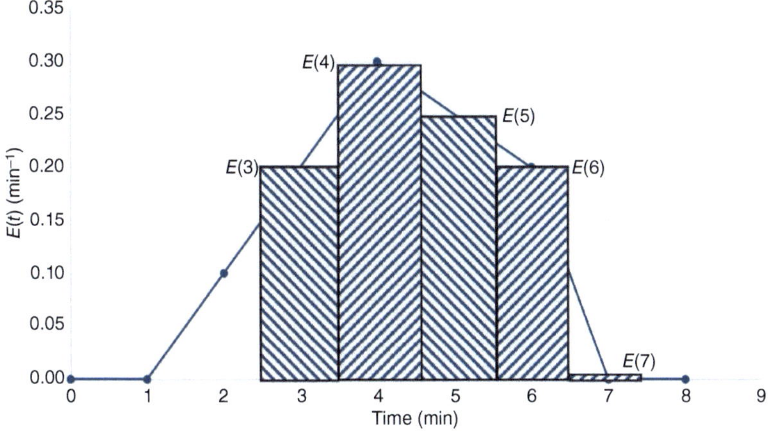

Obviously, this is a rough approximation. A better approximation would involve using smaller time increments, but this is not possible with the obtained data.

More precision is obtained if average value of the $E(t)$ function is used, so:

$$f_{>3} = \frac{E(3) + E(4)}{2} \cdot \Delta t + \frac{E(4) + E(5)}{2} \cdot \Delta t + \frac{E(5) + E(6)}{2} \cdot \Delta t$$
$$+ \frac{E(6) + E(7)}{2} \cdot \Delta t + \frac{E(7) + E(8)}{2} \cdot \Delta t$$
$$= \left( \frac{E(3)}{2} + E(4) + E(5) + E(6) + E(7) + \frac{E(8)}{2} \right) \cdot \Delta t = 0.85$$

In this situation, we are doing the following approximation in the graph:

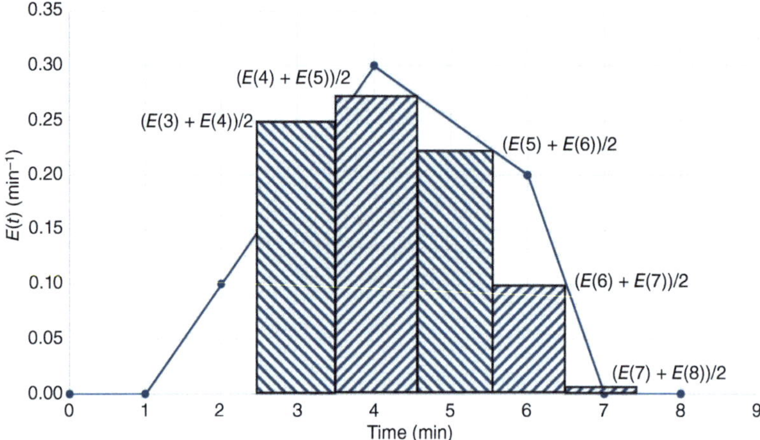


(c) If we look for the fraction of fluid passing between 3 and 4 minutes:

$$f_{3-4} = \int_3^4 E(t)\mathrm{d}t = E(3) \cdot \Delta t + E(4) \cdot \Delta t = 0.5$$

**Problem 1.3** This problem involves a reactor with a flow rate of 10 l/min. Concentration measurements were taken at the outlet during a pulse test at various time intervals. The data obtained are as follows:

| $t$ (min) | $c \times 10^5$ (g/l) | $t$ (min) | $c \times 10^5$ (g/l) |
| --- | --- | --- | --- |
| 0 | 0 | 15 | 238 |
| 0.4 | 329 | 20 | 136 |
| 1.0 | 622 | 25 | 77 |
| 2.0 | 812 | 30 | 44 |
| 3 | 831 | 35 | 25 |
| 4 | 785 | 40 | 14 |

| $t$ (min) | $c \times 10^5$ (g/l) | $t$ (min) | $c \times 10^5$ (g/l) |
| --- | --- | --- | --- |
| 5 | 720 | 45 | 8 |
| 6 | 650 | 50 | 5 |
| 8 | 523 | 60 | 1 |
| 10 | 418 | | |

(a) Determine the volume of the reactor and plot the functions $E(t)$, $F(t)$, and $I(t)$ as a function of time.
(b) Determine the number of tanks required to model the reactor using a TIS system.
(c) Calculate the Peclet number (Pe) if the reactor is modeled using a dispersion model.

**Solution to Problem 1.3**

For details refer the Wiley website at http://www.wiley-vch.de/ISBN9783527354115

(a) This problem is quite similar to Problem 1.1. In a spreadsheet, we calculate:

| $t$ (min) | $C(t)$ (g/l) | $C(t) \cdot \Delta t$ | $E(t)$ | $t \cdot E(t) \cdot \Delta t$ | $(t - t_m)^2 \cdot E(t)$ | $(t - t_m)^2 \cdot E(t) \cdot \Delta t$ |
| --- | --- | --- | --- | --- | --- | --- |
| 0 | 0 | 0 | 0.0000 | 0 | 0 | 0 |
| 0.4 | $3.29 \cdot 10^{-3}$ | $1.32 \cdot 10^{-3}$ | 0.0368 | 0.0059 | 3.4842 | 1.3937 |
| 1 | $6.22 \cdot 10^{-3}$ | $3.73 \cdot 10^{-3}$ | 0.0697 | 0.0418 | 5.7993 | 3.4796 |
| 2 | $8.12 \cdot 10^{-3}$ | $8.12 \cdot 10^{-3}$ | 0.0909 | 0.1819 | 6.0023 | 6.0023 |
| 3 | $8.31 \cdot 10^{-3}$ | $8.31 \cdot 10^{-3}$ | 0.0931 | 0.2792 | 4.7237 | 4.7237 |
| 4 | $7.85 \cdot 10^{-3}$ | $7.85 \cdot 10^{-3}$ | 0.0879 | 0.3516 | 3.2975 | 3.2975 |
| 5 | $7.20 \cdot 10^{-3}$ | $7.20 \cdot 10^{-3}$ | 0.0806 | 0.4031 | 2.1175 | 2.1175 |
| 6 | $6.50 \cdot 10^{-3}$ | $6.50 \cdot 10^{-3}$ | 0.0728 | 0.4367 | 1.2383 | 1.2383 |
| 8 | $5.23 \cdot 10^{-3}$ | $1.05 \cdot 10^{-2}$ | 0.0586 | 0.9371 | 0.2644 | 0.5288 |
| 10 | $4.18 \cdot 10^{-3}$ | $8.36 \cdot 10^{-3}$ | 0.0468 | 0.9362 | 0.0007 | 0.0015 |
| 15 | $2.38 \cdot 10^{-3}$ | $1.19 \cdot 10^{-2}$ | 0.0267 | 1.9989 | 0.6335 | 3.1675 |
| 20 | $1.36 \cdot 10^{-3}$ | $6.80 \cdot 10^{-3}$ | 0.0152 | 1.5230 | 1.4853 | 7.4263 |
| 25 | $7.70 \cdot 10^{-4}$ | $3.85 \cdot 10^{-3}$ | 0.0086 | 1.0779 | 1.9080 | 9.5401 |
| 30 | $4.40 \cdot 10^{-4}$ | $2.20 \cdot 10^{-3}$ | 0.0049 | 0.7391 | 1.9464 | 9.7322 |
| 35 | $2.50 \cdot 10^{-4}$ | $1.25 \cdot 10^{-3}$ | 0.0028 | 0.4899 | 1.7324 | 8.6618 |
| 40 | $1.40 \cdot 10^{-4}$ | $7.00 \cdot 10^{-4}$ | 0.0016 | 0.3136 | 1.3993 | 6.9965 |
| 45 | $8.00 \cdot 10^{-5}$ | $4.00 \cdot 10^{-4}$ | 0.0009 | 0.2016 | 1.0896 | 5.4482 |
| 50 | $5.00 \cdot 10^{-5}$ | $2.50 \cdot 10^{-4}$ | 0.0006 | 0.1400 | 0.8903 | 4.4515 |
| 60 | $1.00 \cdot 10^{-5}$ | $1.00 \cdot 10^{-4}$ | 0.0001 | 0.0672 | 0.2786 | 2.7857 |
| | | $\Sigma(C(t)\mathrm{d}t) = 8.93 \cdot 10^{-2}$ | | $t_m = 10.12$ s | | $\sigma^2 = 80.99$ s$^2$ |

We easily obtain:

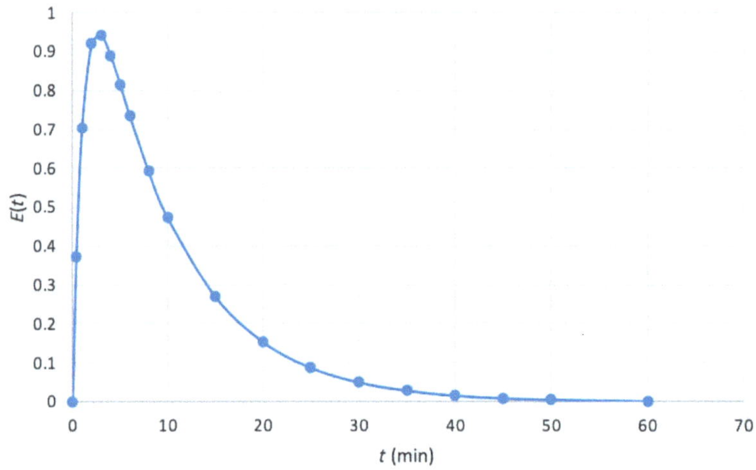

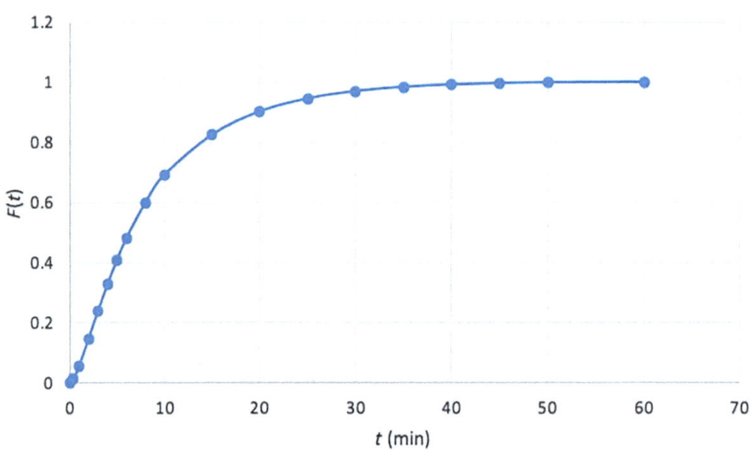

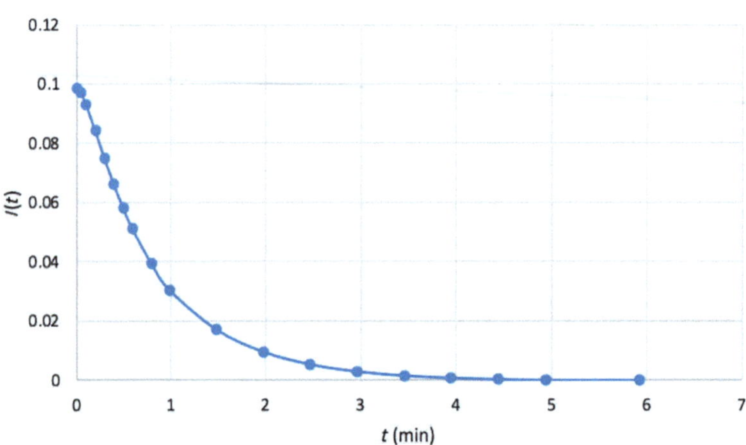

For calculating the volume of the system, we have:

$$\bar{t} = t_m = \frac{V}{Q}; \quad 10.12 = \frac{V}{10}$$

So the volume is 101.2 l.

(b)

$$n_t = \frac{t_m^2}{\sigma^2} = \frac{10.12^2}{80.99} = 1.26 \text{ tanks}$$

(c) As in Problem 1.1, we do not know if the recipient is closed or open. Assuming closed–closed condition, we obtain Bo = 1.33 and Pe = 0.75.

## Problem 1.4

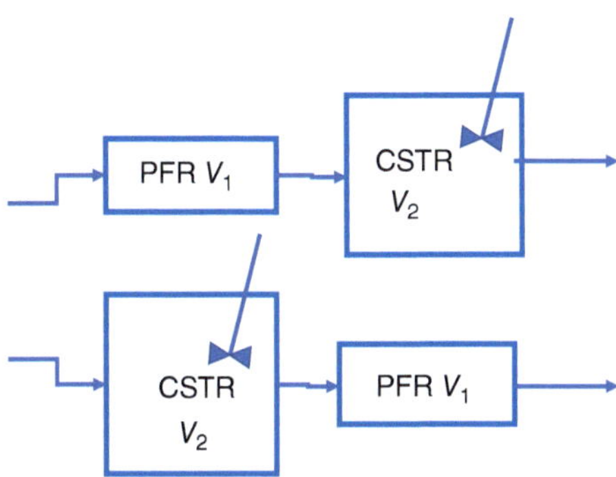


In the given reactor configurations, where the total volume is represented as $V = V_1 + V_2$, and the ratio $V_1/V_2$ is imposed, we need to determine the ratio of the volumes $V_1/V_2$ in the second configuration in order to achieve the same exit conversion, $X_{A2}$, for both a first-order reaction and a second-order reaction.

## Solution to Problem 1.4

For details refer the Wiley website at http://www.wiley-vch.de/ISBN9783527354115

In the first case:

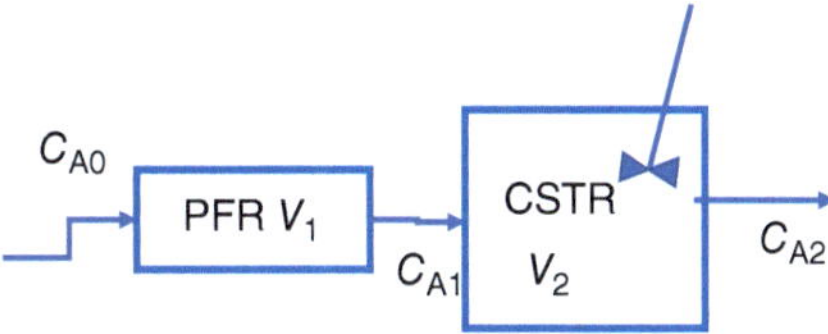


We can write:

$$X_{A1} = 1 - \exp\left(-k \cdot \frac{V_1}{Q_0}\right)$$

and in the CSTR:

$$Q_0 C_{A1} = (-r_A)V_2 + Q_0 C_{A2}$$

$$Q_0 C_{A0}(1 - X_{A1}) = (k \cdot C_{A0}(1 - X_{A2})) \cdot V_2 + Q_0 C_{A0}(1 - X_{A2})$$

$$(1 - X_{A1}) = (k \cdot (1 - X_{A2})) \cdot \frac{V_2}{Q_0} + (1 - X_{A2})$$

And solving for $X_{A2}$:

$$(1 - X_{A2}) = \frac{(1 - X_{A1})}{1 + k \cdot \frac{V_2}{Q_0}} = \frac{\exp\left(-k \cdot \frac{V_1}{Q_0}\right)}{1 + k \cdot \frac{V_2}{Q_0}}$$

In the second configuration:

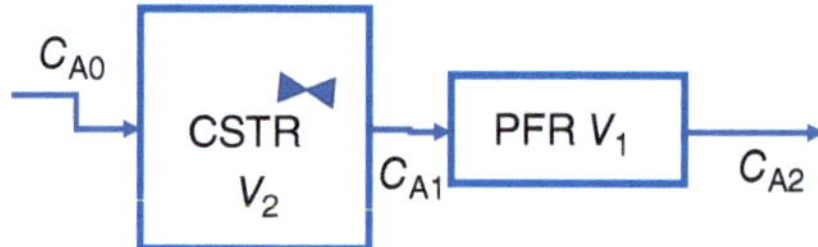


we have, in the CSTR: $Q_0 C_{A0} = (-r_A)V_2 + Q_0 C_{A1}$

So: $X_{A1} = 1 - \frac{1}{1 + k \cdot \frac{V_2}{Q_0}}$. Then, in the PFR:

$$\int_{X_{A1}}^{X_{A2}} \frac{\mathrm{d}X_A}{1 - X_A} = k \int_0^{V_1} \frac{\mathrm{d}V}{Q_0}$$

that is:

$$\ln\left(\frac{1 - X_{A1}}{1 - X_{A2}}\right) = k \cdot \frac{V_1}{Q_0}$$

$$1 - X_{A2} = (1 - X_{A1}) \cdot \exp\left(-k \cdot \frac{V_1}{Q_0}\right)$$

$$= \left(1 - \frac{1}{1 + k \cdot \frac{V_2}{Q_0}}\right) \cdot \exp\left(-k \cdot \frac{V_1}{Q_0}\right) = \frac{\exp\left(-k \cdot \frac{V_1}{Q_0}\right)}{1 + k \cdot \frac{V_2}{Q_0}}$$

We can check that both configurations give the same final conversion.

In the case where a second-order reaction took place, the procedure is similar. In that case, for the first configuration, in the PFR:

$$\mathrm{d}n_A = Q_0 \cdot \mathrm{d}C_A = (r_A) \cdot \mathrm{d}V = -k \cdot C_A^2 \cdot \mathrm{d}V$$

$$-Q_0 \cdot C_{A0} \cdot \mathrm{d}X_A = -k \cdot C_{A0}^2 \cdot (1 - X_A)^2 \cdot \mathrm{d}V$$

$$Q_0 \cdot dX_A = k \cdot C_{A0} \cdot (1 - X_A)^2 \cdot dV$$

$$\int_0^{X_{A1}} \frac{dX_A}{(1 - X_A)^2} = \int_0^{V_1} k \frac{C_{A0}}{Q_0} dV$$

$$-\frac{X_{A1}}{X_{A1} - 1} = k \frac{C_{A0}}{Q_0} V_1$$

$$1 - X_{A1} = \frac{1}{1 + k \cdot C_{A0} \cdot \frac{V_1}{Q_0}}$$

and in the CSTR:

$$Q_0 C_{A0}(1 - X_{A1}) = k \cdot C_{A0}^2 (1 - X_{A2})^2 V_2 + Q_0 C_{A0}(1 - X_{A2})$$

So finally:

$$Q_0 C_{A0} \frac{1}{1 + k \cdot C_{A0} \cdot \frac{V_1}{Q_0}} = k \cdot C_{A0}^2 (1 - X_{A2})^2 V_2 + Q_0 C_{A0}(1 - X_{A2})$$

$$k \cdot C_{A0}(1 - X_{A2})^2 \frac{V_2}{Q_0} + (1 - X_{A2}) - (1 - X_{A1}) = 0$$

$$k \cdot C_{A0} \cdot \overline{t_2} \cdot (1 - X_{A2})^2 + (1 - X_{A2}) - (1 - X_{A1}) = 0$$

This expression is a second-order equation. It is quite complicated to solve the final conversion, but finally, for the first configuration:

$$(1 - X_{A2}) = \frac{-1 + (1 + 4 \cdot k \cdot C_{A0} \cdot \overline{t_2} \cdot (1 - X_{A1}))^{0.5}}{2 \cdot k \cdot C_{A0} \cdot \overline{t_2}}$$

Obviously, the other solution with minus sign does not make sense.
In the second configuration, following the same procedure:

$$Q_0 C_{A0} = k \cdot C_{A0}^2 (1 - X_{A1})^2 V_2 + Q_0 C_{A0}(1 - X_{A1})$$

$$k \cdot C_{A0} \cdot \overline{t_2} \cdot (1 - X_{A1})^2 + (1 - X_{A1}) - 1 = 0$$

$$(1 - X_{A1}) = \frac{-1 + (1 + 4 \cdot k \cdot C_{A0} \cdot \overline{t_2})^{0.5}}{2 \cdot k \cdot C_{A0} \cdot \overline{t_2}}$$

and finally, in the PFR:

$$(1 - X_{A2}) = \frac{1 - X_{A1}}{1 + k \cdot C_{A0} \cdot \overline{t_2} + (1 - X_{A1})}$$

We can check the form of the solution by assuming values of the constants. For example, $k \cdot C_{A0} = 0.1 \text{ mol/l·s}$, $Q_0 = 1 \text{ l/s}$, $V = 100 \text{ l}$, and the following graphs can be calculated. As we can check, in any case, the first configuration (PFR + CSTR) gives a higher conversion than the second configuration (CSTR + PFR). In all cases, $V_1$ refers to the volume of the PFR and $V_2$ to that of CSTR.

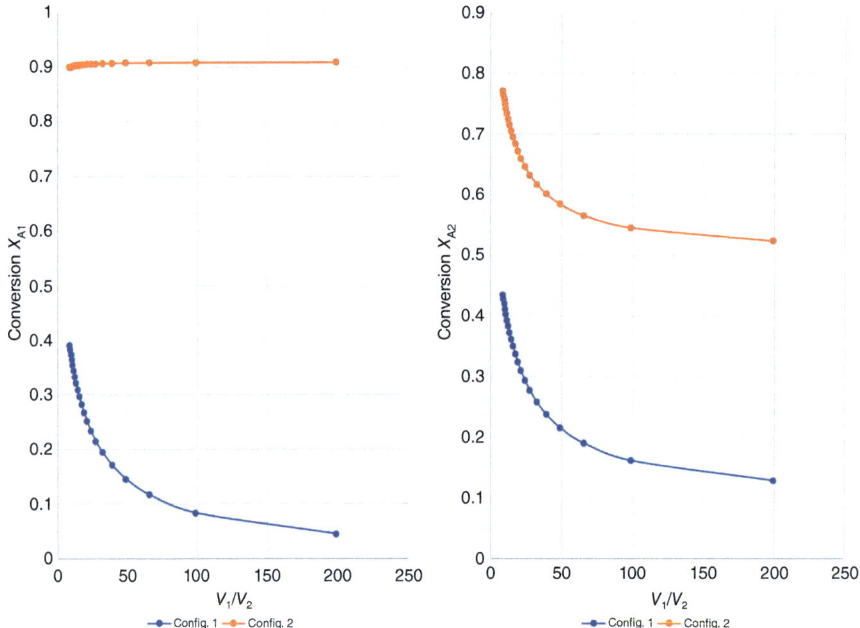

## Problem 1.5

(a) Obtain the cumulative probability function for the following data arising from stable operation with concentration of 0.122 M, followed by a step disturbance with $C = 0.548$ M.

| $t$ (s) | 10 | 20 | 30 | 45 | 60 | 90 | 120 | 200 | 400 | 1000 |
|---|---|---|---|---|---|---|---|---|---|---|
| $C$ (M) | 0.126 | 0.43 | 0.207 | 0.314 | 0.378 | 0.484 | 0.505 | 0.527 | 0.544 | 0.547 |

(b) Obtain a polynomial for its interpolation, and calculate the average residence time using numerical integration by the method of trapezoids.

(c) Calculate the average time using the polynomial.

**Solution to Problem 1.5**

For details refer the Wiley website at http://www.wiley-vch.de/ISBN9783527354115

(a) The resolution of this problem required a spreadsheet. Let us calculate the $E(t)$ as the derivative of the $C_{\text{step}}(t)$ function, i.e.,

$$E(t) = \frac{\mathrm{d}}{\mathrm{d}t}\left[\frac{C(t)}{C_0}\right]_{\text{step input}} = \frac{1}{C_0}\left(\frac{\Delta C}{\Delta t}\right)_{\text{step}}$$

Furthermore, $C_0 = \sum_{\text{all points}} C(t) \cdot \Delta t$. We can do the following calculation, similar to that presented in Problems 1.1 and 1.3:

| $t$ (s) | $C_{step}$ (M) | $C(t)\cdot dt$ | $E(t)$ | $t\cdot E(t)$ | $t\cdot E(t)\cdot dt$ | $(t-t_m)^2\cdot E(t)$ | $(t-t_m)^2\cdot E(t)\cdot dt$ |
|---|---|---|---|---|---|---|---|
| 10 | 0.004 | | 0.0002 | 0.0024 | | 65.133 | |
| 20 | 0.021 | 1.35 | 0.0003 | 0.0055 | 0.04 | 71.085 | 681.09 |
| 30 | 0.085 | 1.75 | 0.0004 | 0.0120 | 0.09 | 98.874 | 849.79 |
| 45 | 0.192 | 3.91 | 0.0006 | 0.0274 | 0.30 | 141.053 | 1799.46 |
| 60 | 0.256 | 5.19 | 0.0007 | 0.0440 | 0.54 | 159.384 | 2253.28 |
| 90 | 0.362 | 12.93 | 0.0009 | 0.0845 | 1.93 | 178.663 | 5070.70 |
| 120 | 0.383 | 14.84 | 0.0010 | 0.1175 | 3.03 | 161.660 | 5104.84 |
| 200 | 0.405 | 41.28 | 0.0010 | 0.2044 | 12.88 | 108.806 | 10 818.61 |
| 400 | 0.422 | 107.10 | 0.0011 | 0.4220 | 62.64 | 16.824 | 12 562.99 |
| 1000 | 0.425 | 327.30 | 0.0011 | 1.0608 | 444.85 | 238.058 | 76 464.73 |
| | | Sum | | | $t_m$ (s) | | $\sigma^2$ |
| | | 515.6375 | | | 526.28 | | 115 605.49 |

Note that the concentration measured has been subtracted from the 0.122 M of the stable operation.

And then:

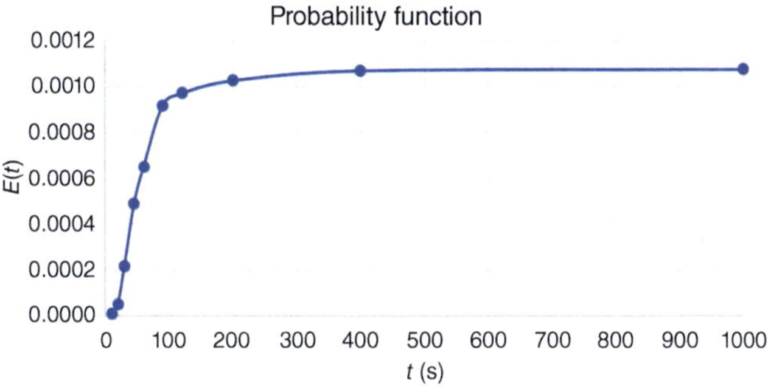


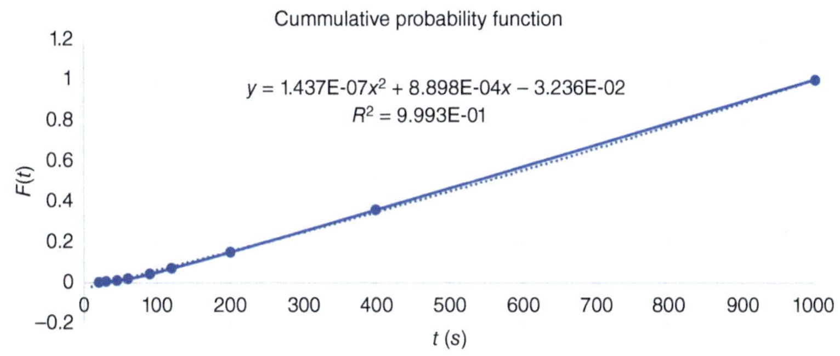

(b) The polynomial of the fitting is shown in the previous figure. Using numerical integration of the data, we obtain $t_m = 532.79$ s.

(c) Using the polynomial shown in the figure, we can do:

$$t_m = \int_0^\infty t \cdot E(t)\mathrm{d}t$$

as:

$$E(t) = \frac{\mathrm{d}F(t)}{\mathrm{d}t} = 2 \cdot 1.437 \cdot 10^{-7}t + 8.898 \cdot 10^{-4}$$

$$t_m = \int_0^\infty t \cdot (2 \cdot 1.437 \cdot 10^{-7}t + 8.898 \cdot 10^{-4}) \cdot \mathrm{d}t$$

This function cannot be integrated because the integral does not converge, as the upper limit is infinity. If we change this limit to, for example, 1000 (high enough to bear in mind all the tracer leaving the system), this can be solved, and:

$$t_m = \int_0^{1000} t \cdot (2 \cdot 1.437 \cdot 10^{-7}t + 8.898 \cdot 10^{-4}) \cdot \mathrm{d}t = 540.7 \text{ s}$$

**Problem 1.6**  A continuous reactor normally operates with a feed of 50 l/s. Under these conditions, the dynamic data for different pulses presented below were obtained.

(a) To what type of reactor do these data surely correspond?
(b) Obtain the residence time distribution.
(c) Estimate the residence time and the volume of the reactor, and
(d) Calculate the total mass of the tracer used in the three runs.

|  | $C_{tracer}$ (mg/l) | | |
| --- | --- | --- | --- |
| $t$ (s) | Run 1 | Run 2 | Run 3 |
| 0 | 0.011 | 0.012 | 0.001 |
| 1 | 0.553 | 0.534 | 0.538 |
| 2 | 0.899 | 0.907 | 0.887 |
| 3 | 0.923 | 0.953 | 0.934 |
| 5 | 0.860 | 0.856 | 0.876 |
| 8 | 0.681 | 0.723 | 0.758 |
| 12 | 0.550 | 0.586 | 0.585 |
| 17 | 0.413 | 0.435 | 0.442 |
| 27 | 0.320 | 0.331 | 0.350 |
| 38 | 0.251 | 0.236 | 0.276 |
| 58 | 0.162 | 0.158 | 0.176 |
| 86 | 0.085 | 0.103 | 0.097 |
| 130 | 0.023 | 0.008 | 0.015 |

**Solution to Problem 1.6**

For details refer the Wiley website at http://www.wiley-vch.de/ISBN9783527354115

(a) and (b) In a spreadsheet, we can calculate numerically:

$$E(t) = \frac{C(t)}{\int_0^\infty C(t) \cdot dt}$$

Let us plot the given data:

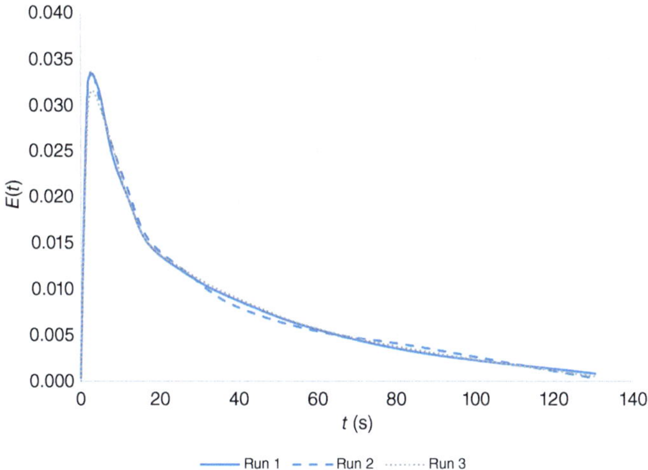


This is quite similar to the RTD of a CSTR, so the reactor behaves as one CSTR.

(c) Mean residence time is calculated following the procedure explained in other problems, obtaining 32.97 (Run 1), 32.53 (Run 2), and 32.98 s (Run 3) (32.83 average value).

Volume of the reactor can be calculated:

$$\bar{t} = \frac{V}{Q} = t_m = \frac{V}{50} = 32.83 \text{ s}$$

resulting $V = 1641.31$

(d) Mass of tracer can be calculated by applying:

$$dN = Q_0 C(t) dt$$

$$N = Q_0 \int_0^\infty C(t) dt$$

The calculation gives $N = 1382$ mg (Run 1), $N = 1412$ mg (Run 2), and $N = 1480$ mg (Run 3).

**Problem 1.7**   The following data correspond to a 2250 l reactor that is fed 150 l/s:

| $t$ (s) | 0 | 2 | 4 | 6 | 8 | 10 | 12 | 14 | 16 | 18 |
|---|---|---|---|---|---|---|---|---|---|---|
| $p(t)$ | 0.00 | 0.02 | 0.03 | 0.05 | 0.1 | 0.19 | 0.32 | 0.47 | 0.6 | 0.71 |

| $t$ (s) | 20 | 22 | 24 | 26 | 28 | 30 | 32 | 34 | 36 | 38 | 40 |
|---|---|---|---|---|---|---|---|---|---|---|---|
| $p(t)$ | 0.80 | 0.86 | 0.90 | 0.93 | 0.95 | 0.97 | 0.98 | 0.98 | 0.99 | 0.99 | 1.00 |

Estimate:

(a) The Peclet number for the dispersion model.
(b) The number of reactors for the model of stirred TIS.

**Solution to Problem 1.7**

For details refer the Wiley website at http://www.wiley-vch.de/ISBN9783527354115

The given function $p(t)$ should correspond to $F(t)$ data, i.e., to a step tracer run, as the value of this function is continuously growing from 0 to 1. These data have been plotted to verify that the trend is as expected. Before beginning to look for the solution that they ask us for, previous calculations must be made:

| $t$ (s) | $p_t(t)$ | $E(t)$ | $E(t)\cdot dt$ | $t\cdot E(t)$ | Integral | $(t - t_m)^2\cdot E(t)$ | Integral |
|---|---|---|---|---|---|---|---|
| 0 | 0 | 0 | 0.02 | 0 | | 0 | |
| 2 | 0.02 | 0.01 | 0.01 | 0.02 | 0.02 | 1.994 | 1.994 |
| 4 | 0.03 | 0.005 | 0.02 | 0.02 | 0.04 | 0.734 | 2.728 |
| 6 | 0.05 | 0.01 | 0.05 | 0.06 | 0.08 | 1.024 | 1.759 |
| 8 | 0.1 | 0.025 | 0.09 | 0.2 | 0.26 | 1.648 | 2.673 |
| 10 | 0.19 | 0.045 | 0.13 | 0.45 | 0.65 | 1.685 | 3.334 |
| 12 | 0.32 | 0.065 | 0.15 | 0.78 | 1.23 | 1.103 | 2.789 |
| 14 | 0.47 | 0.075 | 0.13 | 1.05 | 1.83 | 0.337 | 1.440 |
| 16 | 0.6 | 0.065 | 0.11 | 1.04 | 2.09 | 0.001 | 0.338 |
| 18 | 0.71 | 0.055 | 0.09 | 0.99 | 2.03 | 0.194 | 0.195 |
| 20 | 0.8 | 0.045 | 0.06 | 0.9 | 1.89 | 0.677 | 0.872 |
| 22 | 0.86 | 0.03 | 0.04 | 0.66 | 1.56 | 1.037 | 1.715 |
| 24 | 0.9 | 0.02 | 0.03 | 0.48 | 1.14 | 1.242 | 2.279 |
| 26 | 0.93 | 0.015 | 0.02 | 0.39 | 0.87 | 1.464 | 2.706 |
| 28 | 0.95 | 0.01 | 0.02 | 0.28 | 0.67 | 1.411 | 2.876 |
| 30 | 0.97 | 0.01 | 0.01 | 0.3 | 0.58 | 1.927 | 3.338 |
| 32 | 0.98 | 0.005 | 0 | 0.16 | 0.46 | 1.261 | 3.187 |
| 34 | 0.98 | 0 | 0.01 | 0 | 0.16 | 0.000 | 1.261 |
| 36 | 0.99 | 0.005 | 0 | 0.18 | 0.18 | 1.976 | 1.976 |
| 38 | 0.99 | 0 | 0.01 | 0 | 0.18 | 0.000 | 1.976 |
| 40 | 1 | 0.005 | | 0.2 | 0.2 | 2.851 | 2.851 |
| | | Sum = | 1.00 | | 16.12 | | 42.29 |

The first thing to do is to calculate $E(t)$ by using the data from $F(t)$. This is done by numerical derivation, i.e.,

$$E(t) = \frac{\Delta F(t)}{\Delta t} = \frac{F^{t+1} - F^t}{\Delta t}$$

Once $E(t)$ is known, both $t_m$ and $\sigma^2$ (whose equations are shown in the header of the table) can be calculated.

When we already have the values of both $t_m$ and $\sigma^2$, we can calculate the required values.

The first model that is studied is the TIS. The parameter of such a model is $n_t$:

$$n_t = \frac{t_m^2}{\sigma^2} = 6.14 \text{ tanks}$$

In order to be able to compare the $E(t)$ obtained by the model with the $E(t)$ obtained experimentally, it is also necessary to calculate $\bar{t} = 42.29/6.14 = 2.62\,\text{s}$. On the other hand, to determine the deviation between the points obtained experimentally and with the method, the sum of the squared differences between all the points has been calculated:

$$\text{Deviation} = \sum_{\text{all data}} (E(t)_{\text{calculated}} - E(t)_{\text{measured}})^2$$

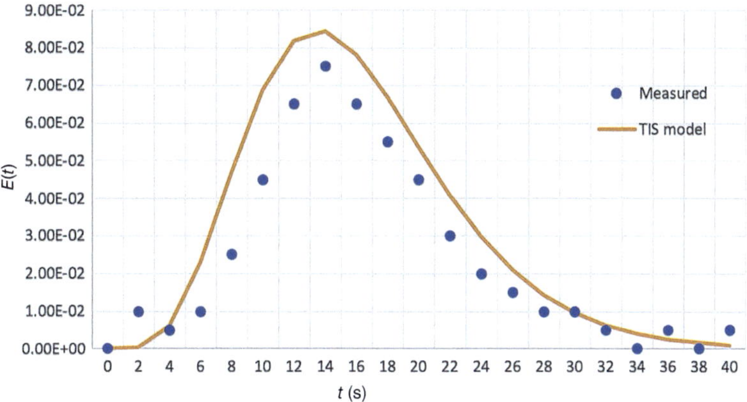


Since the deviation is of the order of $2.36 \cdot 10^{-3}$, it can be concluded that the method can well predict the experimental $E(t)$ values.

For the dispersion model, the first thing to calculate is the Bo number. In this case, the operation that has been done is: $\sigma^2 = 2\text{Bo}$, so $\text{Bo} = 21.14$.

Since Bo is greater than 0.01, a closed–closed system or an o–o system must be chosen. In this exercise, we will use the o–o assumption. The first thing that is done is to calculate the Bo with the equations that govern this o–o dispersion model:

$$t_m = \bar{t} \cdot (1 + 2 \cdot \text{Bo})$$

$$\left(\frac{\sigma}{t_m}\right)^2 = 2 \cdot \text{Bo} + 8 \cdot \text{Bo}^2$$

Obtaining Bo = 0.074. The next step is to calculate $E(t)$ with the closed–closed dispersion model. To do this, the following equation is used:

$$E = \frac{1}{\sqrt{4 \cdot \pi \cdot \text{Bo} \cdot t/\bar{t}}} \cdot \exp\left[ -\frac{\left(1 - \frac{t}{\bar{t}}\right)^2}{4 \cdot t \cdot \text{Bo}/\bar{t}} \right]$$

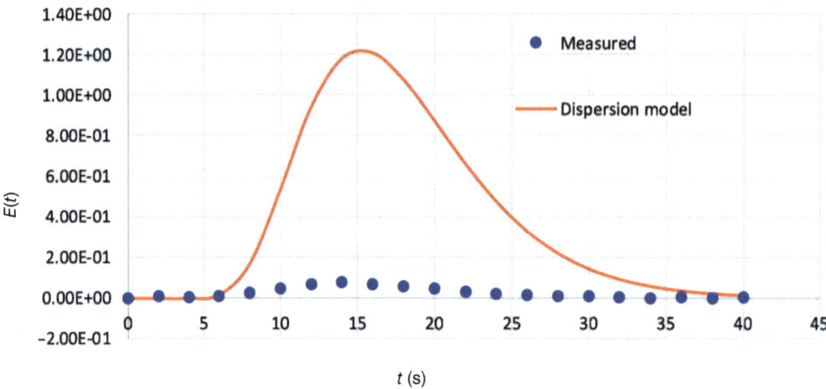


In view of the results, it is clear that the $E(t)$ obtained with the model differs greatly from the $E(t)$ obtained experimentally. Therefore, it can be concluded that this model will not correctly predict this example. On the other hand, when Bo is greater than 0.01 (as in this case), the impulse response is wide, and this gives an asymmetric $E(t)$.

**Problem 1.8**   Consider a continuous reactor fed with a flow $Q_0$ with a composition $C_{A0}$. Both operating variables are held constant. The (total) volume of the reactor is $V_R$ and the space time, $\tau$, is defined based on this volume. The kinetics obeys the expression $(-r_A) = kC_A^a$, and there are no secondary reactions. The reactor is modeled as three identical continuous reactors, perfectly stirred, but it is considered that there is also a certain current "mixed" between the three cameras, which is equal to $\alpha \cdot Q_0$. To visualize this, analyze the following diagram:

Get the dynamic expressions for $\frac{\mathrm{d}C_{A1}}{\mathrm{d}t}$, $\frac{\mathrm{d}C_{A2}}{\mathrm{d}t}$ and $\frac{\mathrm{d}C_{A3}}{\mathrm{d}t}$ as functions of $C_{A0}$, $C_{A1}$, $C_{A2}$, $C_{A3}$, $\alpha$, $\tau$, $a$, and $k$, as required.

**Solution to Problem 1.8**

We will assume well-mixed behavior in all three reactors but consider the accumulation term in the mass balance. In this sense, in the first reactor:

$$\text{Input} + \text{Generation} = \text{Output} + \text{Accumulation}$$

$$n_{A,in} + r_A V = n_{A,out} + V\frac{dC_A}{dt}$$

$$Q_0 C_{A0} + \alpha Q_0 C_{A2} + \left(-kC_{A1}^a\right)V = (1+\alpha)Q_0 C_{A1} + V\frac{dC_{A1}}{dt}$$

Dividing by the flow rate $Q_0$:

$$C_{A0} + \alpha C_{A2}\left(-kC_{A1}^a\right)\bar{t} = (1+\alpha)C_{A1} + \bar{t}\frac{dC_{A1}}{dt}$$

$$\frac{dC_{A1}}{dt} = \frac{C_{A0} + \alpha C_{A2} - (1+\alpha)C_{A1}}{\bar{t}} - \left(kC_{A1}^a\right)$$

In the second reactor:

$$(1+\alpha)Q_0 C_{A1} + \alpha Q_0 C_{A3} + \left(-kC_{A2}^a\right)V = (1+\alpha)Q_0 C_{A2} + \alpha Q_0 C_{A2} + V\frac{dC_{A2}}{dt}$$

$$\frac{dC_{A2}}{dt} = \frac{(1+\alpha)C_{A1} + \alpha C_{A3} - (1+2\alpha)C_{A2}}{\bar{t}} - kC_{A2}^a$$

And in the third one:

$$(1+\alpha)Q_0 C_{A2} + \left(-kC_{A3}^a\right)V = (1+\alpha)Q_0 C_{A3} + V\frac{dC_{A3}}{dt}$$

$$\frac{dC_{A3}}{dt} = \frac{(1+\alpha)C_{A2} - (1+\alpha)C_{A3}}{\bar{t}} - kC_{A3}^a$$

**Problem 1.9**  Draw the $E(t)$ curves for the following systems, with each system represented in a separate sketch:

(a) An ideal PFR with a volume of 2 l and a feed rate of 1 l/min.
(b) An ideal CSTR with a volume of 2 l and a feed rate of 1 l/min.
(c) An ideal PFR with a volume of 2 l, followed by an ideal CSTR with a volume of 2 l. The feed flows in at a rate of 1 l/min.
(d) An ideal CSTR with a volume of 2 l, followed by an ideal PFR with a volume of 2 l. The feed flows in at a rate of 1 l/min.
(e) An ideal CSTR with a volume of 2 l in parallel with an ideal PFR with a volume of 2 l. The feed, with a volume flow rate of 1 l/min, is evenly split between the two reactors: 0.5 l/min to the CSTR and 0.5 l/min to the PFR. After the reactors, the streams are mixed. Provide the $E(t)$ curve for the entire system, not for each individual reactor.

**Solution to Problem 1.9**

(a)  An ideal PFR (volume 2 l, feed 1 l/min):

$$\bar{t} = \frac{2}{1} = 2\,\text{min}$$

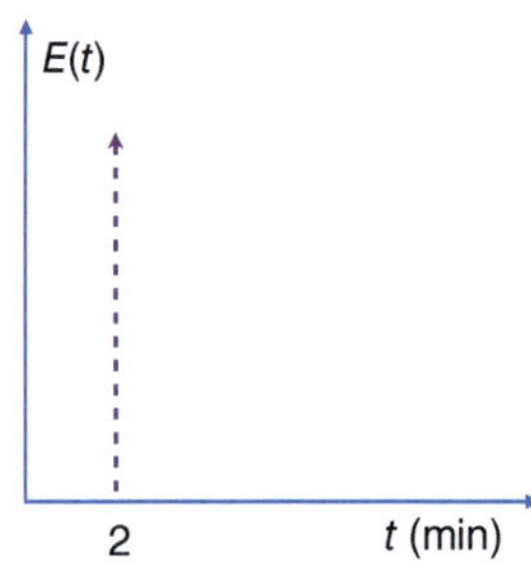


(b)  An ideal CSTR (volume 2 l, feed 1 l/min).

$$\bar{t} = \frac{2}{1} = 2\,\text{min}$$

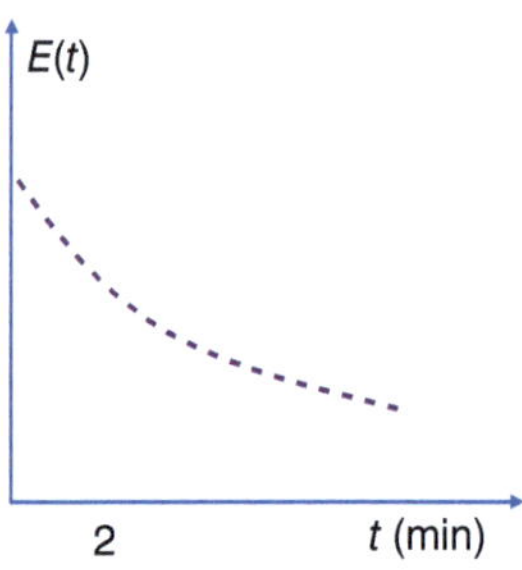


(c)  An ideal PFR (volume 2 l) is followed by an ideal CSTR (volume 2 l). The feed has a volume flow rate of 1 l/min.

$$\overline{t_{\text{PFR}}} = \frac{2}{1} = 2\,\text{min}, \quad \overline{t_{\text{CSTR}}} = \frac{2}{1} = 2\,\text{min}$$

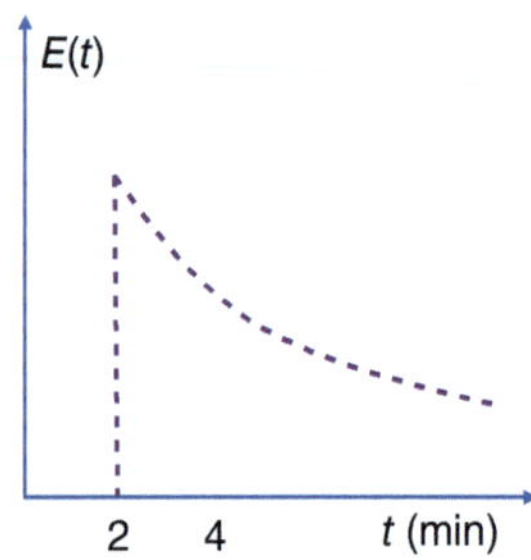

(d) An ideal CSTR (volume 2 l) is followed by an ideal PFR (volume 2 l). The feed has a volume flow rate of 1 l/min.
Same $E(t)$ as in case (c).

(e) An ideal CSTR (volume 2 l) in parallel with an ideal PFR (volume 2 l). The feed (volume flow rate 1 l/min) is equally divided over the reactors: 0.5 l/min to the CSTR and 0.5 l/min to the PFR. After the two reactors, the streams are mixed together. Give the $E(t)$ curve of the complete system, not of each individual reactor.

$$\overline{t_{\text{PFR}}} = \frac{2}{0.5} = 4\,\text{min}, \quad \overline{t_{\text{CSTR}}} = \frac{2}{0.5} = 4\,\text{min}$$

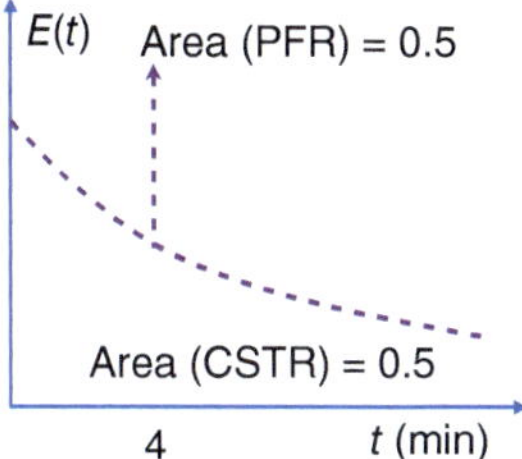


**Problem 1.10** During the "Chemical Reactor Engineering" course, there was a problem that involved sketching the $E(t)$ curve for a system of interconnected ideal reactors. Several answers were provided, and four of the sketches are as follows:

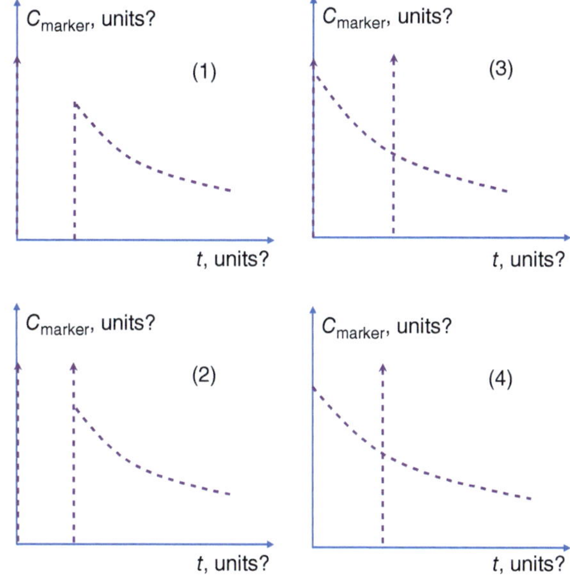

Answer (1) appears to be the correct answer.

(a) Sketch a possible system of interconnected ideal reactors, resulting in sketch 1.
(b) What systems of interconnected reactors correspond to the other sketches?
(c) Sketch the corresponding $F$-curve for each system.
  Denote PFRs in your answer as $P_1, P_2 \ldots$ and CSTRs as $T_1, T_2 \ldots$, with respective volumes $V_{P1}, V_{P2}, \ldots, V_{T1}, V_{T2}, \ldots$. Denote the various volume flow rates as $Q_1, Q_2 \ldots$
(d) Sketch the figure yourself, and put relevant information in it (points of intersection, areas, etc.), expressed in $V_{P1}, V_{P2}, \ldots, V_{T1}, V_{T2}, \ldots, Q_1, Q_2, \ldots$

If the first-order irreversible reaction A $\rightarrow$ products, with rate constant $k$, takes place in this system of reactors, what is the conversion at the outlet of the system? (again, expressed in $V_{P1}, V_{P2}, \ldots, V_{T1}, V_{T2}, \ldots, Q_1, Q_2, \ldots$)?

**Solution to Problem 1.10**

(a) and (b)

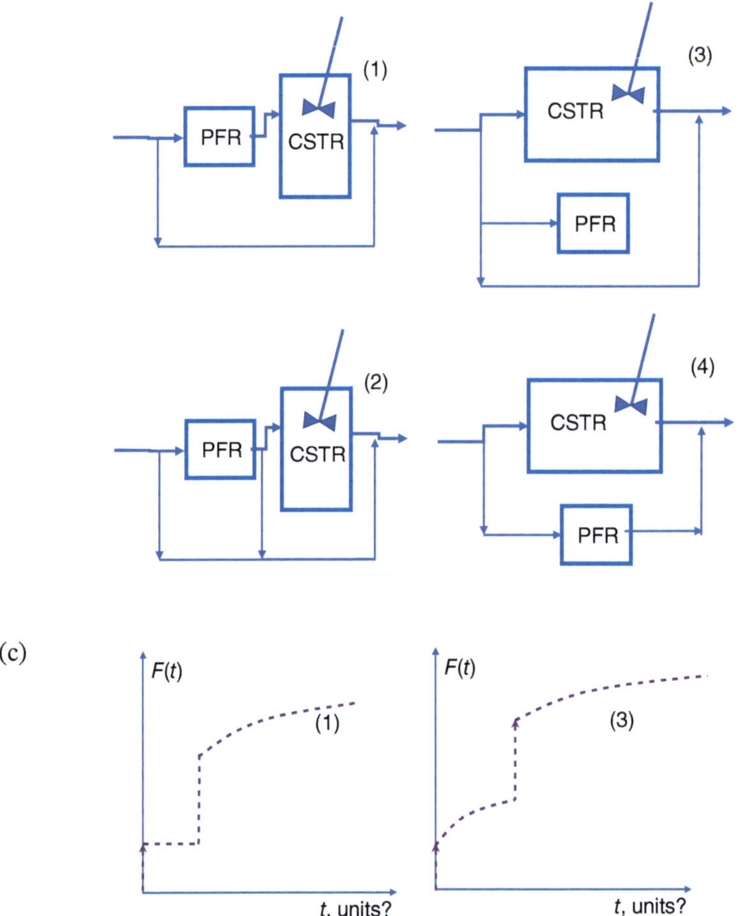

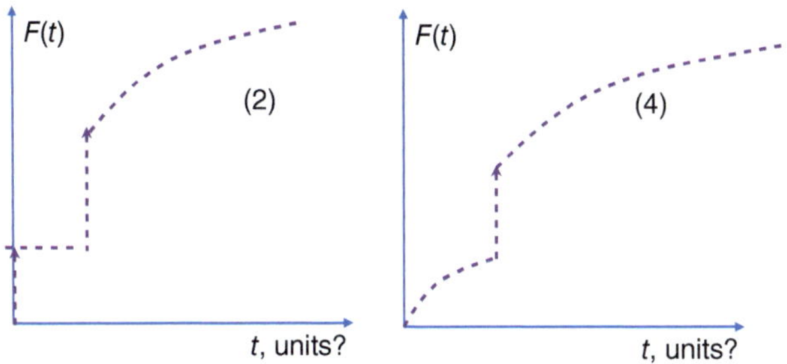



(d)

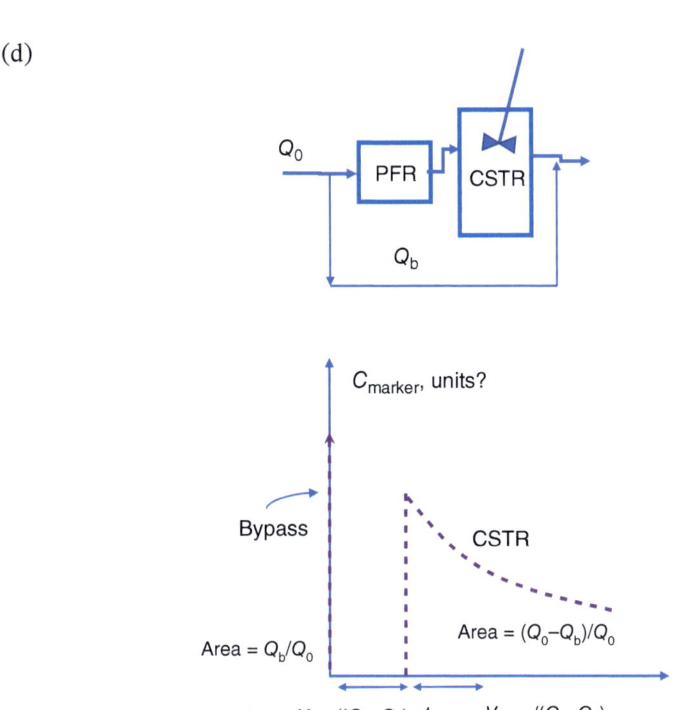



(e) In the PFR:

$$C_{A1} = C_{A0} \cdot \exp(-k\overline{t_{PFR}}) = C_{A0} \cdot \exp\left(-\frac{kV_{PFR}}{Q_0 - Q_b}\right)$$

Then, in the CSTR:

$$C_{A2} = \frac{C_{A1}}{1 + k \cdot \overline{t_{CSTR}}} = \frac{C_{A0} \cdot \exp\left(-\frac{kV_{PFR}}{Q_0 - Q_b}\right)}{1 + k \cdot \frac{V_{CSTR}}{Q_0 - Q_b}}$$

In the mix point, at the end of the system:

$$Q_0 C_{\text{Aout}} = (Q_0 - Q_b) \cdot C_{\text{A2}} + Q_b \cdot C_{\text{A0}}$$

$$C_{\text{Aout}} = \frac{(Q_0 - Q_b)}{Q_0} \cdot \frac{C_{\text{A0}} \cdot \exp\left(-\frac{kV_{\text{PFR}}}{Q_0 - Q_b}\right)}{1 + k \cdot \frac{V_{\text{CSTR}}}{Q_0 - Q_b}} + \frac{Q_b}{Q_0} \cdot C_{\text{A0}}$$

**Problem 1.11**  At time $t = 0$, a step change in tracer concentration occurs in the feed of a reactor with unknown flow behavior. At the reactor outlet, the following $F$-curve is measured:

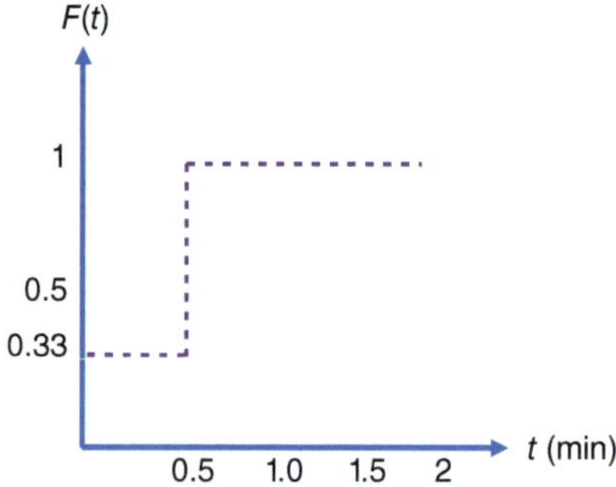

(a) Please sketch the $E$-curve of this unknown reactor, including all relevant information in the sketch.
(b) Which combination of ideal reactors has the same $E$-curve as the unknown reactor?
(c) What is the residence time of the unknown reactor?
(d) If the reaction $A \rightarrow B$, with a rate $(-r_B) = -4C_A$ kmol/m$^3$·min, occurs in the unknown reactor at the residence time determined in part (c), what will be the conversion of A?

**Solution to Problem 1.11**

(a)

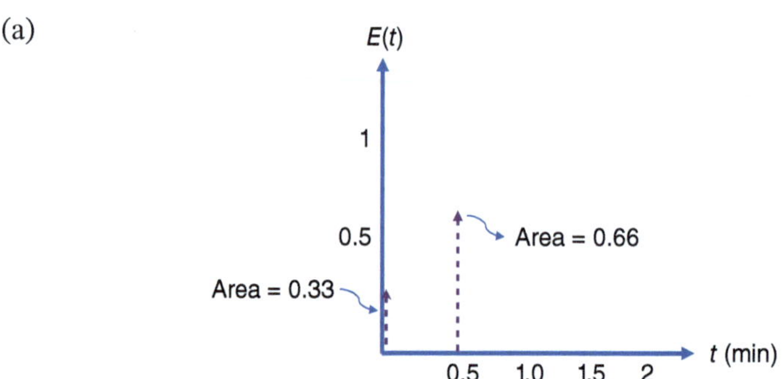

(b)

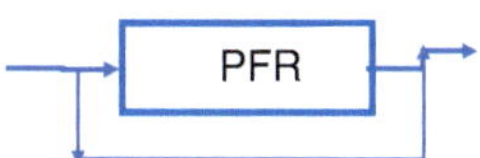

(c) Residence time is: $t = 0.33 \cdot 0 + 0.66 \cdot 0.5 = 0.33$ min

(d) A → B, with $(-r_B) = -4C_A$ kmol/m$^3$·min, takes place in the unknown reactor. What will be the conversion of A?

In a PFR, first-order reaction: $X_A = 1 - \exp(-k\bar{t})$

$$X_A = 1 - \exp(-4 \cdot 0.33) = 0.732$$

As only 66% of the flow passes through the reactor, the correct conversion is:

$$X_{Afinal} = 0.732 \cdot 0.66 = 0.483$$

**Problem 1.12** Three ideal PFRs with volumes of 2, 4, and 6 m$^3$ are interconnected, as shown in the figure below. The water feed, with a volume flow rate $(Q_0)$ of 3 m$^3$/min, is divided between the parallel PFRs as indicated:

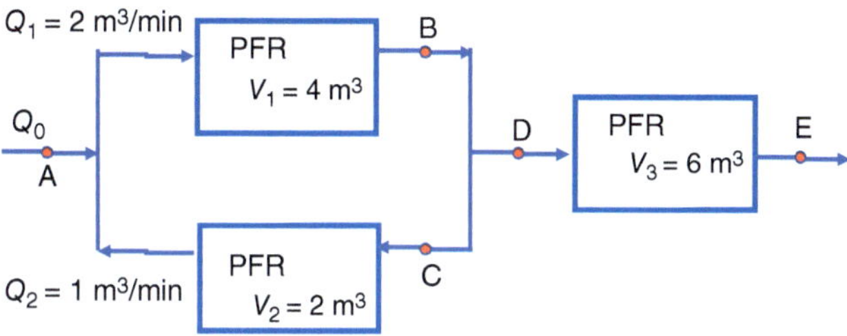


(a) At $t = 0$, a pulse of salt is injected into the feed at point A. Please sketch the salt concentration versus time curves at points B, C, D, and E. Make sure to indicate the scale divisions on the axes.

After conducting the experiment described above, the split ratio of the feed over the two PFRs is changed such that in the new situation $Q_1 = 1$ m$^3$/min and $Q_2 = 1$ m$^3$/min. Once again, at $t = 0$, a pulse of 2 mol of salt is injected into the feed at point A.

(b) Sketch the salt concentration versus time curves at points B, C, D, and E, considering the new split ratio. Indicate the scale divisions on the axes.

(c) Sketch the $F$-curve for the overall system in both situations. Please note that the volume of the tubing connecting the PFRs can be neglected.

## Solution to Problem 1.12

(a)

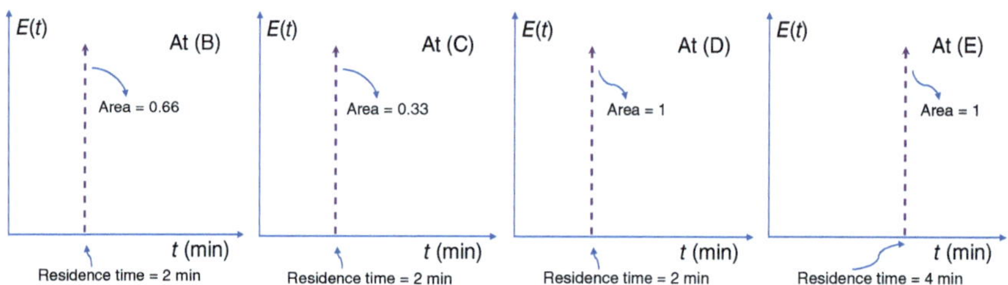

(b)

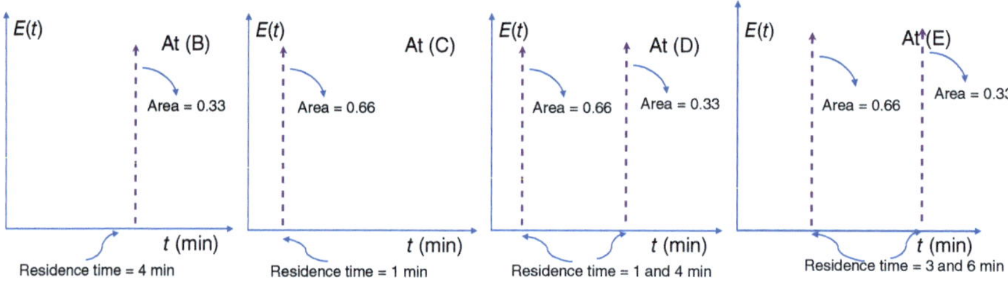

## (c) In the first situation:

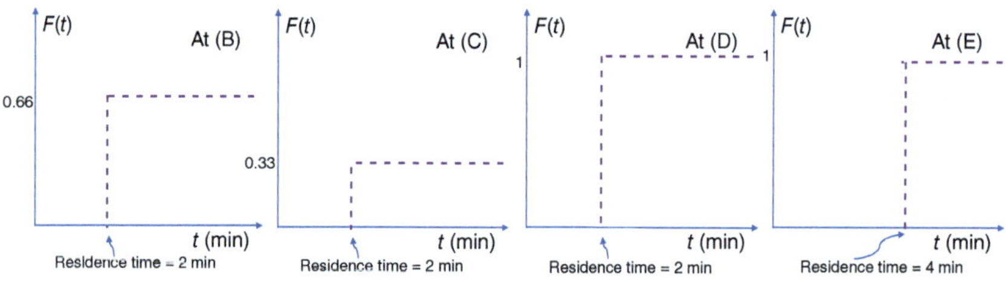

### And in the second:

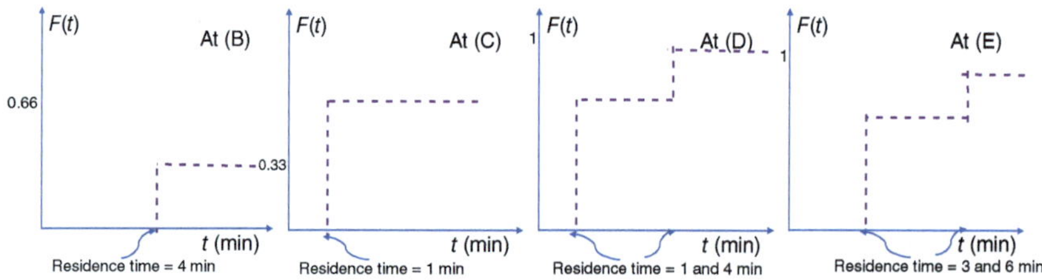

**Problem 1.13**   During the oxidation of phenol to cyclohexanone, assume that the reaction rate is given by $r = k \cdot C_A$ ("A" being phenol) when excess $O_2$ is present. The following parameter values are known:

| | | |
|---|---|---|
| $V = 66\,\mathrm{ml}$ | $Q_0 = 3\,\mathrm{ml/min}$ | $C_{A0} = 0.1\,\mathrm{mol/l}$ |
| $C_A$ (out of the reactor) $= 0.01\,\mathrm{mol/l}$ | $T = 80\,^\circ\mathrm{C}$ | $P = 3\,\mathrm{MPa}$ |

(a) What is the residence time for this reactor? Solve the steady-state PFR material balance. Given the steady-state outlet concentration of A, what is the value of the rate constant $k$?

(b) To perform an RTD step test on the reactor, two inert solvents are used: toluene and cyclohexane. The reactor (PFR) begins with toluene flowing through at steady state. At time $t = 0$, a valve is turned introducing the cyclohexane flow and cutting off the toluene flow. The measured RTD is shown in the figure. What is the mean residence time for the reactor?

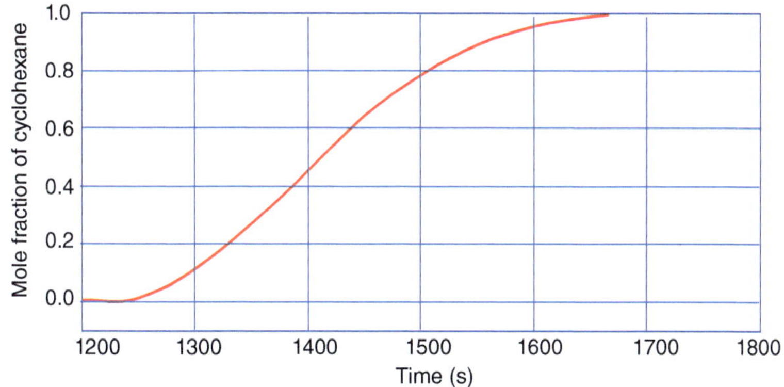


(c) From the data, estimate the dimensionless dispersion number, Bo, describing this reactor.

**Solution to Problem 1.13:**

For details refer the Wiley website at http://www.wiley-vch.de/ISBN9783527354115

(a) Residence time is: $t = V/Q_0 = 66/3 = 22\,\mathrm{min}$

For the PFR in steady state: $dn_A = r_A dV$, which for a first-order reaction gives:

$$C_A = C_{A0} \cdot \exp(-k\bar{t})$$

In the present case:

$$0.01 = 0.1 \cdot \exp(-k \cdot 22)$$

Giving a value of $k = 0.1\,\mathrm{min}^{-1}$

(b) We can read in the plot the time–$F(t)$ data and do the corresponding calculations:

| Time (s) | $F(t)$ | $E(t)$ (s$^{-1}$) | $E(t)\cdot\Delta t$ | $t\cdot E(t)$ | $t\cdot E(t)\cdot\Delta t$ | $(t-t_m)^2\cdot$ $E(t)$ | $(t-t_m)^2\cdot$ $E(t)\cdot\Delta t$ |
|---|---|---|---|---|---|---|---|
| 1200 | 0 | 0.0010 | 0.05 | 1.20 | 61.25 | 101 | 5062.5 |
| 1250 | 0.05 | 0.0010 | 0.05 | 1.25 | 161.25 | 1 | 62.5 |
| 1300 | 0.1 | 0.0040 | 0.2 | 5.20 | 231.25 | 605 | 30 250 |
| 1350 | 0.3 | 0.0030 | 0.15 | 4.05 | 223.75 | 1654 | 82 687.5 |
| 1400 | 0.45 | 0.0035 | 0.35 | 4.90 | 395.00 | 8409 | 840 875 |
| 1500 | 0.8 | 0.0020 | 0.1 | 3.00 | 113.75 | 6503 | 325 125 |
| 1550 | 0.9 | 0.0010 | 0.05 | 1.55 | 58.75 | 4651 | 232 562.5 |
| 1600 | 0.95 | 0.0005 | 0.05 | 0.80 | 40.00 | 6301 | 630 125 |
| 1700 | 1 | 0.0006 | | | | | |
| | | Sum= | 1 | $t_m$ (s)= | 1245 | $\sigma^2$ (s$^2$) | 2 146 750 |
| | | | | $t_m$ (min)= | 20.75 | $\sigma^2$ (min$^2$) | 596.32 |

The value of $E(t)$ is calculated by using a numerical method for deriving the $F(t)$ values read in the graph. In this way:

$$E^t = \frac{F^{t-1} - F^t}{\Delta t}$$

The value of $t_m$ is calculated by integrating (i.e. sum) of the values of $(t\cdot E(t)\cdot\Delta t)$. As we can see, in this system $\bar{t} \neq t_m$. In fact, $\bar{t} > t_m$, so we would expect to get the $E(t)$ signal later. This is probably due to the fact that flow rate is not measured correctly, or that there is a dead volume not contributing to the total residence time in the reactor. In any case, we first try to find the Bo number for applying the dispersion model. Assuming that Bo $< 0.01$, we can do $\sigma^2 = 2 \cdot$ Bo, and we find that Bo $\gg 0.01$, so the assumption is not correct. For Bo $> 0.01$ and assuming o–o conditions:

$$\left(\frac{\sigma}{t_m}\right)^2 = 2 \cdot \text{Bo} + 8 \cdot \text{Bo}^2$$

finding that Bo $= 0.3113$ that is actually quite a high value (Bo $> 0.2$). We can assume that the reactor would be similar to a CSTR. Indeed, if we apply the TIS model, we find that:

$$n_t = \frac{t_m^2}{\sigma^2} = \frac{20.75^2}{596.32} = 0.72 \text{ tanks}$$

**Problem 1.14**   A reaction system is described by the following model:

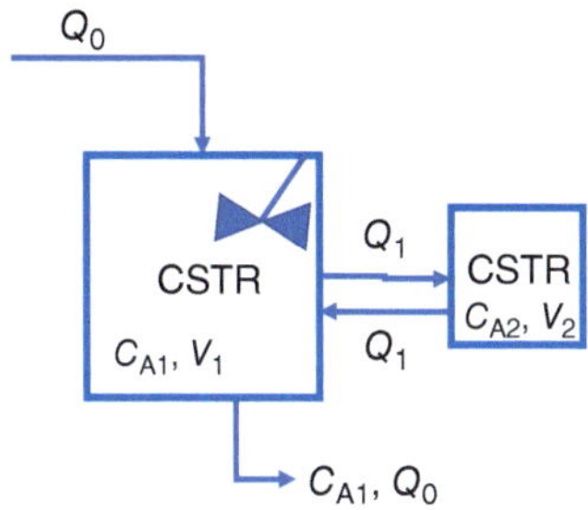


Do a mass balance in this system for a tracer pulse input and determine the residence time distribution function in the case where tank 1 is very small compared to tank 2 and the transfer speed between the reactors is very small.

**Solution to Problem 1.14**

If we propose a molar balance on the tracer, with a pulse injected at $t = 0$ for each of the tanks, we obtain:

$$\text{Accumulation} = \text{Input} - \text{Output} \text{ (there is no tracer generation)}$$

$$V_1 \frac{dC_{T1}}{dt} = Q_1 C_{T2} - (Q_0 C_{T1} + Q_1 C_{T1}) \quad \text{(in the first tank)}$$

$$V_2 \frac{dC_{T2}}{dt} = Q_1 C_{T1} - Q_1 C_{T2} \quad \text{(in the second tank)}$$

where $C_{T1}$ and $C_{T2}$ are, respectively, tracer concentrations in both reactors. These two differential equations are coupled and should be solved simultaneously.

In this model, the two adjustable parameters are the flow rate exchanged ($Q_1$) and the volume of the most agitated region ($V_1$). Remember that the measured volume ($V$) is the sum of $V_1$ and $V_2$. We will call $\beta$ the fraction of the total flow that is transferred between both reactors:

$$Q_1 = \beta \cdot Q_0$$

and $\alpha$ to the fraction of the total volume that corresponds to the most agitated area:

$$V_1 = \alpha \cdot V \rightarrow V_2 = (1 - \alpha) \cdot V$$

On the other hand, the average time ($\bar{t}$) is given by the quotient $V/Q_0$.

The initial conditions ($t = 0$) for this model are: (i) $C_{T1} = (C_{T1})_0$, and (ii) $(C_{T2})_0 = 0$. Analytical solution is possible in the present case and is the following:

$$\left[ \frac{C_{T1}}{(C_{T1})_0} \right]_{\text{pulse}} = \frac{(\alpha m_1 + \beta + 1) \exp\left(\frac{m_2 t}{\bar{t}}\right) - (\alpha m_2 + \beta + 1) \exp\left(\frac{m_1 t}{\bar{t}}\right)}{\alpha (m_1 - m_2)}$$

being:

$$m_1, m_2 = \left[ \frac{1 - \alpha + \beta}{2\alpha(1 - \alpha)} \right] \left[ -1 \pm \sqrt{1 - \frac{4\alpha\beta(1 - \alpha)}{(1 - \alpha + \beta)^2}} \right]$$

Previous equation shows that, if tank 1 is small compared to 2 ($\alpha$ small), and the transfer speed between both reactors is small ($\beta$ small), the second exponential term tends to 1 during the first part of the response to an impulse injection. During the second part, the first exponential term tends to 0. If we represent the logarithm of the tracer concentration versus time, the response curve will tend to a straight line at both ends of the curve, and the parameters will be obtained from the slopes ($m_1$ for $t \to \infty$ and $m_2$ for $t \to 0$) and the cut points of both lines (for $t \to \infty$, the cut point is $-\{\alpha m_2 + \beta + 1\}/\alpha\{m_1 - m_2\}$).

**Problem 1.15**  Figure shows a combination of ideal reactors used to model a real reactor.

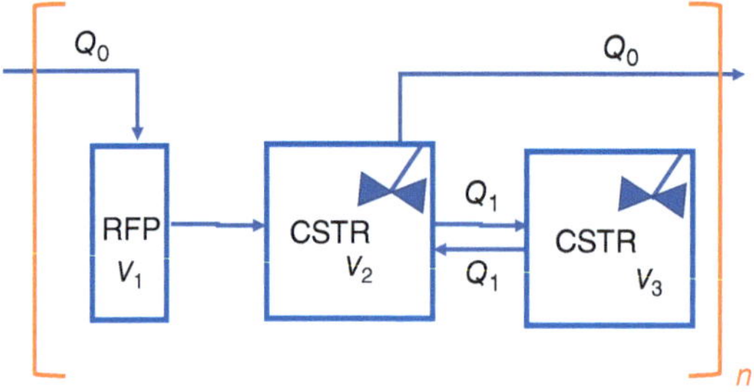



(a) Qualitatively sketch the RTD (i.e. $E(t)$, $F(t)$) that would result if two of the following combinations of ideal reactors were connected in series (i.e. $n = 1$).
(b) How would your results change if two were connected in series (i.e. $n = 2$)?
(c) How about $n = 5$ or 10?

**Solution to Problem 1.15**
For details refer the Wiley website at http://www.wiley-vch.de/ISBN9783527354115

For $n = 1$, we have a signal corresponding to one CSTR with interchange displaced by a value of time corresponding to the RFP residence time. If $n = 2$, the signal is displaced two times:

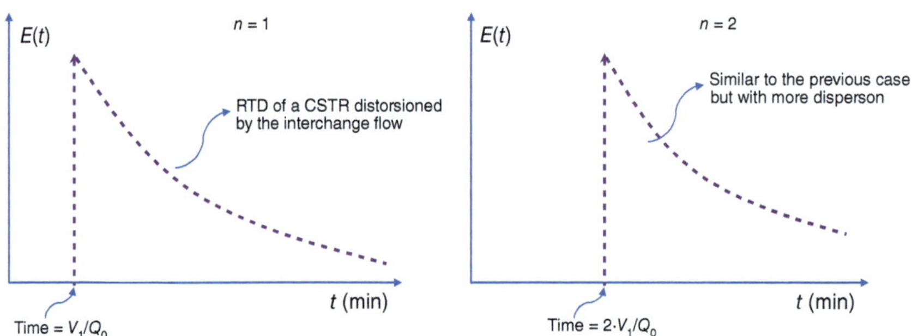

For values of "$n$" higher, the system would behave as a PFR, with residence time $n$ times the residence time of the first reactor.

In Problem 3.2, we will see that the transfer function of this system of reactors is ($n = 1$):

$$E(s) = \frac{(\exp(-\overline{t_1}s)) \cdot \left(\frac{1}{1+\overline{t_2}s}\right)}{1+\beta-\beta\left(\frac{1}{1+\overline{t_3}s}\right)\left(\frac{1}{1+\overline{t_2}s}\right)}$$

being $\beta$ the ratio $Q_1/Q_0$. For a system where $n = 2$:

$$E(s) = \left[\frac{(\exp(-\overline{t_1}s)) \cdot \left(\frac{1}{1+\overline{t_2}s}\right)}{1+\beta-\beta\left(\frac{1}{1+\overline{t_3}s}\right)\left(\frac{1}{1+\overline{t_2}s}\right)}\right]^2$$

The inverse of the Laplace transform is a very complicated function $E(t)$.

We can do a simulation using Matlab®. For a system consisting of two CSTRs with flow rate $Q_1$ between them, expression for the pulse run concentration is that shown in the previous problem, but now $\alpha$ is the ratio $V_2/(V_2 + V_3)$.

For example, for $n = 1$–4, with $V_1 = 0.9\,V$ and $Q_1 = 0.1 \cdot Q_0$, $\overline{t_{RFP}} = 10$ and $\overline{t_{CSTR}} = 20$, we can do:

```
clear all
close all

alfa=0.9;
beta=0.9;

tm_pfr=10;
tm_cstrs=20;

m_1=[(1-alfa+beta)/(2*alfa*(1-alfa))]*[-1+((1-(4*alfa*beta*(1-alfa))/
    (1-alfa+beta)^2))^0.5];
m_2=[(1-alfa+beta)/(2*alfa*(1-alfa))]*[-1-((1-(4*alfa*beta*(1-alfa))/
    (1-alfa+beta)^2))^0.5];

t=0:150;

td=(t-tm_pfr)/tm_cstrs;

E1=((alfa*m_1+beta+1)*exp(m_2*td)-(alfa*m_2+beta+1)*exp(m_1*td))/
    alfa/(m_1-m_2);
E1=(t>tm_pfr).*E1;
subplot(2,2,1), plot(t,E1)
area=sum(E1*1);
E1=E1/area;

tmdistrib=sum(E1.*t*1)

title('n=1')
xlabel('Time')
ylabel('E(t)')
%%%%
```

```
E2=conv(E1,E1);
t2=2*t(1):2*t(end);
area=sum(E2*1);
E2=E2/area;

tmdistrib=sum(E2.*t2*1)

subplot(2,2,2), plot(t2,E2)

title('n=2')
xlabel('Time')
ylabel('E(t)')

%%%%
E3=conv(E2,E1);
t3=3*t(1):3*t(end);
area=sum(E3*1);
E3=E3/area;

tmdistrib=sum(E3.*t3*1)

subplot(2,2,3), plot(t3,E3)

title('n=3')
xlabel('Time')
ylabel('E(t)')

%%%%
E4=conv(E3,E1);
t4=4*t(1):4*t(end);
area=sum(E4*1);
E4=E4/area;

tmdistrib=sum(E4.*t4*1)

subplot(2,2,4), plot(t4,E4)

title('n=4')
xlabel('Time')
ylabel('E(t)')
```

Getting:

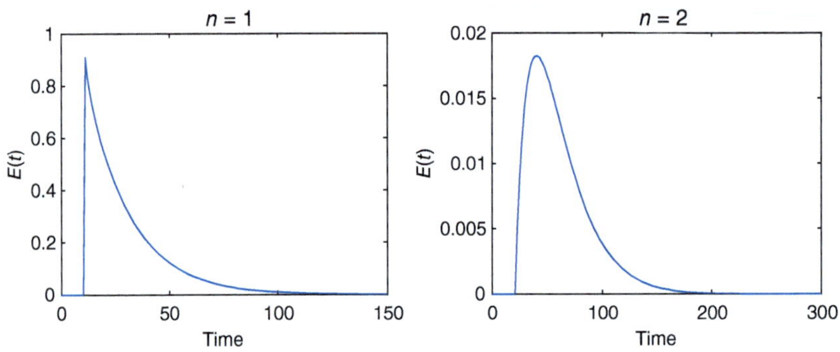

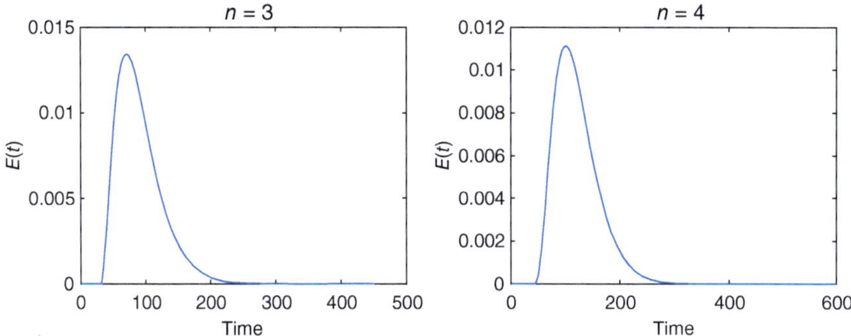

During the calculation, we can check that, for $n = 1$, $t_m = \overline{t_{\text{PFR}}} + \overline{t_{\text{CSTR}}} = 30$.

For $n = 2$, $t_m = 2 \cdot (\overline{t_{\text{PFR}}} + \overline{t_{\text{CSTR}}}) = 60$

For $n = 3$, $t_m = 3 \cdot (\overline{t_{\text{PFR}}} + \overline{t_{\text{CSTR}}}) = 90$

For $n = 4$, $t_m = (\overline{t_{\text{PFR}}} + \overline{t_{\text{CSTR}}}) = 120$

And comparing all four distributions:

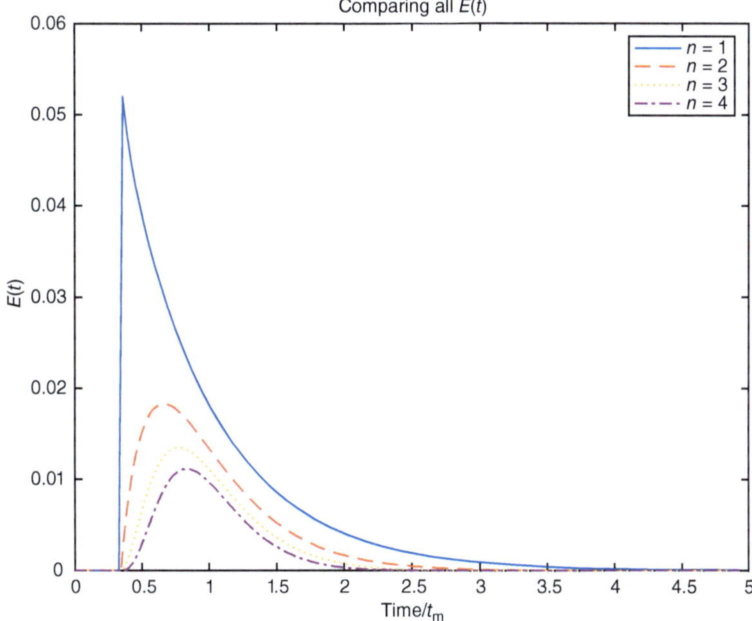

**Problem 1.16** A tracer step input experiment is being performed, feeding pure water, switching to salt water, and analyzing the conductivity at the outlet. The data represented in the figure was obtained.

(a) What will the output signal be like if the tracer had been injected into pulse?
(b) Calculate the mean residence time of the reactor.

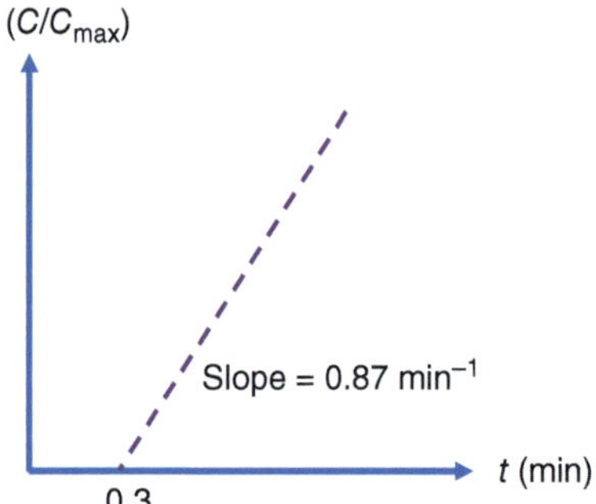



**Solution to Problem 1.16**

(a) What we have obtained is a cumulative RTD (i.e. $F(t)$ curve), and it is easy to find the time that the tracer will exit with $C_{\max}$:

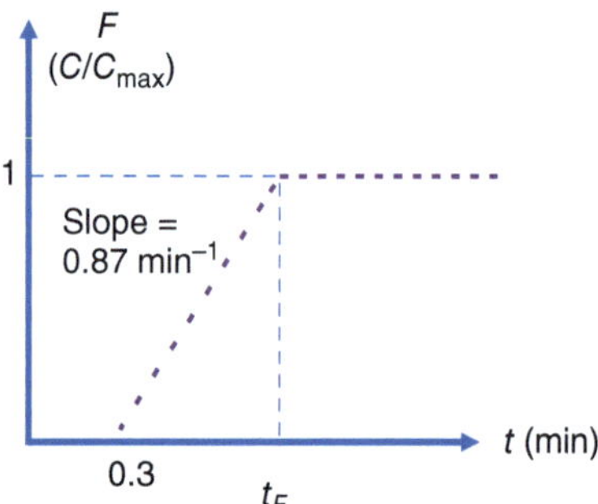



$$F = at + b$$

$$0 = 0.87 \cdot 0.3 + b$$

$$1 = 0.87t + b = 0.87t + (-0.87 \cdot 0.3)$$

$$F = 0.87t - 0.91$$

$$t_F = \frac{1 + 0.261}{0.87} = 1.45 \text{ min}$$

(b) The mean residence time is:

$$t_m = \int_0^\infty t \cdot E(t)\mathrm{d}t$$

And:

$$E(t) = \frac{\mathrm{d}}{\mathrm{d}t}\left[\frac{C(t)}{C_0}\right]_{\text{step input}} = \frac{\mathrm{d}F}{\mathrm{d}t} = \frac{\mathrm{d}(0.87t - 0.91)}{\mathrm{d}t} = 0.87 \text{ min}^{-1}$$

Mean residence time:

$$t_m = \int_0^\infty t \cdot E(t)dt = \int_{0.3}^{t_F} t \cdot 0.87dt = \frac{0.87}{2}(1.45^2 - 0.3^2) = 0.8754 \text{ min}$$

The $E(t)$ function is:

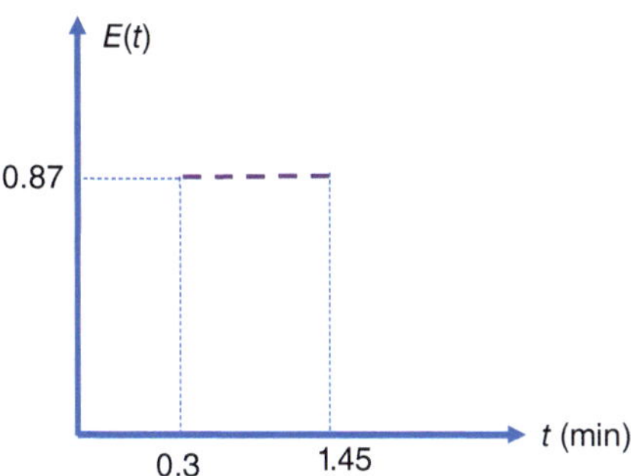


We can check that:

$$\int_0^\infty E(t)dt = \int_{0.3}^{1.45} 0.87dt = 0.87(1.45 - 0.3) = 1$$

**Problem 1.17**   A test is performed with a tracer pulse, and a curve is obtained that follows the expression $C = (t-2)^2$ for $0 \le t \le 2$ and $C = 0$ at any other time. Calculate the mean residence time using the equation for $E(t)$ and $F(t)$.

**Solution to Problem 1.17**

In the reactor, we have a pulse that input gives:

$$C(t) = (t-2)^2$$

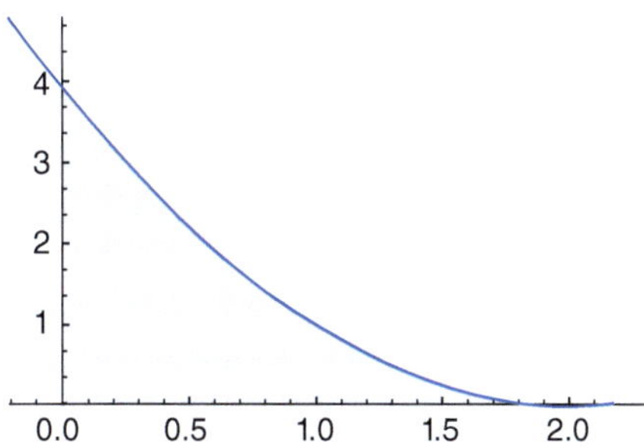


The area under the curve is defined between 0 and 2 (time where $C(t)$ equals to zero). And in this way:

$$\int_0^\infty C(t)dt = \int_0^2 (t-2)^2 dt = 2.6667$$

And so:

$$E(t) = \frac{(t-2)^2}{2.6667}$$

Following the definition of $F(t)$:

$$F(t) = \int_0^t E(t)dt = \int_0^t \frac{(t-2)^2}{2.6667}dt = 0.124\,984t^3 - 0.749\,906t^2 + 1.499\,81t + K$$

being "$K$" an integration constant. By definition, we know that:

$$\lim_{t \to \infty} F(t) = 1 = \lim_{t \to 2} F(t)$$

so the constant "$K$" should be equal to zero.

**Problem 1.18**  A reaction system is described by the following model:

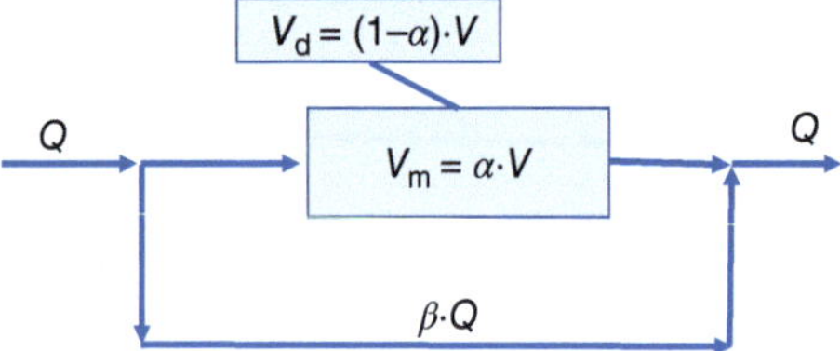


At a given moment, the system presents values of $\alpha = 0.1$ and $\beta = 0.9$. A series of reforms are carried out in the system, and it is achieved that $\alpha = 0.9$ and $\beta = 0.1$.

(a) Qualitatively draw the response of the system to a tracer impulse, comparing the two situations.
(b) Also, draw the corresponding $F(t)$ curves for both situations.
(c) Indicate what reforms had to be made to achieve this change.

**Solution to Problem 1.18**

(a) In this system, the residence time compared to that expected is:

$$\overline{t_{actual}} = \frac{V_m}{Q_{reactor}} = \frac{\alpha V}{(1-\beta)Q} = \frac{\alpha}{(1-\beta)}\overline{t_{expected}}$$

As we have that $\alpha = 0.1$ and $\beta = 0.9$ at the beginning:

$$\overline{t_{actual}} = \frac{0.1}{(1-0.9)}\overline{t_{expected}} = \overline{t_{expected}}$$

In the second case:

$$\overline{t_{actual}} = \frac{0.9}{(1-0.1)}\overline{t_{expected}} = \overline{t_{expected}}$$

So it will be very difficult to observe these changes. Nevertheless, the bypass produces a very important peak at time equal to zero in both situations, but the amount of tracer passing through the bypass is much higher in the first case. We can have something similar to:

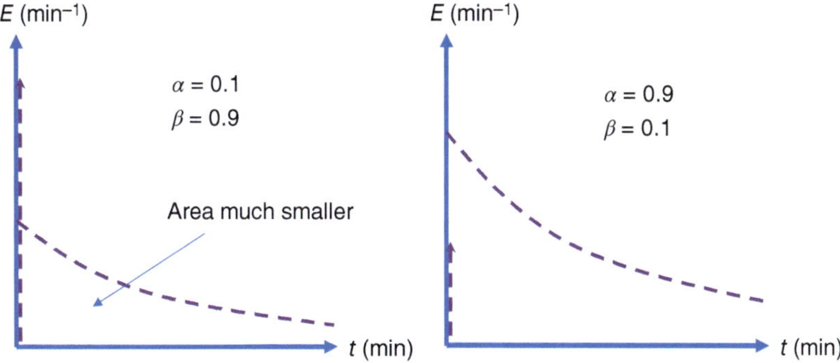


(b)  The cumulative time function would have the following aspect:

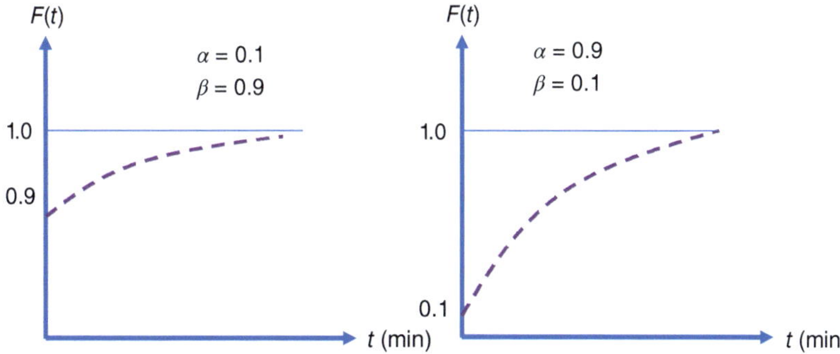


(c)  Usually, a better agitation is procured to eliminate dead volumes. Also, reduction of bypass flow is achieved with better agitation or by changing the reactor filling.

**Problem 1.19**   In a tracer input experiment, a perfect pulse is injected, and the following distribution is obtained:

(a)  Calculate the mean residence time of the reactor.
(b)  What will the output signal be like if the tracer has been injected in step?

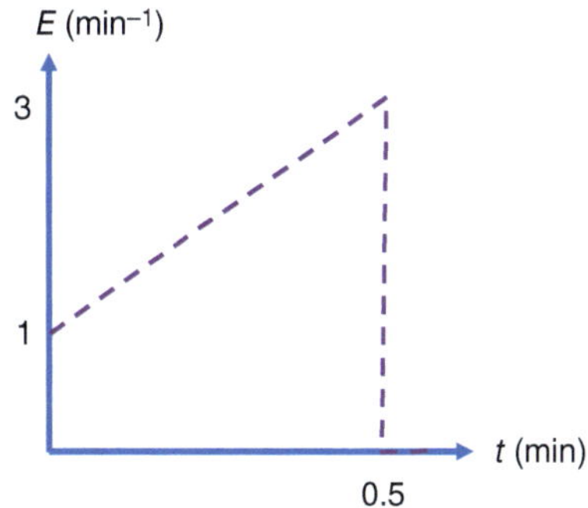

**Solution to Problem 1.19**

(a) In the signal obtained, the total area is $1 \cdot 0.5 + (0.5 \cdot 2/2) = 1$ (just checking).

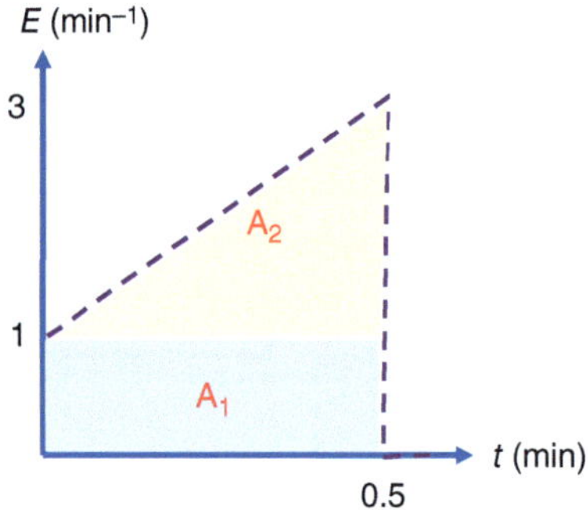


The straight-line connecting points $(0, 1)$ and $(0.5, 3)$ is $E(t) = 1 + 4t$, so:

$$t_m = \int_0^\infty t \cdot E(t)\mathrm{d}t = \int_0^{0.5} t \cdot (1 + 4t)\mathrm{d}t = 0.2917 \text{ min}$$

(b) If a step tracer run is done, we will get:

$$F(t) = \int_0^t E(t)\mathrm{d}t = \int_0^t (1 + 4t)\mathrm{d}t = 2t^2 + t$$

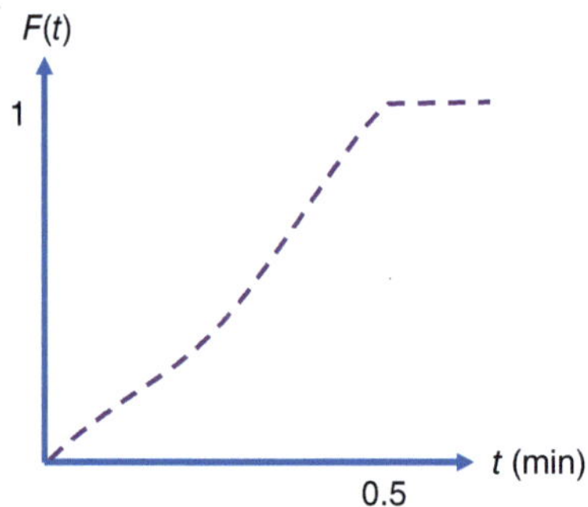


**Problem 1.20** A commercial-scale distillation tray is being studied for residence time using a fiber optic technique. A pulse of a 10 g/l solution of Rhodamine-B dye was injected into the downcomer of the top tray. The response data, summarizing the time (min) and corresponding output voltage (V), are provided as follows:

| Time (min) | Output voltage (V) |
| --- | --- |
| 0.0 | 0.00 |
| 0.1 | 0.00 |
| 0.2 | 2.80 |
| 0.3 | 4.48 |

| Time (min) | Output voltage (V) |
|---|---|
| 0.4 | 3.32 |
| 0.5 | 1.70 |
| 0.6 | 0.84 |
| 0.7 | 0.39 |
| 0.8 | 0.18 |
| 0.9 | 0.11 |
| 1.0 | 0.07 |
| 1.1 | 0.03 |
| 1.2 | 0.04 |
| 1.3 | 0.01 |
| 1.4 | 0.01 |
| 1.5 | 0.01 |
| 1.6 | 0.00 |
| 1.7 | 0.01 |
| 1.8 | 0.00 |
| 1.9 | 0.00 |

Note: The output voltage is proportional to the concentration of the dye.

(a) Determine the mean residence time and the characteristic $F(t)$ curve for this system.
(b) Use the dispersion model and the variance method to determine Bo (Bodenstein number) and Pe (Peclet number).
(c) Estimate the number of equal-sized CSTRs in series that would exhibit a comparable dispersion to the observed experimental data.
(d) Provide comments on the obtained results.

**Solution to Problem 1.20**

For details refer the Wiley website at http://www.wiley-vch.de/ISBN9783527354115

(a) Using similar techniques to those mentioned in other problems, we easily find:

| $C(t)$ | $C(t)\cdot\Delta t$ | $E(t)$ | $E(t)\cdot\Delta t$ | $F(t)$ | $t\cdot E(t)$ | $t\cdot E(t)\cdot\Delta t$ | $(t-t_m)^2\cdot E(t)$ | $(t-t_m)^2\cdot E(t)\cdot\Delta t$ |
|---|---|---|---|---|---|---|---|---|
| 0.000 | 0.000 | 0.000 | 0.000 | 0.000 | 0.000 | 0.000 | 0.000 | 0.000 |
| 0.000 | 0.140 | 0.000 | 0.100 | 0.000 | 0.000 | 0.000 | 0.000 | 0.000 |
| 2.800 | 0.364 | 2.000 | 0.260 | 0.200 | 0.400 | 0.020 | 0.064 | 0.003 |
| 4.480 | 0.390 | 3.200 | 0.279 | 0.520 | 0.960 | 0.068 | 0.020 | 0.004 |
| 3.320 | 0.251 | 2.371 | 0.179 | 0.757 | 0.949 | 0.095 | 0.001 | 0.001 |
| 1.700 | 0.127 | 1.214 | 0.091 | 0.879 | 0.607 | 0.078 | 0.018 | 0.001 |

| $C(t)$ | $C(t)\cdot\Delta t$ | $E(t)$ | $E(t)\cdot\Delta t$ | $F(t)$ | $t\cdot E(t)$ | $t\cdot E(t)\cdot\Delta t$ | $(t-t_m)^2\cdot E(t)$ | $(t-t_m)^2\cdot E(t)\cdot\Delta t$ |
|---|---|---|---|---|---|---|---|---|
| 0.840 | 0.062 | 0.600 | 0.044 | 0.939 | 0.360 | 0.048 | 0.029 | 0.002 |
| 0.390 | 0.029 | 0.279 | 0.020 | 0.966 | 0.195 | 0.028 | 0.029 | 0.003 |
| 0.180 | 0.015 | 0.129 | 0.010 | 0.979 | 0.103 | 0.015 | 0.023 | 0.003 |
| 0.110 | 0.009 | 0.079 | 0.006 | 0.987 | 0.071 | 0.009 | 0.021 | 0.002 |
| 0.070 | 0.005 | 0.050 | 0.004 | 0.992 | 0.050 | 0.006 | 0.019 | 0.002 |
| 0.030 | 0.004 | 0.021 | 0.003 | 0.994 | 0.024 | 0.004 | 0.011 | 0.002 |
| 0.040 | 0.003 | 0.029 | 0.002 | 0.997 | 0.034 | 0.003 | 0.019 | 0.002 |
| 0.010 | 0.001 | 0.007 | 0.001 | 0.998 | 0.009 | 0.002 | 0.006 | 0.001 |
| 0.010 | 0.001 | 0.007 | 0.001 | 0.999 | 0.010 | 0.001 | 0.007 | 0.001 |
| 0.010 | 0.001 | 0.007 | 0.000 | 0.999 | 0.011 | 0.001 | 0.009 | 0.001 |
| 0.000 | 0.000 | 0.000 | 0.000 | 0.999 | 0.000 | 0.001 | 0.000 | 0.000 |
| 0.010 | 0.001 | 0.007 | 0.000 | 1.000 | 0.012 | 0.001 | 0.012 | 0.001 |
| 0.000 | 0.000 | 0.000 | 0.000 | 1.000 | 0.000 | 0.001 | 0.000 | 0.001 |
| 0.000 | | 0.000 | | 1.000 | 0.000 | 0.000 | 0.000 | 0.000 |

$\Sigma(C(t)\mathrm{d}t) = 1.4 \qquad \Sigma(E(t)\mathrm{d}t) = 1.0 \qquad\qquad t_m = 0.38\ \mathrm{min} \qquad \sigma^2 = 0.029\ \mathrm{min}^2$

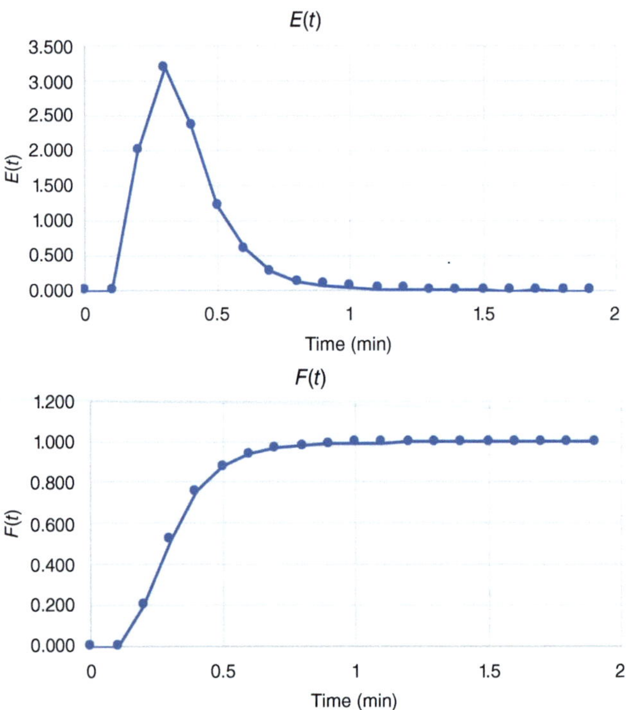

(b) For the dispersion model to be applied, first we should assume $Bo < 0.01$ and then:

$$\sigma^2 = 2 \cdot Bo$$

We obtain $Bo = 0.014$, so this assumption is almost valid. We can do now:

$$Pe = \frac{1}{Bo} = 69.05$$

(c) With the values of $t_m$ and $\sigma^2$, using the TIS model:

$$n_t = \frac{t_m^2}{\sigma^2} = \frac{0.38^2}{0.029} = 4.98 \text{ tanks} \approx 5 \text{ tanks}$$

(d) This reactor has a very small dispersion, and so the number of tanks that should be used in the TIS model is quite high. The system is completely solvable and its RTD is:

$$E = \frac{1}{\bar{t} \cdot \sqrt{4 \cdot \pi \cdot Bo}} \cdot \exp\left[ -\frac{\left(1 - \frac{t}{\bar{t}}\right)^2}{4 \cdot Bo} \right]$$

We can compare the experimental and estimated $E(t)$ curves using the spread-sheet:

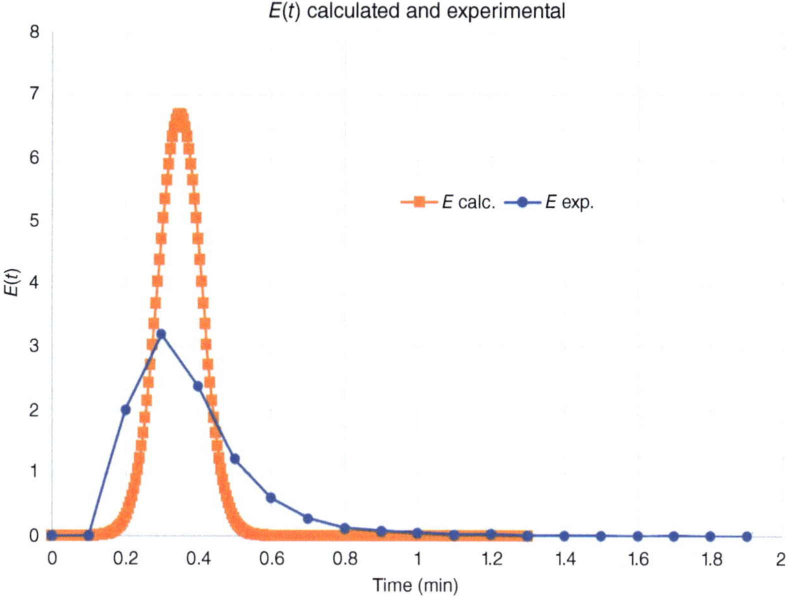


**Problem 1.21**   A common issue encountered in the development of gas and liquid phase chromatographic analysis methods is the phenomenon of "tailing." Tailing makes it challenging to characterize the retention time of a specific species due to the prolonged tail associated with the peak. An approximate representation of tailing

can be described by the time dependence of the effluent concentration of the species as follows:

$$C_A = 0 \qquad \text{for } 0 < t < 8.5\,\text{min}$$

$$C_A = 10^5\,e^{-2t} \quad \text{for } t > 8.5\,\text{min}$$

for $C_A$ in pmol/ml. The curve is generated by injecting a pulse of a sample containing species A into a solvent flowing through the chromatographic column at a constant flow rate.

(a) Determine the mathematical form of the $F(t)$ curve corresponding to the response of the pulse test.
(b) Use the $F(t)$ curve to calculate the average residence time of species A in the column.

**Solution to Problem 1.21**

For details refer the Wiley website at http://www.wiley-vch.de/ISBN9783527354115

(a) The total area of tracer is:

$$\int_0^\infty C(t)dt = \int_0^\infty 10^5 \cdot \exp(-2t)dt = 2.01 \cdot 10^{-3}$$

So, for $t > 8.5$:

$$E(t) = \frac{C(t)}{\int_0^\infty C(t)dt} = 4.83 \cdot 10^7 \cdot \exp(-2t)$$

We can check that:

$$\int_0^\infty E(t)dt = 1$$

In this way:

$$F(t) = \int_0^t E(t)dt = \int_0^t (4.83 \cdot 10^7 \cdot \exp(-2t))dt = 1 - 2.415 \cdot 10^7 \exp(-2t)$$

(b)

$$t_m = \int_0^\infty t \cdot E(t)dt = \int_{8.5}^\infty t \cdot (4.83 \cdot 10^7 \cdot \exp(-2t))dt = 8.89\,\text{min}$$

We can use a numerical method too. In the spreadsheet corresponding to this problem, values of time are given and $C_A$ is calculated. Following the procedure shown before in other problems:

$$\Sigma(C(t)dt) = 0.001\,204\,66$$

$$\Sigma(E(t)dt) = 1.0$$

$$t_m = 9.29\,\text{min}$$

$$\sigma^2 = 0.229\,\text{min}^2$$

Doing the graphical plot:

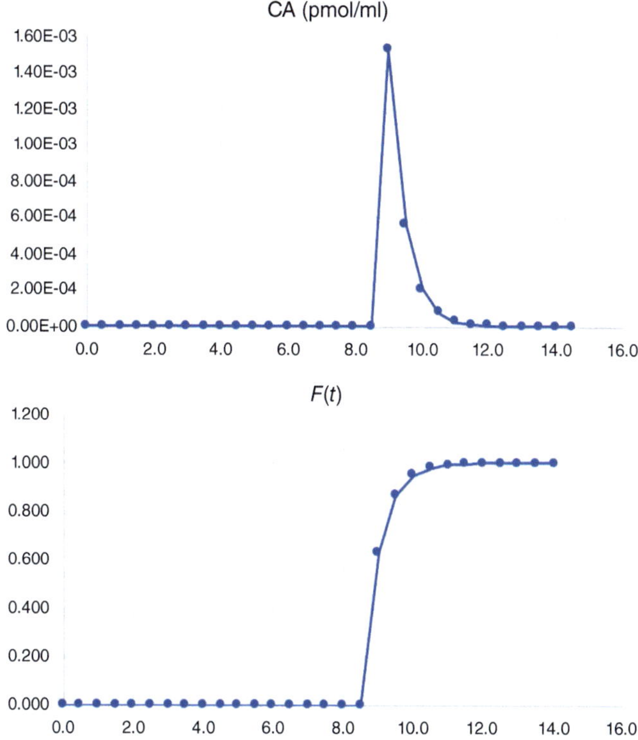


**Problem 1.22**  The techniques discussed in this chapter for analyzing residence time distribution functions can be applied to study flow conditions in streams or rivers, particularly when assessing the dispersion of pollutants from a source. The following data was obtained from a study of the South Platte River. The average flow rate is $15.68\,\text{m}^3/\text{s}$, the length of the reach is $6065\,\text{m}$, and the natural concentration of $K^+$ ions in the stream is $8.2\,\text{mg/l}$.

At time zero, $453.5\,\text{kg}$ of $K_2CO_3$ was dumped into the upstream end of the reach. Samples were periodically collected at the downstream end of the reach and analyzed for $K^+$ ions. The results are presented in the table below.

| $t$ (min) | $K^+$ at downstream (g/m$^3$) |
| --- | --- |
| 0.0 | 8.2 |
| 60.0 | 8.2 |
| 75.0 | 8.2 |
| 90.0 | 8.2 |
| 105.0 | 8.4 |
| 120.0 | 9.6 |
| 130.0 | 13.6 |

| $t$ (min) | $K^+$ at downstream (g/m³) |
| --- | --- |
| 132.5 | 14.8 |
| 134.0 | 14.8 |
| 138.0 | 14.6 |
| 142.5 | 13.2 |
| 150.0 | 12.8 |
| 165.0 | 10 |
| 180.0 | 9.2 |
| 195.0 | 8.2 |
| 210.0 | 8.2 |
| 300.0 | 8.2 |

(a) Calculate the fraction of the tracer ($K_2CO_3$) that was recovered.

(b) Based on the recovered amount of tracer, calculate the $F(t)$ curve. Additionally, compute the average residence time and plot the effluent concentration of $K^+$ ions exceeding the background level.

**Solution to Problem 1.22**

For details refer the Wiley website at http://www.wiley-vch.de/ISBN9783527354115

(a) Subtracting the natural $K^+$ level, we find:

| $t$ (min) | $C(t)$ | $C(t)\cdot\Delta t$ | $E(t)$ | $E(t)\cdot\Delta t$ | $F(t)$ | $t\cdot E(t)\cdot\Delta t$ | $(t-t_m)^2\cdot$ $E(t)\cdot\Delta t$ |
| --- | --- | --- | --- | --- | --- | --- | --- |
| 0.0 | 0 | 0.00 | 0.000 | 0.000 | 0.000 | 0.000 | 0.000 |
| 60.0 | 0 | 0.00 | 0.000 | 0.000 | 0.000 | 0.000 | 0.000 |
| 75.0 | 0 | 0.00 | 0.000 | 0.000 | 0.000 | 0.000 | 0.000 |
| 90.0 | 0 | 1.50 | 0.000 | 0.006 | 0.000 | 0.000 | 0.000 |
| 105.0 | 0.2 | 12.00 | 0.001 | 0.051 | 0.013 | 0.666 | 9.368 |
| 120.0 | 1.4 | 34.00 | 0.006 | 0.144 | 0.101 | 5.992 | 33.747 |
| 130.0 | 5.4 | 15.00 | 0.023 | 0.063 | 0.330 | 18.389 | 36.857 |
| 132.5 | 6.6 | 9.90 | 0.028 | 0.042 | 0.399 | 8.331 | 9.322 |
| 134.0 | 6.6 | 26.00 | 0.028 | 0.110 | 0.441 | 5.577 | 4.366 |
| 138.0 | 6.4 | 25.65 | 0.027 | 0.108 | 0.550 | 14.945 | 6.567 |
| 142.5 | 5.0 | 36.00 | 0.021 | 0.152 | 0.645 | 15.178 | 1.840 |
| 150.0 | 4.6 | 48.00 | 0.019 | 0.203 | 0.791 | 22.234 | 3.212 |
| 165.0 | 1.8 | 21.00 | 0.008 | 0.089 | 0.905 | 31.294 | 32.824 |
| 180.0 | 1 | 7.50 | 0.004 | 0.032 | 0.968 | 15.124 | 68.928 |
| 195.0 | 0 | 0.00 | 0.000 | 0.000 | 0.968 | 5.707 | 42.389 |
| 210.0 | 0 | 0.00 | 0.000 | 0.000 | 0.968 | 0.000 | 0.000 |
| 300.0 | 0 | | 0.000 | | 0.968 | 0.000 | 0.000 |

$$\Sigma(C(t)dt) = 236.55$$

$$\Sigma(E(t)dt) = 1.0$$

$$t_m = 143.44 \text{ min}$$

$$\sigma^2 = 249.42 \text{ min}^2$$

We can see that:

$$\sum(C(t)dt) = 236.55 \frac{\text{g} \cdot \text{min}}{\text{m}^3}$$

Taking into account that the flow rate is $15.68 \text{ m}^3/\text{s}$, we have that the amount recovered is:

$$M = 236.55 \text{ g} \cdot \frac{\text{min}}{\text{m}^3} \cdot 15.68 \frac{\text{m}^3}{\text{s}} \cdot 60 \frac{\text{s}}{\text{min}} = 222\,546.24 \text{ g} = 222.5 \text{ kg}$$

This means that only 49% of the $K^+$ dumped is recovered.

(b)

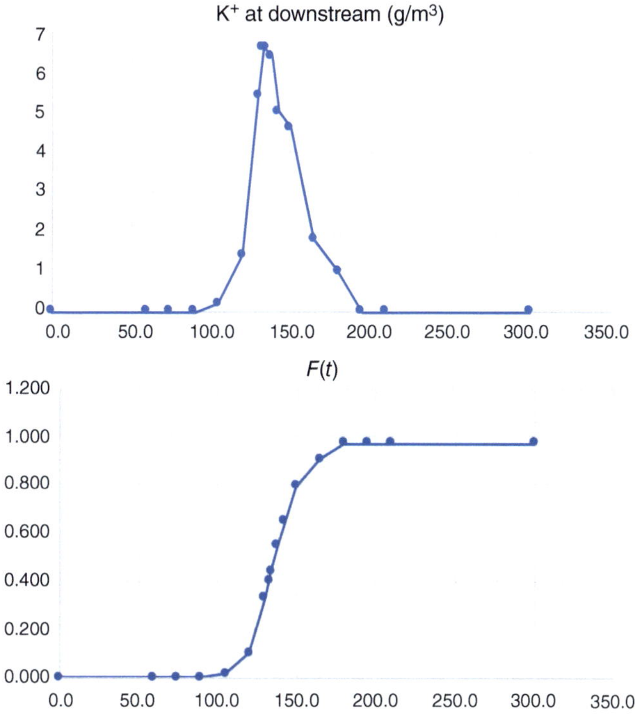


**Problem 1.23**  By injecting brine into an experimental continuous reactor, a signal is obtained at the exit given by:

| $t$ (min) | 1 | 3 | 5 | 7 | 9 | 11 | 13 |
|---|---|---|---|---|---|---|---|
| $C$ (g/l) | 0 | 2 | 3 | 4 | 2 | 1 | 0 |

The flow rate at the inlet is $10\,\text{l/min}$.

(a) How much salt was injected?
(b) Calculate $E(t)$, $F(t)$, mean residence time, and variance.

**Solution to Problem 1.23**

For details refer the Wiley website at http://www.wiley-vch.de/ISBN9783527354115

To solve this problem, we should use a spreadsheet. First, data given in the statement is introduced, and then we should do the following calculations:

| $t$ (min) | $C(t)$ | $C(t){\cdot}\Delta t$ | $E(t)$ | $F(t)$ | $t{\cdot}E(t){\cdot}\Delta t$ | $(t-t_m)^2{\cdot}$ $E(t)$ | $(t-t_m)^2{\cdot}$ $E(t){\cdot}\Delta t$ |
|---|---|---|---|---|---|---|---|
| 1 | 0.00 | 0.00 | 0.0000 | 0.0000 | 0 | 0 | 0 |
| 3 | 2.00 | 4.00 | 0.0833 | 0.1667 | 0.5000 | 1.0208 | 2.0417 |
| 5 | 3.00 | 6.00 | 0.1250 | 0.4167 | 1.2500 | 0.2813 | 0.5625 |
| 7 | 4.00 | 8.00 | 0.1667 | 0.7500 | 2.3333 | 0.0417 | 0.0833 |
| 9 | 2.00 | 4.00 | 0.0833 | 0.9167 | 1.5000 | 0.5208 | 1.0417 |
| 11 | 1.00 | 2.00 | 0.0417 | 1.0000 | 0.9167 | 0.8438 | 1.6875 |
| 13 | 0.00 | 0.00 | 0.0000 | 1.0000 | 0.0000 | 0.0000 | 0.0000 |
| | $\Sigma(C(t)\mathrm{d}t)=24.00$ | | | | $t_m=6.50\,\text{min}$ | | $\sigma^2=5.42\,\text{min}^2$ |

The total area of the $C(t)$ curve is:

$$\text{Area} = \int_0^\infty C(t)\mathrm{d}t = \sum_{\text{all points}} C(t)\Delta t = 24\,\frac{\text{g}}{\text{l}}\cdot\text{min}$$

The total amount of tracer introduced is:

$$M_0 = Q_0\int_0^\infty C(t)\mathrm{d}t = 10\,\frac{\text{l}}{\text{min}}\cdot 24\,\frac{\text{g}\cdot\text{min}}{\text{l}} = 240\,\text{g}$$

For calculating the RTD:

$$E(t) = \frac{C(t)}{\int_0^\infty C(t)\cdot\mathrm{d}t} = \frac{C(t)}{24}$$

And:

$$t_m = \int_0^\infty t\cdot E(t)\mathrm{d}t = \sum_{\text{all points}} t\cdot E(t)\cdot\Delta t = 6.5\,\text{min}$$

In a similar way, variance is calculated:

$$\sigma^2 = \sum_{\text{all points}} (t-t_m)^2\cdot E(t)\cdot\Delta t = 5.41\,\text{min}^2$$

Finally, we can calculate $F(t)$ by attending to its definition:

$$F(t) = \sum_{\text{all points at time}<t} E(t)\cdot\Delta t$$

We can check the form of the graphs showing the distributions:

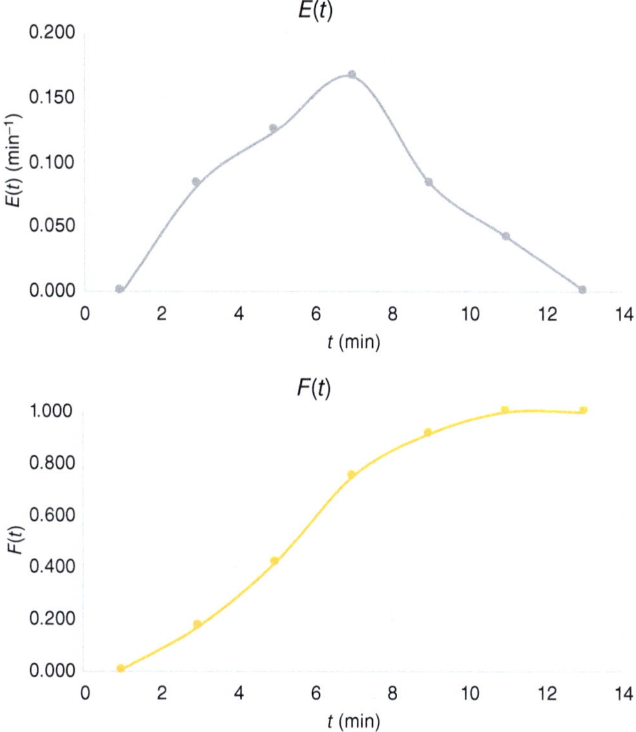

# 2

# Chemical Reaction in Non-ideal Reactors

## Summary of Most Important Models

### Calculation of Conversion

#### Tanks-in-series (TIS) Model and Chemical Reaction
First-order: reaction:

$$X_{\mathrm{A}} = 1 - \frac{1}{(1 + \bar{t}_i \cdot k)^{n_t}}$$

Other kinetics: sequential molar balances must be made in each reactor.

#### Dispersion Model and Chemical Reaction
$n$-th order: kinetics:

$$1 - X_{\mathrm{A}} = \frac{4a \cdot \exp\left(\frac{1}{2\mathrm{Bo}}\right)}{(1 + a)^2 \cdot \exp\left(\frac{a}{2\mathrm{Bo}}\right) - (1 - a)^2 \cdot \exp\left(-\frac{a}{2\mathrm{Bo}}\right)}$$

$$a = [1 + 4 \cdot \mathrm{Da} \cdot \mathrm{Bo}]^{1/2}$$

$$\mathrm{Da} = k \cdot \bar{t} \cdot C_{\mathrm{A0}}^{n-1}$$

Conversion can also be calculated graphically, with the aid, for example, of the following graphs (please consult spreadsheet "dispersion model conversion.xlsx" for more details):

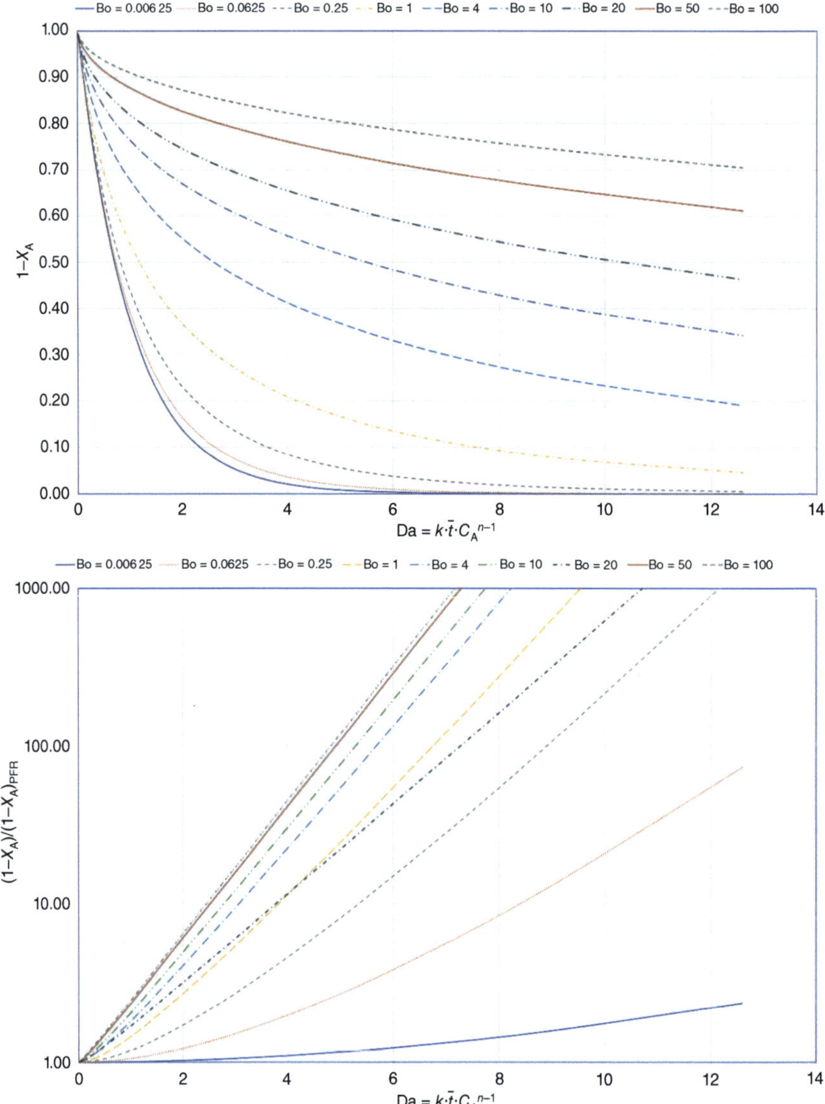


### From RTD Runs

*First-order kinetics:* the information of $E(t)$ is enough to calculate the conversion, e.g., using the segregation model:

$$\overline{X_A} = \int_0^\infty X_A(t)E(t)\mathrm{d}t$$

*Other kinetics:* A model of the reactor is needed to estimate the conversion:

o Dispersion model
o Tanks-in-series model
o Combination of ideal reactors: Once the model is fitted, the corresponding mass balances will be applied to calculate actual conversion.

## Mass Balance in Ideal Reactors

In the PFR, the mass balance is:

$$\mathrm{d}n_A = r_A \mathrm{d}V$$

$$Q_0 \mathrm{d}C_A = r_A \mathrm{d}V$$

For first-order kinetics:

$$Q_0 \mathrm{d}C_A = -kC_A \mathrm{d}V$$

$$\int_{C_{A0}}^{C_A} \frac{\mathrm{d}C_A}{C_A} = -k \int_0^V \frac{\mathrm{d}V}{Q_0}$$

$$\ln\left(\frac{C_A}{C_{A0}}\right) = -k\frac{V}{Q_0} = -k\bar{t}$$

$$C_A = C_{A0} \cdot \exp(-k\bar{t})$$

And equally:

$$X_A = 1 - \exp(-k\bar{t})$$

If the reaction is second-order:

$$Q_0 \mathrm{d}C_A = -kC_A^2 \mathrm{d}V$$

$$-C_{A0}Q_0 \mathrm{d}X_A = -kC_{A0}^2(1 - X_A)^2 \mathrm{d}V$$

$$X_A = \frac{kC_{A0}\bar{t}}{1 + kC_{A0}\bar{t}}$$

In the CSTR:

$$n_{A0} + (r_A)V = n_A$$

$$Q_0 C_{A0} - (kC_A)V = Q_0 C_A$$

$$C_{A0} - (kC_A)\bar{t} = C_A$$

$$C_A = \frac{C_{A0}}{1 + k\bar{t}}$$

And:

$$X_A = 1 - \frac{1}{1 + k\bar{t}} = \frac{k\bar{t}}{1 + k\bar{t}}$$

In a batch reactor:

$$\text{Input} + \text{Generation} = \text{Output} + \text{Accumulation}$$

$$0 + (r_A) \cdot V = 0 + \frac{\mathrm{d}C_A}{\mathrm{d}t}$$

$$(r_A) \cdot V = V \cdot \frac{\mathrm{d}N_A}{\mathrm{d}t}$$

$$r_A = \frac{\mathrm{d}N_A}{\mathrm{d}t}$$

## Arrhenius Law for Kinetic Constants

$$k = k_0 \exp\left(-\frac{E}{R_g T}\right)$$

At two different temperatures:

$$k_{T1} = k_{T_2} \exp\left(-\frac{E}{R_g}\left(\frac{1}{T_1} - \frac{1}{T_2}\right)\right)$$

And also:

$$\frac{E}{R_g} = \frac{\ln\left(\frac{k_{T2}}{k_{T1}}\right)}{\frac{1}{T_1} - \frac{1}{T_2}}$$

For more details, please consult Conesa (2019).

**Problem 2.1** For the $E(t)$ distribution and modeling of Problem 1.3, calculate the reactor conversion for a given reaction of first-order with a constant ($k$) of 0.1 min$^{-1}$ for the following models: the TIS model, segregation model, dispersion model, plug flow, and completely stirred tank reactor.

**Solution to Problem 2.1**

For details refer the Wiley website at http://www.wiley-vch.de/ISBN9783527354115

In Problem 1.2 we calculated:

$$\bar{t} = t_m = 10.12 \text{ s}$$

$$\sigma^2 = 80.99 \text{ s}^2$$

$$n_t = \frac{t_m^2}{\sigma^2} = \frac{10.12^2}{80.99} = 1.26 \text{ tanks}$$

Using the dispersion model, as it occurs in Problem 1.1, we do not know if the recipient is closed or open. Assuming closed–closed condition, we obtain Bo = 1.33 and Pe = 0.75.

For calculating the different conversions, we use the corresponding mass balances. In the TIS model, we have:

$$X_A = 1 - \frac{1}{(1 + \bar{t}_i \cdot k)^{n_t}} = 1 - \frac{1}{\left(1 + \frac{10.12}{1.26} \cdot 0.1\right)^{1.26}} = 0.5247$$

Using the segregation model:

$$\overline{X_A} = \int_0^\infty X_A(t) E(t) dt$$

We can use a spreadsheet to calculate the conversion. $X_A(t)$ is given by the conversion obtained in a batch reactor that for a first-order reaction can be written:

$$X_A(t) = 1 - \exp(-k \cdot t)$$

| Time (min) | $X(t)$ | $X(t) \cdot E(t) \cdot \Delta t$ |
|---|---|---|
| 0 | 0.0000 | 0.0000 |
| 0.4 | 0.0392 | 0.0006 |
| 1 | 0.0952 | 0.0040 |
| 2 | 0.1813 | 0.0165 |
| 3 | 0.2592 | 0.0241 |
| 4 | 0.3297 | 0.0290 |
| 5 | 0.3935 | 0.0317 |
| 6 | 0.4512 | 0.0328 |
| 8 | 0.5507 | 0.0645 |
| 10 | 0.6321 | 0.0592 |
| 15 | 0.7769 | 0.1035 |
| 20 | 0.8647 | 0.0658 |
| 25 | 0.9179 | 0.0396 |
| 30 | 0.9502 | 0.0234 |
| 35 | 0.9698 | 0.0136 |
| 40 | 0.9817 | 0.0077 |
| 45 | 0.9889 | 0.0044 |
| 50 | 0.9933 | 0.0028 |
| 60 | 0.9975 | 0.0011 |
| $X_{\text{mean}}$ | | 0.524 |

Second column is calculated using the previous equation, and the last one by multiplying $X(t)$ by $E(t)\Delta t$. Finally, summing all values in the last column:

$$\overline{X_{\text{A}}} = \sum_{\text{all points}} X_{\text{A}}(t)E(t)\Delta t = 0.524$$

Using the dispersion model:

$$1 - X_{\text{A}} = \frac{4a \cdot \exp\left(\frac{1}{2\text{Bo}}\right)}{(1+a)^2 \cdot \exp\left(\frac{a}{2\text{Bo}}\right) - (1-a)^2 \cdot \exp\left(-\frac{a}{2\text{Bo}}\right)}$$

with

$$a = [1 + 4(\bar{t} \cdot k) \cdot \text{Bo}]^{1/2}$$

With the data of this problem:

$$a = [1 + 4(10.12 \cdot 0.1) \cdot 1.33]^{1/2} = 2.529$$

obtaining: $X_A = 0.5291$.

If the system is a simple FPR, a mass balance would give:

$$dn_A = Q_0 \cdot dC_A = (r_A) \cdot dV = -k \cdot C_A \cdot dV$$

Using the relationships $C_A = C_{A0}(1 - X_A)$ and $dC_A = C_{A0} \cdot (-dX_A)$ we can obtain:

$$\int_0^{X_A} \frac{dX_A}{1 - X_A} = k \int_0^V \frac{dV}{Q_0}$$

and:

$$\bar{t} = \frac{V}{Q_0} = \ln\left(\frac{1}{1 - X_A}\right)$$

so:

$$X_A = 1 - \exp(-k \cdot \bar{t}) = 1 - \exp(-0.1 \cdot 10.12) = 0.636$$

In a CSTR:

$$Q_0 C_{A0} = r_A \cdot V + Q_0 \cdot C_A$$

that gives:

$$X_A = 1 - \frac{1}{1 + k \cdot \bar{t}} = 0.503$$

**Problem 2.2**   A 5000-l packed reactor is used to dimerize A following the second-order reaction 2A $\rightarrow$ B. The constant for the rate, referred to as mole of reaction, is $8.7 \times 10^{-3}$ l/mol·s. Feed consists of 15 l/s of a 0.8 M solution of A. The reactor has an internal diameter of 88 cm, and the axial dispersion coefficient is 500 cm$^2$/s. The packed bed porosity is 0.45, and the density of the packed catalyst is 0.8 g/cm$^3$. Density cm$^3$ refers to packed bed volume.

(a) Obtain the homogeneous expression of the reaction rate and calculate the conversion fraction at the outlet of an ideal plug flow reactor (PFR).
(b) Compare the result with the conversion that would be obtained considering the dispersion model.

**Solution to Problem 2.2**

(a) The expression for the kinetics would be:

$$(-r_A) = kC_A^2$$

The mass balance for the PFR is (differential form):

$$\text{Input} - \text{Output} + \text{Generation} = \text{Accumulation}$$

$$n_A - (n_A + dn_A) + (r_A) \cdot dV = 0$$

$$dn_A = (r_A) \cdot dV$$

$$Q_0 \cdot dC_A = \left(-kC_A^2\right) \cdot dV$$

$$\int_{C_{A0}}^{C_A} \frac{dC_A}{\left(-C_A^2\right)} = k\frac{V}{Q_0}$$

$$-\frac{C_{A0}dX_A}{-C_{A0}^2(1-X_A)^2} = k\bar{t}$$

Or, using the conversion $X_A$:

$$\int_0^{X_A} \frac{dX_A}{(1-X_A)^2} = kC_{A0}\frac{V}{Q_0} = kC_{A0}\bar{t}$$

$$\frac{X_A}{(1-X_A)} = kC_{A0}\bar{t}$$

And then:

$$X_A = \frac{kC_{A0}\bar{t}}{1+kC_{A0}\bar{t}} = \frac{8.7\times10^{-3}\cdot0.8\cdot5000/15}{1+8.7\times10^{-3}\cdot0.8\cdot5000/15} = 0.699$$

(b) The data we have are the following:

- $d_t = 88\,\text{cm}$
- $D = 500\,\text{cm}^2/\text{s}$
- $\varepsilon = 0.45$
- $\rho = 0.8\,\text{g/cm}^3$

The parameter we should know for calculating conversion with the dispersion model is Bo, i.e., the dispersion number, which for a packed bed is:

$$\text{Bo} = \frac{\varepsilon D}{d_p \cdot u}$$

Value of superficial velocity is $u = \frac{Q_0}{(\pi \cdot d_t^2)\cdot\varepsilon} = 5.48\ \frac{\text{cm}}{\text{s}}$ and particle diameter, $d_p$, can be estimated using the correlation:

$$\varepsilon = 0.4 + 0.05\left(\frac{d_p}{d_t}\right) + 0.412\left(\frac{d_p}{d_t}\right)^2$$

obtaining $d_p = 0.25\,\text{m}$ and Bo $= 1.59$.

With value of $kC_{A0}\bar{t} = 2.32$ and Bo $= 1.59$, we can check with the corresponding equation or graphically that conversion is $X_A = 0.60$:

$$1 - X_A = \frac{4a \cdot \exp\left(\frac{1}{2\text{Bo}}\right)}{(1+a)^2 \cdot \exp\left(\frac{a}{2\text{Bo}}\right) - (1-a)^2 \cdot \exp\left(-\frac{a}{2\text{Bo}}\right)}$$

$$a = [1 + 4 \cdot \text{Da} \cdot \text{Bo}]^{1/2}$$

$$\text{Da} = k \cdot \bar{t} \cdot C_A^{n-1}$$

**Problem 2.3**  Given the following cumulative function of residence times:

$P(t) = t/20$ for $0 < t < 10$

$P(t) = t/10 - 0.5$ for $10 < t < 15$

$P(t) = 1$ for $t > 15$

(a) Determine the average residence time, and
(b) Calculate the conversion that would be obtained for a reaction with a single reactant, with no change in density and a rate constant equal to $0.1\,\text{s}^{-1}$.

**Solution to Problem 2.3**

For details refer the Wiley website at http://www.wiley-vch.de/ISBN9783527354115

(a) From the data of $P(t)$, we can calculate the derivative that would be $E(t)$. Then, following the same procedure shown in Problems 1.1 and 1.3:

| $t$ (s) | $P(t)$ | $E(t)$ | $t \cdot E(t)$ | $t \cdot E(t) \cdot \Delta t$ |
|---|---|---|---|---|
| 0 | 0 | | 0 | |
| 1 | 0.05 | 0.05 | 0.05 | 0.025 |
| 2 | 0.1 | 0.05 | 0.1 | 0.075 |
| 3 | 0.15 | 0.05 | 0.15 | 0.125 |
| 4 | 0.2 | 0.05 | 0.2 | 0.175 |
| 5 | 0.25 | 0.05 | 0.25 | 0.225 |
| 6 | 0.3 | 0.05 | 0.3 | 0.275 |
| 7 | 0.35 | 0.05 | 0.35 | 0.325 |
| 8 | 0.4 | 0.05 | 0.4 | 0.375 |
| 9 | 0.45 | 0.05 | 0.45 | 0.425 |
| 10 | 0.5 | 0.05 | 0.5 | 0.475 |
| 11 | 0.6 | 0.1 | 1.1 | 0.8 |
| 12 | 0.7 | 0.1 | 1.2 | 1.15 |
| 13 | 0.8 | 0.1 | 1.3 | 1.25 |
| 14 | 0.9 | 0.1 | 1.4 | 1.35 |
| 15 | 1 | 0.1 | 1.5 | 1.45 |
| | $\Sigma(E(t)\text{d}t) = 1.0$ | | $t_m = 8.5\,\text{s}$ | |

Note that $E(t)$ is calculated by numerical derivation of the values of $p(t)$.

(b) As $k = 0.1\,\text{s}^{-1}$, the reaction is first-order. In this case, we can use:

$$\overline{X_A} = \int_0^\infty X_A(t)E(t)\text{d}t$$

With $X_A(t) = 1 - \exp(-kt)$ that represents the conversion obtained in a batch reactor. Doing the corresponding calculation, we obtain:

| $t$ (s) | $X_A(t)$ | $X_A(t) \cdot E(t)$ | $X_A(t) \cdot E(t) \cdot \Delta t$ |
|---|---|---|---|
| 0 | 0.000 | 0.000 | 0.002 |
| 1 | 0.095 | 0.005 | 0.007 |
| 2 | 0.181 | 0.009 | 0.011 |
| 3 | 0.259 | 0.013 | 0.015 |
| 4 | 0.330 | 0.016 | 0.018 |
| 5 | 0.393 | 0.020 | 0.021 |
| 6 | 0.451 | 0.023 | 0.024 |
| 7 | 0.503 | 0.025 | 0.026 |
| 8 | 0.551 | 0.028 | 0.029 |
| 9 | 0.593 | 0.030 | 0.031 |
| 10 | 0.632 | 0.032 | 0.049 |
| 11 | 0.667 | 0.067 | 0.068 |
| 12 | 0.699 | 0.070 | 0.071 |
| 13 | 0.727 | 0.073 | 0.074 |
| 14 | 0.753 | 0.075 | 0.077 |
| 15 | 0.777 | 0.078 | |
| | | $\Sigma(X_A(t) \cdot E(t) \cdot dt) = 0.523$ | |

So the value of $\overline{X_A}$ is 0.523.

**Problem 2.4** A reactor working in continuous mode is being fed with a solution of A of concentration 2 M. The reaction rate is given by:

$$(-r_A) = kC_A^2$$

with $k = 2\,1/\text{mol} \cdot \text{s}$. Calculate the mean residence time ($t_m$), the conversion fraction corresponding to a time equal to $t_m$, and the conversion fraction given the following residence time distributions:

$$E(t) = \frac{1}{2.506} \exp(-0.5t^2 + 8t - 32)$$
$$E(t) = 0.25t \exp(-0.5t)$$

**Solution to Problem 1.8**

For details refer the Wiley website at http://www.wiley-vch.de/ISBN9783527354115

(a) For calculating mean residence time, with the $E(t)$:

$$t_m = \int_0^\infty t \cdot E(t) \cdot dt = \int_0^\infty \frac{t}{2.506} \exp(-0.5t^2 + 8t - 32)dt = 8.00 \, \text{min}$$

We can also estimate value of variance:

$$\sigma^2 = \int_0^\infty (t-8)^2 \cdot \frac{1}{2.506} \exp(-0.5t^2 + 8t - 32)dt = 1 \text{ min}^2$$

If a TIS model is assumed:

$$n_t = \frac{t_m^2}{\sigma^2} = 64$$

This means that this reactor is more similar to a PFR than to a CSTR (high value of the number of tanks). Assuming now a dispersion model, Bo can be obtained:

$$\left(\frac{\sigma}{t_m}\right)^2 = 2 \cdot \text{Bo} - 2 \cdot \text{Bo}^2 \cdot \left[1 - \exp\left(-\frac{1}{\text{Bo}}\right)\right]$$

resulting Bo = 0.022. Now, using the equation or the figures corresponding to the dispersion model (shown in the initial part of this chapter), with Da = $kC_{A0}\,\bar{t} = 2 \cdot 2 \cdot 8 = 32$, a conversion of $X_A = 0.988$ is obtained.

(b) Now:

$$t_m = \int_0^\infty t \cdot E(t) \cdot dt = \int_0^\infty 0.25t^2 \exp(-0.5t)dt = 4.00 \text{ min}$$

and:

$$\sigma^2 = \int_0^\infty (t-4)^2 \cdot 0.25t \exp(-0.5t)dt = 8 \text{ min}^2$$

$$n_t = \frac{t_m^2}{\sigma^2} = 2$$

In this case, only two CSTRs are needed to explain $E(t)$ curve. In the first reactor:

$$kC_{A1}^2\bar{t} + C_{A1} - C_{A0} = 0$$

and in the second one:

$$kC_{A2}^2\bar{t} + C_{A2} - C_{A1} = 0$$

We can check $C_{A1} = 0.442$ M and $C_{A2} = 0.181$ M, so the conversion is $(1 - 0.181/2) = 0.91$.

**Problem 2.5** In the liquid phase, contaminant A decomposes according to the kinetics:

$$(-r_A) = kC_A^{1.5}$$

with $k = 4.38\,1^{0.5}/\text{mol}^{0.5} \cdot \text{s}$.

The flow is represented by the following cumulative probability data:

| $t$ (s) | 0 | 15 | 30 | 45 | 60 | 75 | 90 | 105 | 120 | 135 | 150 |
|---|---|---|---|---|---|---|---|---|---|---|---|
| $p(t)$ | 0 | 0.02 | 0.05 | 0.13 | 0.3 | 0.55 | 0.72 | 0.85 | 0.93 | 0.97 | 1 |

The supply contains 1.5 M of A. Directly using the accumulated function and integration by trapezoids, we obtain:

(a) The average residence time and the concentration that would be obtained at the outlet if all the fluid had exactly this residence time.
(b) The actual average outlet concentration.

**Solution to Problem 2.5**

For details refer the Wiley website at http://www.wiley-vch.de/ISBN9783527354115

(a) Bearing in mind that the data of $p(t)$ represent the function $F(t)$, we can calculate, as in previous problems:

| $t$ (s) | $F(t)$ | $E(t)$ | $t \cdot E(t)$ | $t \cdot E(t) \cdot dt$ | $C(t)$ (mol/l) | $(t-t_m)^2 \cdot E(t)$ | $(t-t_m)^2 \cdot E(t) \cdot dt$ |
|---|---|---|---|---|---|---|---|
| 0 | 0 | 0.0000 | 0 | 0.15 | 0 | 0.000 | 42.185 |
| 15 | 0.02 | 0.0013 | 0.02 | 0.6 | 0.03 | 5.625 | 79.610 |
| 30 | 0.05 | 0.0020 | 0.06 | 2.25 | 0.075 | 4.990 | 86.285 |
| 45 | 0.13 | 0.0053 | 0.24 | 6.9 | 0.195 | 6.515 | 82.690 |
| 60 | 0.3 | 0.0113 | 0.68 | 14.475 | 0.45 | 4.511 | 36.893 |
| 75 | 0.55 | 0.0167 | 1.25 | 17.025 | 0.825 | 0.408 | 11.648 |
| 90 | 0.72 | 0.0113 | 1.02 | 14.475 | 1.08 | 1.145 | 49.373 |
| 105 | 0.85 | 0.0087 | 0.91 | 11.625 | 1.275 | 5.438 | 104.948 |
| 120 | 0.93 | 0.0053 | 0.64 | 7.5 | 1.395 | 8.555 | 124.770 |
| 135 | 0.97 | 0.0027 | 0.36 | 4.95 | 1.455 | 8.081 | 134.215 |
| 150 | 1 | 0.0020 | 0.3 | 0 | 1.5 | 9.814 | |
| | | | $t_m$ (s) = 79.95 | | | $\sigma^2$ (s$^2$) = 752.62 | |

If a PFR with residence time of $79.95\,$s is considered, we would do a mass balance:

$$\text{Input} - \text{Output} + \text{Generation} = \text{Acummulation}$$

$$n_A - (n_A + dn_A) + (r_A)dV = 0$$

$$dn_A = (r_A) \cdot dV$$

$$Q_0 \cdot dC_A = \left(-kC_A^{1.5}\right) \cdot dV$$

$$\int_{C_{A0}}^{C_A} \frac{dC_A}{\left(-C_A^{1.5}\right)} = k\frac{V}{Q_0}$$

$$\frac{2}{\sqrt{C_{A0}}} - \frac{2}{\sqrt{C_A}} = k\bar{t}$$

$$C_A = \left(\frac{2}{\frac{2}{\sqrt{C_{A0}}} + k\bar{t}}\right)^2$$

Introducing the data, we obtain $C_A = 0.0717\,$mol/l.

(b) We can use, for example, the TIS model.

$$n_t = \frac{t_m^2}{\sigma^2} = 8.49 \approx 9$$

Note that, as we are working with a n-th order different from one, we should do balances on an integer number of reactors. In each of these:

$$n_{Ai-1} + r_A V = n_A$$

$$Q_0 C_{Ai-1} + \left(-kC_{Ai}^{1.5}\right) V = Q_0 C_{Ai}$$

$$C_{Ai} = \frac{C_{Ai-1}}{1 + ktC_{Ai}^{0.5}}$$

This equation can be solved numerically. In that way, we obtain (in mol/l):

$C_{A1} = 0.3405$

$C_{A2} = 0.1145$

$C_{A3} = 0.0497$

$C_{A4} = 0.0257$

$C_{A5} = 0.0150$

$C_{A6} = 0.0095$

$C_{A7} = 0.00649$

$C_{A8} = 0.00464$

$C_{A9} = 0.00346$

**Problem 2.6**  In a continuous reactor, several reactions are taking place in the liquid phase, which are summarized in the following expressions:

$$A \rightarrow 2B$$

$$A \rightarrow 2C$$

The reaction rates per mole of reaction are given by $r_1 = k_1 \sqrt{C_A}$ and $r_2 = k_2 \sqrt{C_A}$. The reaction rate constants are 0.020 and 0.005 $mol^{0.5}/min \cdot l^{0.5}$, respectively. The species A is being fed with a concentration of 0.65 M. The residence time distribution is:

$E(t) = -0.037 + 0.008\,78t - 3.07 \times 10^{-4}t^2 + 2.93 \times 10^{-6}t^3$ for $5 < t < 50\,min$
$E(t) = 0$ for other values of time.

Determine:

(a) The average residence time.
(b) The conversion and selectivity to B at the outlet of the real reactor.

**Solution to Problem 2.6**

For details refer the Wiley website at http://www.wiley-vch.de/ISBN9783527354115

(a) First of all, let us see the form of the $E(t)$ and calculate their momenta. As in previous problems, we construct the following table:

| $t$ (min) | $E(t)$ | $t \cdot E(t)$ | $t \cdot E(t) \cdot \Delta t$ | $(t - t_m)^2 \cdot E(t) \cdot \Delta t$ |
| --- | --- | --- | --- | --- |
| 0 | 0 | 0 | | 0 |
| 2 | 0 | 0 | 0 | 0 |
| 4 | 0 | 0 | 0 | 0 |
| 6 | 0.005 | 0.063 | 0.063 | 3.213 |
| 8 | 0.015 | 0.241 | 0.305 | 7.228 |
| 10 | 0.023 | 0.230 | 0.472 | 8.362 |
| 12 | 0.029 | 0.351 | 0.581 | 7.693 |

| *t* (min) | *E(t)* | *t · E(t)* | *t · E(t) · Δt* | *(t − t$_m$)$^2$ · E(t) · Δt* |
|---|---|---|---|---|
| 14 | 0.034 | 0.473 | 0.824 | 6.066 |
| 16 | 0.037 | 0.590 | 1.063 | 4.122 |
| 18 | 0.039 | 0.696 | 1.286 | 2.317 |
| 20 | 0.039 | 0.785 | 1.481 | 0.947 |
| 22 | 0.039 | 0.853 | 1.638 | 0.169 |
| 24 | 0.037 | 0.897 | 1.750 | 0.021 |
| 26 | 0.035 | 0.916 | 1.814 | 0.450 |
| 28 | 0.032 | 0.909 | 1.826 | 1.330 |
| 30 | 0.029 | 0.876 | 1.785 | 2.488 |
| 32 | 0.026 | 0.819 | 1.696 | 3.722 |
| 34 | 0.022 | 0.741 | 1.560 | 4.828 |
| 36 | 0.018 | 0.645 | 1.386 | 5.620 |
| 38 | 0.014 | 0.536 | 1.181 | 5.953 |
| 40 | 0.011 | 0.421 | 0.957 | 5.746 |
| 42 | 0.007 | 0.306 | 0.727 | 5.004 |
| 44 | 0.005 | 0.201 | 0.507 | 3.840 |
| 46 | 0.002 | 0.113 | 0.314 | 2.499 |
| 48 | 0.001 | 0.055 | 0.168 | 1.379 |
| 50 | 0.001 | 0.038 | 0.093 | |
| | | | $t_m = 23.47\,\mathrm{min}$ | $\sigma^2 = 83.00\,\mathrm{min}^2$ |

Graphically, the $E(t)$ is:

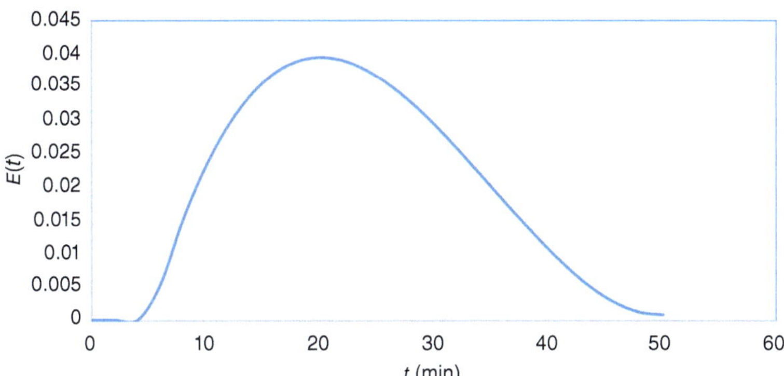


As we can check, average residence time is 23.47 min.

(b) From this data, using the TIS model:

$$n_t = \frac{23.47^2}{83} = 6.64 \,\text{tanks}$$

In a batch reactor, we would have:

$$\frac{dC_A}{dt} = -k_1\sqrt{C_A} - k_2\sqrt{C_A} = -(k_1 + k_2)\sqrt{C_A}$$

$$\frac{dC_B}{dt} = +2k_2\sqrt{C_A}$$

and so:

$$\int_{C_{A0}}^{C_A} \frac{dC_A}{\sqrt{C_A}} = -(k_1 + k_2)t$$

i.e.,

$$2\sqrt{C_A} - 2\sqrt{C_{A0}} = -(k_1 + k_2)t$$

hence:

$$\sqrt{C_A} = \frac{-t(k_1 + k_2) + 2\sqrt{C_{A0}}}{2} \quad \therefore C_A = \left(\frac{-(k_1 + k_2)t + 2\sqrt{C_{A0}}}{2}\right)^2$$

From B:

$$\frac{dC_B}{dt} = +2k_2\sqrt{C_A} = k_2\left(-(k_1 + k_2)t + 2\sqrt{C_{A0}}\right)$$

$$dC_B = k_2\left(-t(k_1 + k_2) + 2\sqrt{C_{A0}}\right)dt \therefore C_B = \frac{1}{2}k_2 t\left(-(k_1 + k_2)t + 4\sqrt{C_{A0}}\right)$$

Assuming a segregation model:

$$\overline{C_A} = \int_0^\infty C_A(t)E(t)dt = \int_0^\infty \left(\frac{(k_1 + k_2)t + 2\sqrt{C_{A0}}}{2}\right)^{0.5} E(t)dt$$

$$\overline{C_B} = \int_0^\infty C_B(t)E(t)dt = \int_0^\infty \left(\frac{1}{2}k_2 t\left((k_1 + k_2)t + 4\sqrt{C_{A0}}\right)\right) E(t)dt$$

In the spreadsheet, we will do the numerical integration:

| $t$ (min) | $E(t)$ | $C_A$ (mol/l) | $C_B$ (mol/l) | $C_A(t) \cdot E(t)dt$ | $C_B(t) \cdot E(t)dt$ |
|---|---|---|---|---|---|
| 0 | 0.0000 | 0.6500 | 0.0000 | 0.0000 | 0.0000 |
| 2 | 0.0000 | 0.6103 | 0.0159 | 0.0000 | 0.0000 |
| 4 | 0.0000 | 0.5719 | 0.0312 | 0.0000 | 0.0000 |
| 6 | 0.0053 | 0.5347 | 0.0461 | 0.0056 | 0.0005 |
| 8 | 0.0151 | 0.4988 | 0.0605 | 0.0151 | 0.0018 |
| 10 | 0.0230 | 0.4641 | 0.0744 | 0.0214 | 0.0034 |
| 12 | 0.0292 | 0.4306 | 0.0877 | 0.0252 | 0.0051 |
| 14 | 0.0338 | 0.3984 | 0.1006 | 0.0269 | 0.0068 |
| 16 | 0.0369 | 0.3675 | 0.1130 | 0.0271 | 0.0083 |
| 18 | 0.0387 | 0.3378 | 0.1249 | 0.0261 | 0.0097 |
| 20 | 0.0392 | 0.3094 | 0.1362 | 0.0243 | 0.0107 |
| 22 | 0.0388 | 0.2822 | 0.1471 | 0.0219 | 0.0114 |

| $t$ (min) | $E(t)$ | $C_A$ (mol/l) | $C_B$ (mol/l) | $C_A(t) \cdot E(t)dt$ | $C_B(t) \cdot E(t)dt$ |
|---|---|---|---|---|---|
| 24 | 0.0374 | 0.2563 | 0.1575 | 0.0192 | 0.0118 |
| 26 | 0.0352 | 0.2316 | 0.1674 | 0.0163 | 0.0118 |
| 28 | 0.0325 | 0.2081 | 0.1767 | 0.0135 | 0.0115 |
| 30 | 0.0292 | 0.1860 | 0.1856 | 0.0109 | 0.0108 |
| 32 | 0.0256 | 0.1650 | 0.1940 | 0.0084 | 0.0099 |
| 34 | 0.0218 | 0.1453 | 0.2019 | 0.0063 | 0.0088 |
| 36 | 0.0179 | 0.1269 | 0.2092 | 0.0045 | 0.0075 |
| 38 | 0.0141 | 0.1097 | 0.2161 | 0.0031 | 0.0061 |
| 40 | 0.0105 | 0.0938 | 0.2225 | 0.0020 | 0.0047 |
| 42 | 0.0073 | 0.0791 | 0.2284 | 0.0012 | 0.0033 |
| 44 | 0.0046 | 0.0657 | 0.2337 | 0.0006 | 0.0021 |
| 46 | 0.0025 | 0.0535 | 0.2386 | 0.0003 | 0.0012 |
| 48 | 0.0011 | 0.0425 | 0.2430 | 0.0001 | 0.0006 |
| 50 | 0.0008 | 0.0328 | 0.2469 | | |
| | | | Sum = | 0.2799 | 0.1478 |

So the concentration of A at the exit is 0.2799 M (i.e. the conversion is 0.57). The selectivity to B is:

$$S_B = \frac{2C_B}{C_{A0} - C_A} = 3.66$$

If the reactor were a PFR with residence time equal to 23.47 min:

$$C_A = \left( \frac{-(k_1 + k_2)\bar{t} + 2\sqrt{C_{A0}}}{2} \right)^2 = 0.263 \text{ M} \rightarrow X_A = 0.6$$

**Problem 2.7**  A first-order reaction occurs in a tubular-geometry reactor. The distribution of residence times is given by:

$E(t) = 0.04t$ for $0 < t < 5$ min
$E(t) = 0.4 - 0.04t$ if $5 < t < 10$ min
$E(t) = 0$ if $t > 10$ min

At 50 °C, the reaction rate constant referred to as mole of reactant is 0.1833 min$^{-1}$. The activation energy for the reaction being produced is 25 cal/mol.

(a) Calculate the conversion fraction at the actual reactor outlet if it is operated at 50 °C.
    On the other hand, the possibility of carrying out the conversion in an ideal PFR that has the same residence time is being studied. Calculate:
(b) The conversion fraction at the output.
(c) The uncertainty in that conversion fraction if there is an uncertainty of $\pm 1$ °C in the operating temperature.

The goal of this problem is to raise concern about whether deviations from ideality in flow patterns are worth considering when there are significant uncertainties in other parameters.

**Solution to Problem 2.7**

For details refer the Wiley website at http://www.wiley-vch.de/ISBN9783527354115

(a) In the PFR with first-order reaction:

$$X_A = 1 - \exp(-k\bar{t})$$

With:

$$\bar{t} = t_m = \int_0^\infty t \cdot E(t)\,dt$$

| | | | | Segregation Model | |
|---|---|---|---|---|---|
| $t$ (min) | $E(t)$ (min$^{-1}$) | $t \cdot E(t)$ | $(t \cdot E(t))$ dt | $\exp(-kt) \cdot E(t)$ | $(\exp(-kt) \cdot E(t))$ dt |
| 0 | 0 | 0 | 0.020 | 0 | 0.017 |
| 1 | 0.040 | 0.040 | 0.100 | 0.033 | 0.044 |
| 2 | 0.080 | 0.160 | 0.260 | 0.055 | 0.062 |
| 3 | 0.120 | 0.360 | 0.500 | 0.069 | 0.073 |
| 4 | 0.160 | 0.640 | 0.820 | 0.077 | 0.078 |
| 5 | 0.200 | 1 | 0.980 | 0.080 | 0.067 |
| 6 | 0.160 | 0.960 | 0.900 | 0.053 | 0.043 |
| 7 | 0.120 | 0.840 | 0.740 | 0.033 | 0.026 |
| 8 | 0.080 | 0.640 | 0.500 | 0.018 | 0.013 |
| 9 | 0.040 | 0.360 | 0.180 | 0.008 | 0.004 |
| 10 | 0 | 0 | 0 | 0 | 0 |
| 11 | 0 | 0 | 0 | 0 | 0 |
| 12 | 0 | 0 | 0 | 0 | 0 |
| 13 | 0 | 0 | | 0 | |
| | | Sum = | 5 | | 0.428 |

So $\bar{t} = t_m = 5$ min and the conversion calculated using segregation model is 0.428.

(b) In the ideal PFR:

$$X_A = 1 - \exp(-k\bar{t}) = 0.6$$

Using the segregation model for the non-ideal flow:

$$\overline{X_A} = \int_0^\infty X_A(t)E(t)\,dt$$

$$\overline{X_A} = \int_0^\infty (1 - \exp(-kt))E(t)\,dt$$

$$= \int_0^\infty E(t)\mathrm{d}t - \int_0^\infty \exp(-kt)E(t)\mathrm{d}t$$

$$= 1 - \int_0^\infty \exp(-kt)E(t)\mathrm{d}t$$

Using a numerical method, we can calculate: $\overline{X_A} = 0.572$

(c) For evaluating the effect of the uncertainty of the temperature, we should calculate the values of $k$ at 51 and 49 °C. Bearing in mind that:

$$k = k_0 \exp\left(-\frac{E}{R_g T}\right)$$

$$0.1833 = k_0 \exp\left(-\frac{25}{1.987\,(273 + 50)}\right)$$

and $k_0 = 1.54 \cdot 10^{16}$ min$^{-1}$.

Now:

$$k_{49\,°C} = 0.162 \text{ min}^{-1}$$

$$k_{51\,°C} = 0.206 \text{ min}^{-1}$$

Doing a similar calculation to that shown previously:

$$X_{A,49\,°C} = 1 - \exp(-k_{49\,°C}\bar{t}) = 0.556$$

$$X_{A,51\,°C} = 1 - \exp(-k_{51\,°C}\bar{t}) = 0.644$$

**Problem 2.8**  To decompose a contaminant A, a photoreactor is used . For simplicity, let us consider that the reactor is tubular and is made of transparent glass at the wavelength necessary to excite the photocatalyst. This tube that forms the reactor has a length of 8 m and an internal diameter of 2.21 cm. The outside of the glass tube is uniformly illuminated by an ultraviolet light source. The reaction occurs in the presence of UV light and can be schematized as:

$$A + \tfrac{1}{2}B \to \text{Safe products}$$

where B is an oxidizing agent. The reaction rate in homogeneous units is given by:

$$(-r_A) = kC_B^{0.8}C_A = 0.84\,\frac{\text{L}^{0.8}}{\text{mol}^{0.8}\cdot \text{s}}\,C_B^{0.8}C_A$$

where the catalyst concentration and light intensity are included within the rate constant. The suspended catalyst is very fine particles of $TiO_2$, whose volume must be assumed to be negligible for calculation purposes. The feed consists of an aqueous solution diluted with 0.05 M of A and 0 03 M of B. Determine the conversion fraction achieved if the volumetric flow fed is: (a) 7.5 l/min and (b) 1.5 l/min.

**Solution to Problem 2.8**

For details refer the Wiley website at http://www.wiley-vch.de/ISBN9783527354115

In the PFR:

$$\mathrm{d}n_A = Q_0 \cdot \mathrm{d}C_A = r_A \cdot \mathrm{d}V$$

$$r_A = -k \cdot C_A \cdot C_B^{0.8}$$

So:

$$Q_0 \cdot dC_A = -k \cdot C_A \cdot C_B^{0.8} \cdot dV$$

And:

$$n_B = n_{B0} + n_{A0} \cdot (-1/2) \cdot X_A$$

With:

$$X_A = (n_{A0} - n_A)/n_{A0}$$

i.e.,

$$n_B = n_{B0} - 1/2 \cdot (n_{A0} - n_A)$$

As we are treating a liquid, flow rate is constant, and:

$$C_B = C_{B0} - 1/2 \cdot C_{A0} + 1/2 \cdot C_A$$

and we have:

$$-dV \cdot \frac{k}{Q_0} = \frac{dC_A}{C_A \cdot \left(C_{B0} - \frac{1}{2}C_{A0} + \frac{1}{2}C_A\right)^{0.8}}$$

$$-V \cdot \frac{k}{Q_0} = \int_{C_{A0}}^{C_A} \frac{dC_A}{C_A \cdot \left(C_{B0} - \frac{1}{2}C_{A0} + \frac{1}{2}C_A\right)^{0.8}}$$

(a) The left part of the equation equals $-20.62$ using the data given. Right part should be integrated until equals this number. In this case, a good solution is $C_A = 0.019 \, \text{mol/l}$, so the conversion is $X_A = 0.603$:

| $C_A$ (mol/l) | $1/(C_A \cdot [C_{B,in} - 0.5 \cdot C_{A,in} + 0.5 \cdot C_A]^{0.8})$ $(\text{l/mol})^{1.8}$ | Integral $(\text{l/mol})^{0.8}$ |
|---|---|---|
| 0.0500 | 330.6 | −0.201 |
| 0.0494 | 337.4 | −0.205 |
| 0.0488 | 344.3 | −0.210 |
| 0.0482 | 351.5 | −0.214 |
| 0.0476 | 358.9 | −0.219 |
| 0.0470 | 366.6 | −0.223 |
| 0.0464 | 374.5 | −0.228 |
| 0.0458 | 382.7 | −0.233 |
| 0.0452 | 391.2 | −0.238 |
| … | … | … |
| 0.0211 | 1326.4 | −0.817 |
| 0.0205 | 1387.0 | −0.855 |
| 0.0199 | 1452.0 | |
| | Sum → | −20.622 |

(b) Now, $-V \cdot k/Q_0 = -103.11$, and finally, $C_A = 0.003 \, \text{mol/l}$, $X_A = 0.9399$.

**Problem 2.9**    A continuous reactor can be modeled as eight stirred tank reactors connected in series. The reaction kinetics can be considered first-order with a rate constant, referred to as mole of reactant, of $0.25\,\text{min}^{-1}$. $625\,\text{l/min}$ are fed, and the reactor volume is $1.275\,\text{l}$. Calculate the output conversion fraction and analyze which ideal continuous reactors performance is closest.

**Solution to Problem 2.9**

In the TIS model:

$$X_A = 1 - \frac{1}{(1 + \bar{t_i} \cdot k)^{n_t}}$$

but

$$\bar{t_i} = \frac{\bar{t}}{n_t} = \frac{\left(\frac{Q_0}{V}\right)}{n_t} = \frac{\frac{1275}{625}}{8} = 0.255\,\text{min}$$

And the conversion is:

$$X_A = 1 - \frac{1}{(1 + 0.255 \cdot 0.25)^8} = 0.390$$

If it were a PFR:

$$X_A = 1 - \exp(-k\bar{t}) = 1 - \exp\left(-0.25 \cdot \frac{1275}{625}\right) = 0.399$$

If, on the contrary, the reactor were a single CSTR:

$$X_A = \frac{k\bar{t}}{1 + k\bar{t}} = 0.337$$

So, undoubtedly, the reactor is more like a PFR.

**Problem 2.10**    A 2000-l tubular reactor is fed with $20\,\text{mol/s}$ of which $0.2\,\text{mol/s}$ is A and the rest is an inert I. The operating conditions are $200\,°\text{C}$ and $2\,\text{atm}$. The reaction order is unity, and the rate constant referred to as mole of A is $0.2\,\text{s}^{-1}$. The internal diameter of the reactor is $20\,\text{cm}$. If the axial dispersion is $50\,\text{m}^2/\text{s}$, calculate the output conversion and compare it with that which would be estimated based on the ideal plug flow model.

**Solution to Problem 2.10**

For details refer the Wiley website at http://www.wiley-vch.de/ISBN9783527354115

In the dispersed plug flow, considering a first-order reaction:

$$1 - X_A = \frac{4a \cdot \exp\left(\frac{1}{2\text{Bo}}\right)}{(1 + a)^2 \cdot \exp\left(\frac{a}{2\text{Bo}}\right) - (1 - a)^2 \cdot \exp\left(-\frac{a}{2\text{Bo}}\right)}$$

$$a = [1 + 4(\bar{t} \cdot k) \cdot \text{Bo}]^{1/2}$$

In the reactor $S = \pi\left(\frac{d_{\text{int}}}{2}\right)^2 = 0.0314\,\text{m}^2$, and the length can be calculated:

$$L = \frac{V}{S} = \frac{2\,\text{m}^3}{0.0314\,\text{m}^2} = 63.66\,\text{m}$$

Using the ideal gas equation, the flow rate and superficial velocity can also be calculated:

$$P \cdot Q_0 = n_0 \cdot R_g T$$

$$Q_0 = 20 \cdot 0.082 \cdot 473/2 = 377.86 \; \frac{l}{s}$$

$$u = Q_0/S = 12.35 \; \frac{m}{s}$$

We can calculate:

$$\mathrm{Bo} = \frac{D_e}{u \cdot L} = 0.063$$

The value of residence time is:

$$\bar{t} = V/Q_0 = 5.15 \text{ s}$$

and the "$a$" value is 1.12, finally giving a conversion of $X_A = 0.6226$.

If the reactor were a PFR, the conversion would be:

$$X_A = 1 - \exp(-k\bar{t}) = 0.6434$$

**Problem 2.11**  The decomposition of a compound that is produced with the following reaction is being studied:

$$A \rightarrow 3B$$

Experimental observations were carried out in an ideal isothermal batch reactor with a constant volume. The concentration profile was measured, starting from an initial composition of 75% by volume of A and 25% by volume of inert substance.

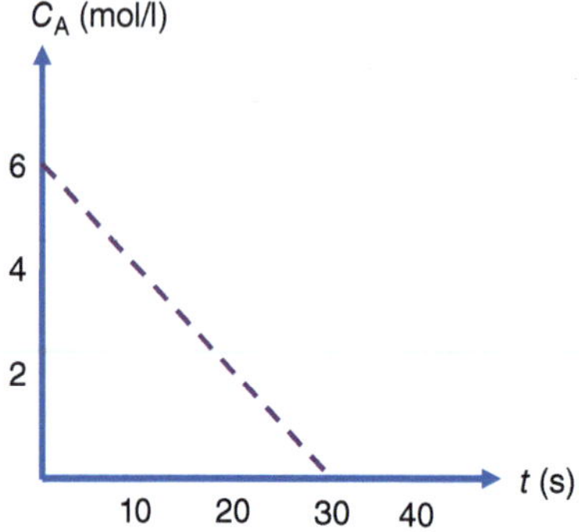

In a context of uniform initial conditions and temperatures, an ideal PFR, an ideal continuous stirred tank reactor (CSTR), and an ideal batch reactor (currently operating at constant pressure) are contrasted.

In the case of continuous reactors, spatial time is analyzed, while for batch reactors, it is necessary to calculate the precise time to achieve the complete reaction.

**Solution to Problem 2.11**
For details refer the Wiley website at http://www.wiley-vch.de/ISBN9783527354115

In the batch reactor:

$$N_A = N_{A0}(1 - X_A)\,(\text{mol})$$

$$N_B = N_{B0} + 3N_{A0}X_A = 3N_{A0}X_A\,(\text{mol})$$

$$N_{\text{total}} = N_A + N_B = N_{A0}(1 + 2X_A)$$

The expansion factor in this reaction is $\varepsilon_{\text{exp}} = 2$, so the volume expands as follows:

$$V = V^0 \frac{P^0}{P} \frac{T}{T^0}(1 + 2X_A) = V^0(1 + 2X_A)$$

where the superscript "0" indicates the beginning of the reaction, and the effect of pressure and temperature is eliminated by isothermal and isochore operations. The concentration of the reactant then follows:

$$C_A = \frac{N_A}{V} = \frac{N_{A0}(1 - X_A)}{V^0(1 + 2X_A)} = C_{A0}\frac{(1 - X_A)}{(1 + 2X_A)}$$

The mass balance in the reactor is:

$$\frac{dN_A}{dt} = r_A \cdot V$$

The behavior shown in the figure is only possible if kinetics is zero-order $(-r_A) = k$, then:

$$-N_{A0}\frac{dX_A}{dt} = -k \cdot V^0(1 + 2X_A)$$

$$C_{A0}\int_0^{X_A} \frac{dX_A}{(1 + 2X_A)} = k \cdot \int_0^t dt$$

$$C_{A0}\frac{\ln(2X_A + 1)}{2} = kt$$

In the figure, $C_{A0} = 6\,\text{M}$ and $t = 30\,\text{s}$ for $X_A = 1$, so $k = 0.109\,\text{s}^{-1}$.
In the batch reactor, when $X_A = 1$, $t = 30\,\text{s}$. For one PFR:

$$\frac{dn_A}{dV} = r_A$$

$$dV = \frac{dn_A}{r_A} = Q\frac{dC_A}{r_A} = Q\frac{dC_A}{-k}$$

$$V = Q \cdot (C_A - C_{A0})/-k$$

$$C_A = C_{A0} - k \cdot \bar{t}$$

For $C_A = 0$, we found that $\bar{t} = 30\,\text{s}$.
In a CSTR:

$$r_A = \frac{n_A - n_{A0}}{V} = -k$$

We find that:

$$C_A = C_{A0} - k \cdot \bar{t}$$

So again, the answer is 30 seconds.

**Problem 2.12**   To investigate the liquid-phase reaction A → products, a series of experiments were conducted in a small experimental ideal CSTR. The experimental data obtained are as follows:

| $t$ (s) | $C_{A0}$ (kmol/m$^3$) | $C_A$ (kmol/m$^3$) |
|---------|----------------------|--------------------|
| 44  | 5 | 4.0 |
| 52  | 8 | 6.0 |
| 58  | 8 | 5.5 |
| 70  | 8 | 7.0 |
| 78  | 8 | 5.0 |
| 166 | 5 | 2.0 |
| 225 | 1 | 0.5 |
| 300 | 2 | 1.0 |
| 468 | 5 | 2.0 |

(a) Avoiding determining the reaction order, the space time (residence time) needed to convert a feed containing 6 kmol/m$^3$ of A can be calculated in such a way that the conversion is 2/3 in an ideal CSTR.

(b) How much space time would be needed to achieve a comparable level of conversion in an ideal PFR?

**Solution to Problem 2.12**

For details refer the Wiley website at http://www.wiley-vch.de/ISBN9783527354115

(a) The conversion can be calculated using the data:

$$X_A = \frac{C_{A0} - C_A}{C_{A0}}$$

and also the reaction rate, by doing a mass balance:

$$Q_0 C_{A0} + r_A V = Q_0 C_A$$

$$-r_A = Q_0(C_{A0} - C_A)/V = X_A C_{A0}/\bar{t}$$

where $t_m$ is mean residence time (equals $\bar{t}$) in the data. We can obtain:

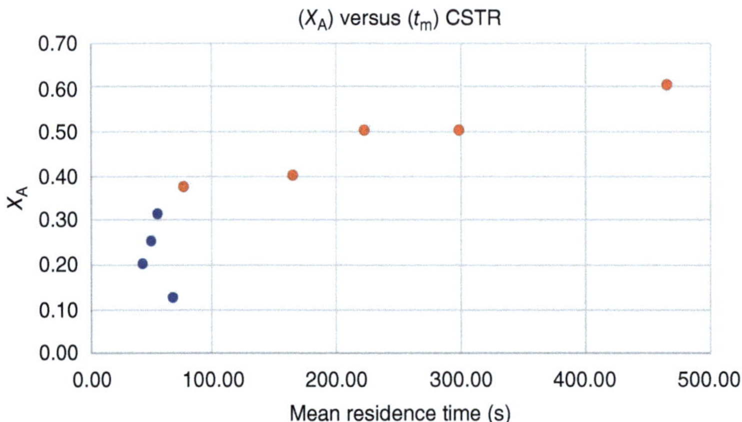

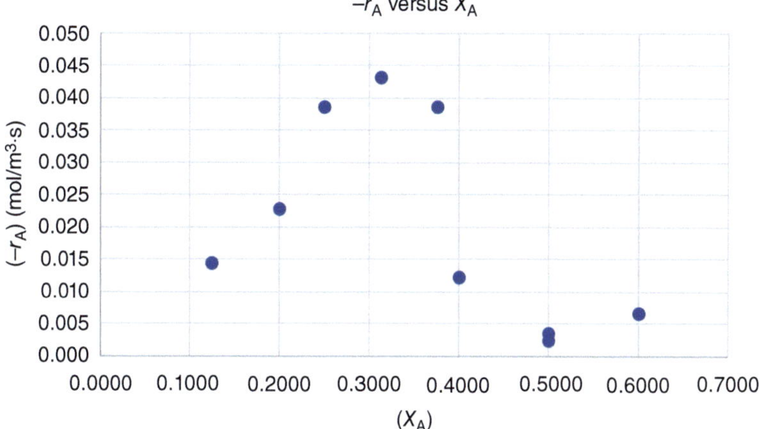


As the desired conversion is $X_A = 0.66$, we should extrapolate the data. Doing a first-order extrapolation, we have that the corresponding $(-r_A)$ is $0.00843\,\text{mol/m}^3 \cdot \text{s}$ so:

$$0.008\,43\,\frac{\text{mol}}{\text{m}^3 \cdot \text{s}} = 0.66 C_{A0}/t_m$$

and $t_m$ should be 580 seconds.

(b) If the reactor were a PFR, the mass balance would be:

$$-\frac{C_{A0}\mathrm{d}X_A}{r_A} = \frac{\mathrm{d}V}{Q_0}$$

And so:

$$C_{A0}\int_0^{X_A} \frac{\mathrm{d}X_A}{(-r_A)} = \frac{V}{Q_0} = t_m$$

For doing the integral, we should use a numerical method. We have:

| Time (s) | $(C_{A,0})$ (kmol/m³) | $(C_A)$ (kmol/m³) | Conversion $(X_A)$ | Reaction rate A $(-r_A)$ (mol/m³ · s) | $1/(-r_A)$ (m³ · s/mol) | Integral (m³ · s/mol) |
|---|---|---|---|---|---|---|
| 44.00 | 5.000 | 4.000 | 0.2000 | 0.022 73 | 44.00 | 0.00 |
| 52.00 | 8.000 | 6.000 | 0.2500 | 0.038 46 | 26.00 | 1.75 |
| 58.00 | 8.000 | 5.500 | 0.3125 | 0.043 10 | 23.20 | 1.54 |
| 70.00 | 8.000 | 7.000 | 0.1250 | 0.014 29 | 70.00 | −8.74 |
| 78.00 | 8.000 | 5.000 | 0.3750 | 0.038 46 | 26.00 | 12.00 |
| 166.0 | 5.000 | 3.000 | 0.4000 | 0.012 05 | 83.00 | 1.36 |
| 225.0 | 1.000 | 0.5000 | 0.5000 | 0.002 222 | 450.00 | 26.65 |
| 300.0 | 2.000 | 1.000 | 0.5000 | 0.003 333 | 300.00 | 0.00 |
| 468.0 | 5.000 | 2.000 | 0.6000 | 0.006 410 | 156.00 | 22.80 |
| PFR ideal | 6.000 | | 0.6667 | 0.008 43 | 118.58 | 9.15 |
| | | | | | Mean time $(t_m)$ (s) | 399.1 |

Note that tm for $X_A = 0.667$ (with $C_{A0} = 6\,\text{kmol/m}^3$) is lower than that necessary for $X_A = 0.6$ (with $C_{A0} = 5\,\text{kmol/m}^3$). This is only possible if kinetics is not first-order.

**Problem 2.13**  In a laboratory-scale isothermal ideal CSTR, substance A undergoes conversion into substance B. Each mole of A results in the formation of one mole of B. The CSTR is supplied with a stream of pure A, which has previously experienced partial conversion into B. The feed composition is as follows: $C_{A,\text{feed}} = 10\,\text{kmol/m}^3$ and $C_{B,\text{feed}}/C_{A,\text{feed}} = 10/990$. The following table illustrates the observed correlation between space time and feed conversion:

| $t$ (min) | Conversion |
| --- | --- |
| 0 | 0.01 |
| 0.030 | 0.2 |
| 0.041 | 0.4 |
| 0.049 | 0.5 |
| 0.062 | 0.6 |
| 0.124 | 0.8 |

(a) Establish the kinetics of this reaction.
(b) When does the maximum rate of disappearance of species A occur in terms of space time?

**Solution to Problem 2.13**

For details refer the Wiley website at http://www.wiley-vch.de/ISBN9783527354115

(a) In this situation, we have:

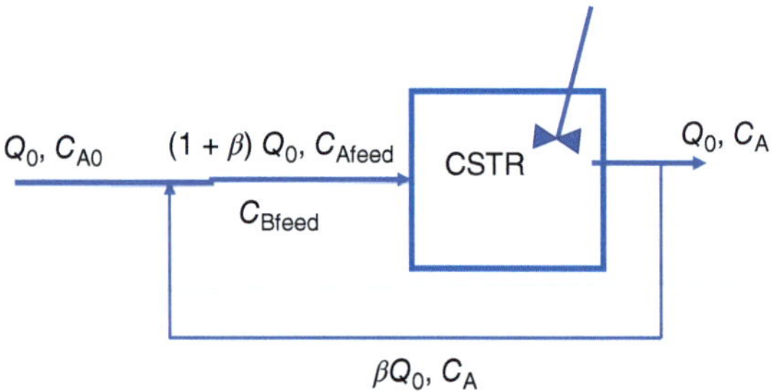


Let us try to find the kinetic expression. We will assume, as a first guess, a second-order kinetics:

$$(-r_A) = kC_A C_B$$

A balance of "A" in the reactor would be:

$$n_{A,\text{feed}} + (r_A)V = n_A$$
$$(1 + \beta)Q_0 C_{A,\text{feed}} - kC_A C_B V = (1 + \beta)Q_0 C_A$$

Let the quotient $\frac{V}{Q_0}$ be the space time in the reactor $(\bar{t})$:

$$C_{A,feed} - kC_A C_B \frac{\bar{t}}{(1+\beta)} = C_A$$

With:

$$C_A = C_{A,feed}(1 - X_A)$$
$$C_B = C_{B,feed} + C_{A,feed}X_A$$

So:

$$C_{A,feed} - kC_{A,feed}(1 - X_A)(C_{B,feed} + C_{A,feed}X_A)\frac{\bar{t}}{(1+\beta)} = C_{A,feed}(1 - X_A)$$

Solving for the kinetic constant:

$$k = \frac{-C_{A,feed}(1 - X_A) + C_{A,feed}}{C_{A,feed}(1 - X_A)(C_{B,feed} + C_{A,feed}X_A)\frac{\bar{t}}{(1+\beta)}}$$

$$= \frac{C_{A,feed}X_A}{C_{A,feed}(1 - X_A)(C_{B,feed} + C_{A,feed}X_A)\frac{\bar{t}}{(1+\beta)}}$$

Let us try to check if all the data in the reactor give the same value of kinetic constant:

| Time (min) | $X_A$ | $C_A$ | $C_B$ | $k$ (l/min $\cdot$ mol) | $r_A$ (mol/min $\cdot$ l) |
|---|---|---|---|---|---|
| 0 | 0.01 | 9.9 | 0.1 | | |
| 0.03 | 0.2 | 8 | 2 | 4.006 | 64.096 |
| 0.041 | 0.4 | 6 | 4 | 4.005 | 96.110 |
| 0.049 | 0.5 | 5 | 5 | 4.041 | 101.021 |
| 0.062 | 0.6 | 4 | 6 | 4.005 | 96.124 |
| 0.124 | 0.8 | 2 | 8 | 4.022 | 64.349 |

As we can check, the "$k$" value is the same in the different runs, so probably the kinetic law is second-order.

(b) In the previous table, it was also indicated that the value of $(-r_A) = kC_A C_B$, reaching a maximum at 0.049 min.

**Problem 2.14** The transformation of substance A into substance B, following the reaction A → B, occurs across three series-connected ideal CSTRs. These CSTRs possess volumes of 4, 10, and 20 l, respectively. The feed comprises a flow of pure A at a rate of 100 l/min, featuring a concentration of $C_A = 10$ mol/l.

The reaction rate equation is defined as $(-r_A) = kC_A^2$, where $k = 2$ l/mol $\cdot$ min.

(a) Our initial task involves determining the exit concentrations of the first, second, and third tanks.

(b) Subsequently, we will construct a graph of $1/(-r_A)$ versus $C_A$, delineating the regions corresponding to the volumes of the first, second, and third tanks.

(c) Within the graph from part (b), we will demarcate the region representing the volume of a PFR necessary to achieve an equivalent conversion to the three CSTRs in series.

(d) Estimation of the PFR volume from part (c) will be conducted using area ratios and the specified tank volumes. We will then juxtapose this estimated volume with the volume of a PFR derived from the molar balance for conversion in a PFR from $C_{A0}$ to $C_{A3}$ (the concentration of A in the third CSTR).

**Solution to Problem 2.14**

For details refer the Wiley website at http://www.wiley-vch.de/ISBN9783527354115

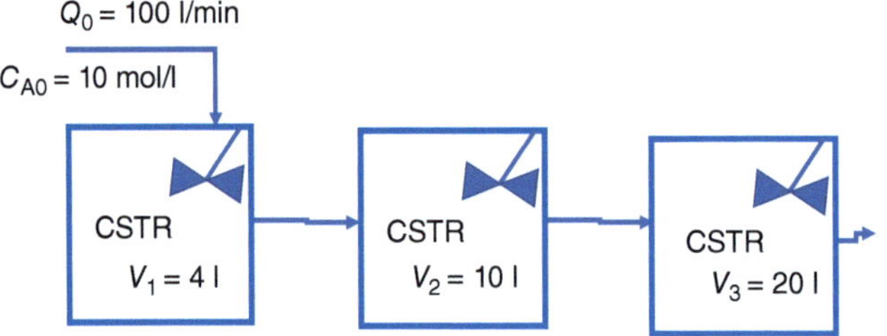


(a) The mean residence times in each reactor are 0.04, 0.1, and 0.2 min, respectively. In the first reactor, the mass balance is:

$$Q_0 C_{A0} - \left(k C_{A1}^2\right) V = Q_0 C_{A1}$$
$$C_{A1}^2 k \overline{t_1} + C_{A1} - C_{A0} = 0$$

With the data, a value of 6.55 M is found for $C_{A1}$, representing a conversion of 0.34.

Repeating the calculation with all three reactors, we have:

| | $C_A$ (mol/l) | $X_A$ | $r_A$ (mol/l·min) | $1/r_A$ (min·l/mol) |
|---|---|---|---|---|
| Input | 10.00 | 0.0000 | 100.0 | 0.010 00 |
| Reactor 1 | 6.56 | 0.3441 | 43.0 | 0.023 25 |
| Reactor 2 | 5.00 | 0.5000 | 25.0 | 0.040 00 |
| Reactor 3 | 3.90 | 0.6096 | 15.2 | 0.065 62 |

And we can do the needed figures:

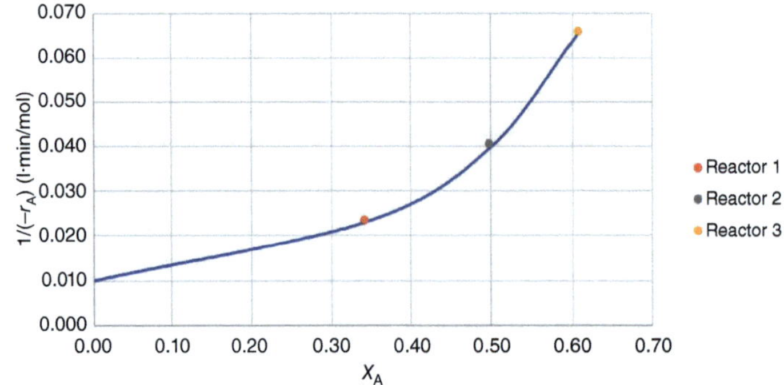

(b)

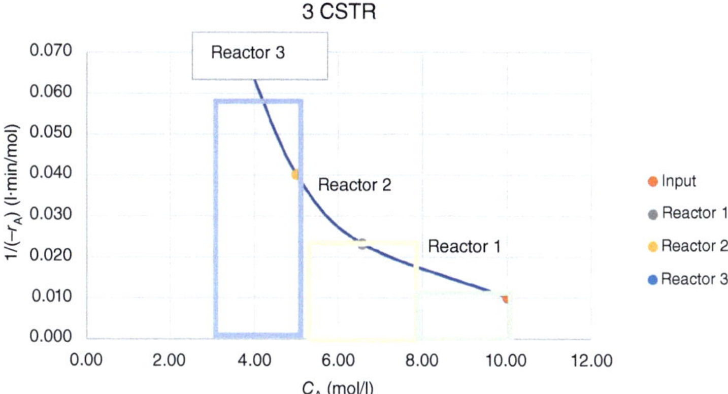

(c)

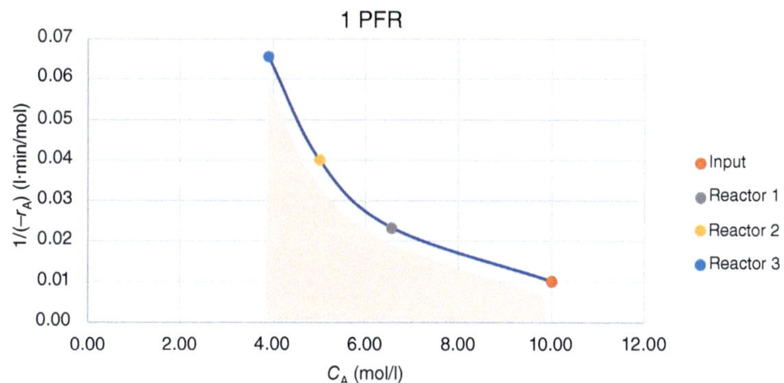

Using the relation:

$$\bar{t}_{PFR} = \int_{C_{A0}}^{C_A} \frac{dC_A}{(-r_A)} = \sum_{C_{A0}}^{C_A} \frac{(C_{A,i+1} - C_{A,i})}{(-r_{A,i})}$$

And doing a numerical integration, we find: $\bar{t}_{PFR} = 7.82\,l$.

(d) On the other hand, because of the mass balance:

$$Q_0 dC_A = -kC_A^2 dV$$

$$-Q_0 C_{A0} dX_A = -k(C_{A0}(1 - X_A))^2 dV$$

And we obtain:

$$t_{PFR} = \frac{X_A}{1 - X_A} \frac{1}{k \cdot C_{A0}} = 7.81\,l$$

As we can see, the result is quite similar to the graphical method.

**Problem 2.15**  Species B is derived from species A through a first-order irreversible reaction A → B. The system configuration involves two parallel CSTRs, as depicted in the accompanying figure. The volumes of the CSTRs are labeled as $V_1$ and $V_2$, with $V_1$ being greater than $V_2$.

(a) In order to achieve an identical level of conversion $X_{A,out}$ as $X_{A1} = X_{A2}$, we need to determine the volume of a CSTR capable of converting the molar flow $Q_0 \cdot C_{A0}$.

(b) To maximize conversion when the CSTRs are connected in series, it is necessary to determine the optimal sequence of the two reactors, $V_1$ or $V_2$, to be placed first.

(c) Assuming the total volume of the two CSTRs remains constant while allowing flexibility in choosing the volume of each individual CSTR, we can compute the optimal $V_1/V_2$ ratio. This ratio will facilitate achieving the highest possible conversion when the two CSTRs are arranged in series.

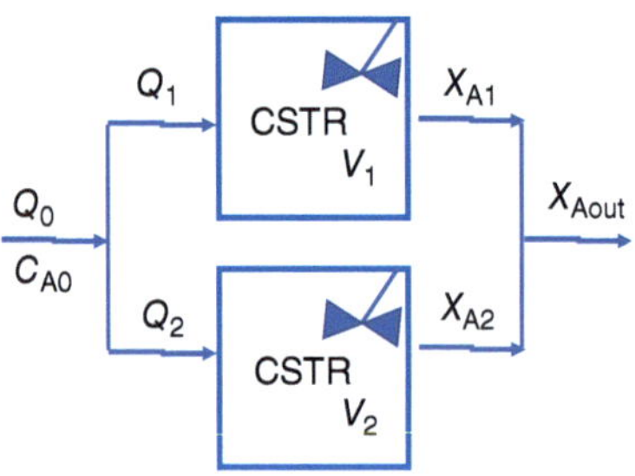


**Solution to Problem 2.15**

(a) A balance in the mix point is:

$$Q_1 X_{A1} + Q_2 X_{A2} = Q_0 X_{Aout}$$

If $X_{A1} = X_{A2}$:

$$(Q_1 + Q_2)X_{A1} = Q_0 X_{Aout}$$
$$(Q_0)X_{A1} = Q_0 X_{Aout}$$
$$X_{A1} = X_{Aout}$$

So, in this way, $V = V_1 + V_2$.

(b) With two CSTRs connected in series:

$$X_{Aout} = 1 - \frac{1}{(1 + \overline{t_1} \cdot k)(1 + \overline{t_2} \cdot k)}$$

As we have that $\overline{t} = \overline{t_1} + \overline{t_2}$, the combined conversion is not dependent on the volume of each reactor.

(c) We have that:

$$X_{Aout} = 1 - \frac{1}{\left(1 + \frac{V_1}{Q_0} \cdot k\right)\left(1 + \frac{V_{total} - V_1}{Q_0} \cdot k\right)}$$

Doing the derivative:

$$\frac{dX_{Aout}}{dV_1} = \frac{\left(\frac{k}{Q_0}\right)^2 (V_{total} - 2V_1)}{\left(\frac{k}{Q_0} \cdot V_1 + 1\right)^2 \left(\frac{k}{Q_0} \cdot (V_{total} - V_1) + 1\right)^2}$$

Equaling to zero:

$$\left(\frac{k}{Q_0}\right)^2 (V_{\text{total}} - 2V_1) = 0$$

So the maximum is found at $V_{\text{total}} = 2V_1$, so the solution is $V_1 = V_2$.

**Problem 2.16**  The production of product R is set to take place within a reactor system featuring an ideal CSTR and an ideal PFR operating simultaneously (refer to the figure).

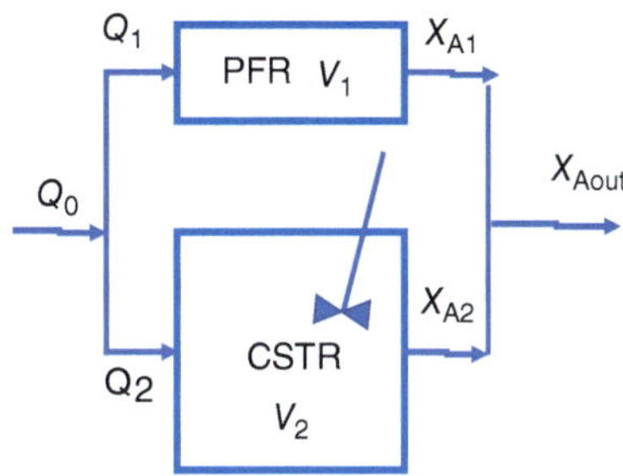


**Key Data:**
- Reaction equation: A → R, with a reaction rate constant of $k = 1\,\text{kmol/m}^3 \cdot \text{s}$
- $V_1 = 2\,\text{m}^3$
- $V_2 = 3\,\text{m}^3$
- Initial concentration of A $(C_{A0}) = 1\,\text{kmol/m}^3$
- Original flow rate $(Q_0) = 10\,\text{m}^3/\text{s}$

The objective is to determine the optimal division of the original flow $Q_0$ into flows $Q_1$ and $Q_2$ to maximize the overall conversion of the system.

**Solution to Problem 2.16**
In the PFR, we have:

$$X_{A1} = 1 - \exp\left(-\frac{kV_1}{Q_1}\right)$$

And in the CSTR:

$$X_{A2} = \frac{\frac{kV_2}{Q_2}}{1 + \frac{kV_2}{Q_2}}$$

In the mix point:

$$Q_1 X_{A1} + Q_2 X_{A2} = Q_0 X_{Aout}$$

$$Q_1\left(1 - \exp\left(-\frac{kV_1}{Q_1}\right)\right) + Q_2\left(\frac{\frac{kV_2}{Q_2}}{1 + \frac{kV_2}{Q_2}}\right) = Q_0 X_{Aout}$$

With the corresponding data, and taking into account that $Q_2 = Q_0 - Q_1$:

$$Q_1\left(1 - \exp\left(-\frac{3}{Q_1}\right)\right) + (10 - Q_1)\left(\frac{\frac{2}{10-Q_1}}{1 + \frac{2}{10-Q_1}}\right) = 10 X_{Aout}$$

$$Q_1\left(1 - \exp\left(-\frac{3}{Q_1}\right)\right) + \left(\frac{2}{1 + \frac{2}{10 - Q_1}}\right) = 10X_{\text{Aout}}$$

Deriving:

$$\frac{dX_{\text{Aout}}}{dQ_1} = \frac{1}{10} \cdot \left(\frac{Q_1^2 - 24Q_1 + 140}{(Q_1 - 12)^2} - \frac{\exp(-3Q_1)(Q_1 + 3)}{Q_1}\right)$$

Equalizing to zero, we find that the maximum is located at $Q_1 = 5.6588\,\text{m}^3/\text{s}$ and $Q_2 = 4.3412\,\text{m}^3/\text{s}$. This point gives a conversion of $X_{\text{Aout}} = 0.388$.

**Problem 2.17**   In a commercial setup, the manufacturing of product R from substance A proceeds via a first-order irreversible liquid-phase reaction: $A \rightarrow R$. The reactor utilized in this procedure demonstrates non-ideal characteristics. Based on experimental findings, the age distribution of volume elements exiting the reactor can be characterized by the equation:

$$E(t) = E(0) - 0.1t$$

We need to determine:

(a)  The value of $E(0)$, represents the initial age distribution in the reactor.
(b)  The conversion attained in the commercial reactor is such that an ideal CSTR with an equivalent average residence time distribution achieves a 60% conversion rate.
(c)  Furthermore, in the scenario where the reaction $A \rightarrow R$ is not a first-order reaction but rather a second-order reaction, we are tasked with drawing conclusions solely based on the residence time distribution. Specifically, we need to determine whether the conversion will be lower, equal to, or higher than the observed conversion in reality. The justification for the answer should be provided.

**Solution to Problem 2.17**

(a)  The cuts with axis are:
$t = 0 \rightarrow E(t) = E(0)$
$E(t) = 0 \rightarrow t_f = E(0)/0.1$
As we know, all distributions $\int_0^\infty E(t)dt = 1$
The area under $E(t)$ is in this case: Area $= \frac{E(0)}{0.1} \cdot \frac{E(0)}{2} = 1$, so $E(0) = \sqrt{0.2} = 0.447\,\text{s}^{-1}$
The value of time where the distribution is zero is $t_f = E(0)/0.1 = 4.47\,\text{s}$
From this distribution, we can also calculate:

$$t_m = \int_0^{4.47} t \cdot (0.447 - 0.1t)dt = 1.48\,\text{s}$$

(b)  We can use the segregation model:

$$\overline{X_A} = \int_0^\infty X_A(t)E(t)dt$$

With:

$$X_A = 1 - \exp(-kt)$$

$$\overline{X_A} = \int_0^{4.47} (1 - \exp(-kt))(0.447 - 0.1t)dt$$

We know that in a CSTR with $\bar{t} = 1.48$ s the conversion is 0.6, so:

$$0.6 = \frac{k \cdot 1.48}{1 + k \cdot 1.48}$$

Obtaining $k = 1.01$ s$^{-1}$. Then:

$$\overline{X_A} = \int_0^{4.47} (1 - \exp(-1.01t))(0.447 - 0.1t)dt = 0.654$$

(c) In the case of a second-order reaction:

$$X_A = \frac{kC_{A0}t}{1 + kC_{A0}t}$$

And, using the segregation model:

$$\overline{X_A} = \int_0^{4.47} \left(\frac{kC_{A0}t}{1 + kC_{A0}t}\right)(0.447 - 0.1t)dt$$

Taking $kC_{A0} = 1.01$ s$^{-1}$, we have $\overline{X_A} = 0.518$.

**Problem 2.18** In a continuous reaction system, the process of suspension polymerization unfolds. This intricate system involves suspending small monomer droplets within oil with a volume flow rate of 1 m$^3$/min. Besides, the system's design may influence factors like temperature control, initiator concentration, and mixing efficiency, which play pivotal roles in the reaction kinetics. The reaction rate of the monomer is governed by the following equation:

$$(-r_{\text{monomer}}) = 0.1 C_{\text{monomer}}^2 \cdots \left(\frac{\text{kmol}}{\text{m}^3 \cdot \text{min}}\right)$$

The system is considered to be fully segregated. Initially, the concentration of the monomer in the droplets is 1 kmol/m$^3$. Based on the analysis of residence time distribution measurements, it is determined that within the reaction system operating at a flow rate of 1 m$^3$/min, no volume elements remain within the system for durations exceeding 10 minutes. Furthermore, it is noteworthy that, for time intervals less than 10 minutes, the age distribution $E(t)$ remains uniform. However, this observation may overlook potential variations in temperature, pressure, and the presence of catalysts, all of which can significantly influence reaction kinetics and product yields.

We need to calculate:

(a) The level of conversion achieved at the reactor outlet.
(b) The reactor's volume.

**Solution to Problem 2.18**

(a) With the monomer being the reactant species:

$$(-r_A) = 0.1 C_A^2$$

For this second-order reaction:

$$C_A = \frac{C_{A0}}{1 + k \cdot t}$$

From the data, we have that:

$$\int_0^{10} E dt = 1$$

So $E = 0.1\ \text{min}^{-1} = \text{constant}$.

Using the segregation model:

$$\overline{C_A} = \int_0^{\infty} C_A(t)E(t)dt = \int_0^{10} \frac{C_{A0}}{1 + k \cdot t} E(t)dt$$

$$= \int_0^{10} \frac{1}{1 + 0.1 \cdot t} 0.1 dt = 0.693\ \frac{\text{mol}}{1}$$

$$X_A = 0.307$$

(b) The volume of the reactor can be calculated using mean residence time:

$$t_m = \int_0^{\infty} t \cdot E(t)dt = \int_0^{10} t \cdot 0.1\ dt = 5$$

$$\bar{t} = t_m = \frac{V}{Q}$$

Obtaining: $V = 0.1\ m^3/min \cdot 5\ min = 0.5\ m^3$

**Problem 2.19**  During the investigation of the first-order irreversible reaction $A \rightarrow P$ within an ideal CSTR, it was found that a conversion of 40% is achieved within an average residence time $\bar{t}$.

Subsequently, the same reaction is conducted in a non-ideal PFR under identical average residence time and temperature conditions. In this non-ideal PFR, injecting a marker pulse at $t = 0$ at the reactor inlet results in the following concentration profile at the outlet:

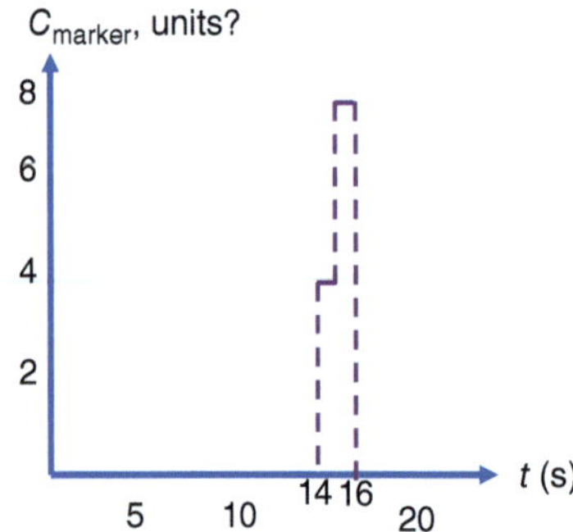


Address the given questions:

(a) Sketch and formulate the relationship between residence time distribution and time for this non-ideal PFR, which involves describing the concentration profile at the outlet. The specific relationship can be derived from the concentration profile graph.

(b) Calculate the residence time $\bar{t}$ used for the non-ideal PFR, which should be the same as in the ideal CSTR investigation, as stated in the problem.

(c) The conversion obtained with the non-ideal PFR can be determined by comparing the concentration of A at the inlet and outlet of the reactor based on the concentration profile obtained from the pulse test. The conversion can be calculated using the appropriate formula.

**Solution to Problem 2.19**

(a) The area under the $C(t)$ curve is $(4 \cdot 1 + 8 \cdot 1) = 12$ so:

$$E(t) = \begin{cases} 0, & t < 14 \\ \dfrac{4}{12}, & 14 \leq t < 15 \\ \dfrac{8}{12}, & 15 \leq t \leq 16 \end{cases}$$

This distribution has a mean residence time of:

$$t_m = \int_0^\infty t \cdot E(t)\mathrm{d}t = \int_{14}^{15} t \frac{4}{12}\mathrm{d}t + \int_{15}^{16} t \frac{8}{12}\mathrm{d}t = 15.16 \text{ s}$$

(b) and (c) In the ideal CSTR:

$$X_A = 1 - \frac{1}{1 + k\bar{t}} = \frac{k\bar{t}}{1 + k\bar{t}} = 0.4$$

And in the non-ideal PFR:

$$\overline{X_A} = \int_0^\infty (1 - \exp(-kt))E(t)\mathrm{d}t$$

Obviously $\bar{t} = 15.16$ s and then:

$$0.4 = \frac{k \cdot 15.16}{1 + k \cdot 15.16}$$

Obtaining $k = 0.044 \text{ s}^{-1}$.

And in the real reactor:

$$\overline{X_A} = \int_0^\infty (1 - \exp(-0.044t))E(t)\mathrm{d}t$$

$$\overline{X_A} = \int_{14}^{15} (1 - \exp(-0.044t))\frac{4}{12}\mathrm{d}t + \int_{15}^{16} (1 - \exp(-0.044t))\frac{8}{12}\mathrm{d}t$$

$$= 0.157 + 0.329 = 0.486$$

**Problem 2.20** In a continuous reactor, the following reaction system is carried out in liquid phase:

$$2A \rightleftharpoons B$$

$$2B \rightarrow \text{By-products}$$

The rate expressions are:

$$r_1 = k_1 C_A^2 - k_1' C_B$$

$$r_2 = k_2 C_B$$

being $k_1$ the kinetic constant of the forward reaction for producing B and $k_1'$ that of the reverse equilibrium. The feed contains 1.7 mol/l of A. The constants for the first reaction are 0.073 and 0.027, respectively, and 0.013 corresponds to the second reaction ($k_2$). The constants are in units consistent with l, mol, and min. The residence time distribution is:

$$E(t) = 0.01334\,t - 8.89 \cdot 10^{-4}t^2 + 1.48 \cdot 10^{-5}t^3$$

for values of time between 0 and 30 min.

Determine the conversion and selectivity to B:

(a) At the outlet of the real reactor.
(b) If it were a PFR.
(c) If it were an ideal stirred tank reactor.

**Solution to Problem 2.20**

For details refer the Wiley website at http://www.wiley-vch.de/ISBN9783527354115

(a) For calculating conversion in the actual reactor, with its residence time distribution, we need either to model the reactor using one of the known models for non-ideal flow or propose a combination of ideal reactors giving the corresponding $E(t)$. Let us check how this distribution is in the reactor. By giving values to time in the $E(t)$ expression, we get the following:

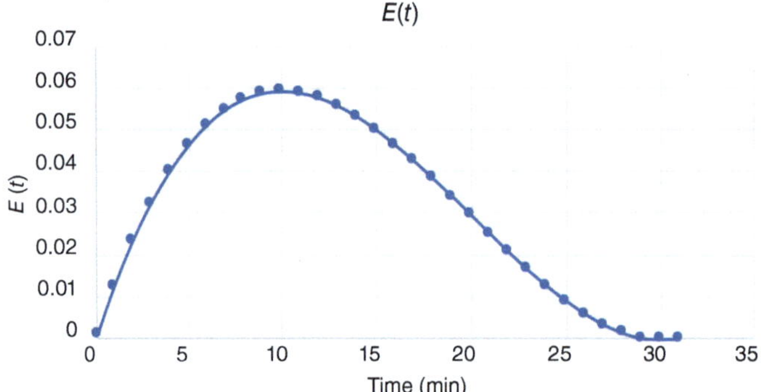


At first sight, this distribution can be modeled using the dispersion model. Let us calculate the corresponding parameters. Using the procedure explained in Problems 1.1 and 1.2, we can calculate:

$$t_m = 11.96\,\text{min}$$

$$\sigma^2 = 35.59\,\text{min}^2$$

And using the relationship:

$$\left(\frac{\sigma}{t_m}\right)^2 = 2 \cdot \text{Bo} - 2 \cdot \text{Bo}^2 \cdot \left[1 - \exp\left(-\frac{1}{\text{Bo}}\right)\right]$$

We can estimate Bo = 0.125.

But in this problem, since two independent reactions are involved and an analytical solution is apparently not possible, at least in a simple way, then let us think of a numerical solution. We begin by stating the reaction rates for two independent species (A and B) and apply the design equations for a liquid-phase batch reactor:

$$\frac{dC_A}{dt} = -2k_1 C_A^2 + 2k_1' C_B$$

$$\frac{dC_B}{dt} = k_1 C_A^2 - k_1' C_B - 2k_2 C_B$$

Since the probability that a fluid element spends more than 30 min in the reactor is zero, the integration is carried out up to this time. Using the finite differences method, integration of $C_A$ can be done by:

$$\frac{C_A^{t+1} - C_A^t}{\Delta t} = -2k_1 \left(C_A^t\right)^2 + 2k_1' C_B^t$$

and:

$$C_A^{t+1} = C_A^t + \Delta t \cdot \left(-2k_1 \left(C_A^t\right)^2 + 2k_1' C_B^t\right)$$

That is subjected to the initial value $C_A\ (t = 0) = 1.7\ \text{M}$. In the same way, $C_B$ can be obtained:

$$\frac{C_B^{t+1} - C_B^t}{\Delta t} = k_1 \left(C_A^t\right)^2 - k_1' C_B^t - 2k_2 C_B^t$$

$$C_B^{t+1} = \left(k_1 \left(C_A^t\right)^2 - k_1' C_B^t - 2k_2 C_B^t\right) \Delta t + C_B^t$$

During the integration of both differential equations by means of the finite difference method, we can also calculate the average values of each concentration by using the following relation:

$$\frac{d\overline{C_i}}{dt} = E(t) \cdot C_i(t)$$

A simple graph is obtained:

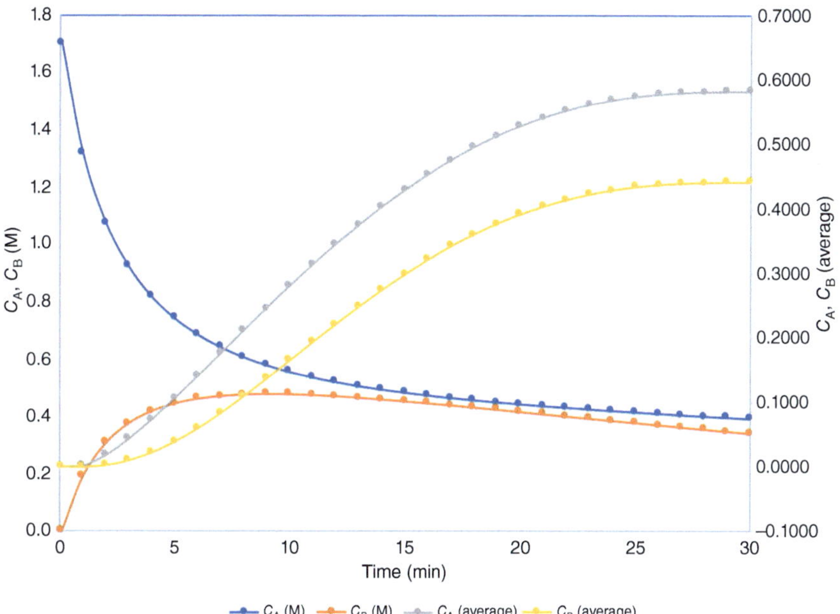

The concentrations of species A and B at the outlet of the system correspond to the average concentrations resulting from integrating the previous equations for up to 30 minutes. In the figure, we can check that:

$$\overline{C_A} = 0.582 \text{ M}$$

$$\overline{C_B} = 0.440 \text{ M}$$

We must be aware that the "profiles" for $\overline{C_A}$ and $\overline{C_B}$ lack practical utility and that we are only interested in the final results of the integration. Both profiles are shown flat for $25 < t < 30$ min; this is because the probability that fed elements come out in this period is low and there is a very little contribution to the average value.

Note that the following procedure is equivalent to applying the segregation model:

$$\overline{C_A} = \int_0^\infty C_A(t)E(t)dt$$

but values of $C_A(t)$ are known numerically instead as a definite mathematical function.

Finally, the conversion of species A at the exit of the reactor is:

$$X_A = 1 - \frac{C_A}{C_{A0}} = \frac{C_{A0} - \overline{C_A}}{C_{A0}} = 0.657$$

And the selectivity is:

$$S_B = \frac{2\overline{C_B}}{C_{A0} - \overline{C_A}} = 0.788$$

(b) Data do not include reactor volume or feed volumetric flow rate; therefore, the space time, which because it is about liquids is equal to the residence time, must be estimated based on the distribution of residence times. As we have already mentioned, $t_m = 11.96$ min. From the data used to generate former figures, we obtain that for a time of approximately 12 min, $C_A = 0.3448$ M and $C_B = 0.2202$ M. We can use these results because, for a PFR, the design equations would be the same, with the only difference being that instead of d$t$ we would have d$\bar{t}$. In this way:

$$X_A\big]_{\text{plug flow}} = 1 - \frac{C_A}{C_{A0}} = 1 - \frac{0.3448}{1.7} = 0.797$$

$$S_B\big]_{\text{plug flow}} = \frac{2C_B}{C_{A0} - C_A} = 0.325$$

(c) For a CSTR, the mass balance of each species gives:

$$\bar{t} = \frac{C_{A0} - C_{A1}}{-2k_1 C_{A1}^2 + 2k_1' C_{B1}}$$

$$\bar{t} = \frac{C_{B1}}{k_1 C_{A1}^2 - k_1' C_{B1} - 2k_2 C_{B1}}$$

With average residence time of 12 min, solving we obtain $C_A = 0.693$ M and $C_B = 0.257$ M, so:

$$X_A\big]_{\text{CSTR}} = 1 - \frac{C_A}{C_{A0}} = 1 - \frac{0.693}{1.7} = 0.592$$

$$S_B\big]_{\text{CSTR}} = \frac{2C_B}{C_{A0} - C_A} = 0.510$$

Inspection of the shape of $E(t)$, in particular by comparing the curve for $E(t)$ with those shown for the ideal reactors, reveals that the real reactor does not have a clear behavior tending toward any of the ideal continuous reactors. The selectivity of this real reactor is much higher than that for a PFR, while its conversion is in the middle of the values for ideal reactors. The wording of the exercise does not specify whether it is a tubular reactor or a stirred tank reactor. The evidence presented indicates that it may be a tubular reactor with considerable mixing or a stirred tank with poor mixing.

**Problem 2.21**  A restaurant intends to install a continuous electric fryer for the preparation of French fries. The fryer functions by continuously introducing a blend of fresh oil and raw fries, with the outgoing oil and cooked fries being separated by a conveyor belt equipped with perforations. The configuration is depicted in the accompanying figure.

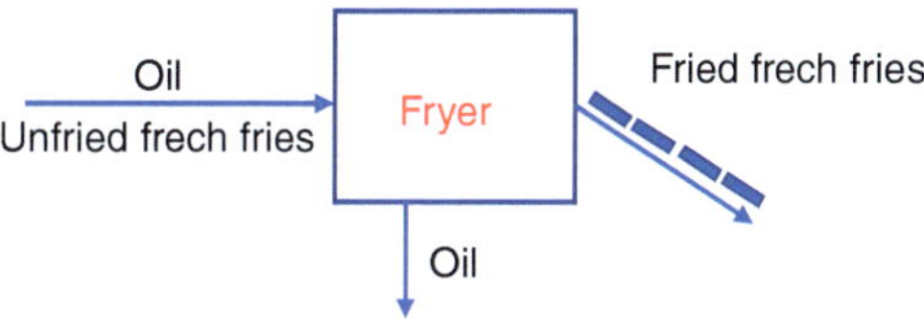


By conducting representative tracer experiments, involving the injection of a pulse of 1 mol of colored oil at the entrance of the fryer at $t = 0$, the ensuing concentration profile at the fryer's exit has been documented.

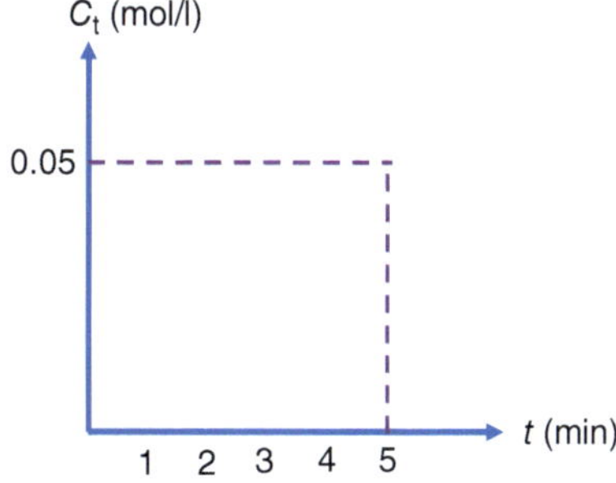


(a) Determine the volume of the fryer.

Given the mass balance equation for the regular batch electric fryer:

$$\frac{\mathrm{d}B}{\mathrm{d}t} = k(1 - B)$$

$B$ represents the degree of browning of the French fries, and the conditions are:

| | |
|---|---|
| $t = 0$ | $B = 0$ |
| $t = 3\,\text{min}$ | $B = 0.5$ |
| $t$ approaches infinity | $B = 1$ (burnt fries) |

(b) Determine the average degree of browning for the continuous fryer, assuming that the residence time distributions of oil and fries in the fryer are the same.

**Solution to Problem 2.21**

From the figure, we have:

$$C(t) = 0.05 \quad 5 \leq t \leq 0$$

The area of this square is $0.25\,\text{mol} \cdot \text{min/l}$, and so:

$$E(t) = \frac{0.05}{0.25} = 0.2 \quad 5 \leq t \leq 0$$

(a) Mean residence time is calculated using:

$$t_m = \int_0^\infty E(t)\mathrm{d}t = \int_0^5 0.2\mathrm{d}t = 2.5\,\text{min}$$

As we are using 1 mol of colored solution:

$$M_0 = \int_0^\infty Q \cdot C(t)\mathrm{d}t$$

$$1 = Q \int_0^5 0.05\mathrm{d}t$$

So $Q = 4\,\text{l/min}$. As we have that $\bar{t} = t_m = V/Q$, $V = 10\,\text{l}$.

(b) Assuming that:

$$\frac{\mathrm{d}B}{\mathrm{d}t} = k(1 - B)$$

we can do:

$$\frac{\mathrm{d}B}{1 - B} = k\mathrm{d}t$$

and finally:

$$B = \exp(-kt)$$

Using the data: $k = 0.231\,\text{min}^{-1}$

And applying the complete segregation scheme:

$$\bar{B} = \int_0^\infty B(t)E(t)\mathrm{d}t = \int_0^5 \exp(-0.231t)0.2\mathrm{d}t = 0.593$$

**Problem 2.22** Within a continuous, non-ideal reactor, the transformation of A into B takes place via a first-order irreversible liquid-phase reaction, A → B. Measurements of the residence time distribution led to the acquisition of a concentration profile at the reactor's outlet following the injection of a pulse marker M at the reactor's inlet:

| Time (min) | 0 | 1 | 2 | 3 | 4 | 5 | 6 | 7 | 8 | 9 | 10 |
|---|---|---|---|---|---|---|---|---|---|---|---|
| $M$ conc. (mol/l) | 0 | 0 | 2 | 3 | 6 | 7 | 5 | 2 | 0 | 0 | 0 |

Previous kinetic measurements conducted in an ideal CSTR had already indicated that at a residence time equivalent to the average residence time in the non-ideal reactor, the conversion reaches 90%.

Based on this information, an estimation can be made regarding the expected conversion in the real system.

**Solution to Problem 2.22**

For details refer the Wiley website at http://www.wiley-vch.de/ISBN9783527354115

As in previous problems, we can calculate:

| $t$ (min) | $C$ (mol/l) | $C(t) \cdot dt$ | $E(t)$ | $t \cdot E(t)$ | $t \cdot E(t) \cdot dt$ | $(t - t_m)^2 \cdot E(t)$ | $(t - t_m)^2 \cdot E(t) \cdot dt$ |
|---|---|---|---|---|---|---|---|
| 0 | 0 | | 0 | 0 | | 0 | |
| 1 | 0 | 0 | 0 | 0 | 0 | 0 | 0 |
| 2 | 2 | 1 | 0.08 | 0.16 | 0.08 | 0.557 568 | 0.278 784 |
| 3 | 3 | 2.5 | 0.12 | 0.36 | 0.26 | 0.322 752 | 0.440 160 |
| 4 | 6 | 4.5 | 0.24 | 0.96 | 0.66 | 0.098 304 | 0.210 528 |
| 5 | 7 | 6.5 | 0.28 | 1.40 | 1.18 | 0.036 288 | 0.067 296 |
| 6 | 5 | 6 | 0.20 | 1.20 | 1.30 | 0.369 920 | 0.203 104 |
| 7 | 2 | 3.5 | 0.08 | 0.56 | 0.88 | 0.445 568 | 0.407 744 |
| 8 | 0 | 1 | 0 | 0 | 0.28 | 0 | 0.222 784 |
| 9 | 0 | 0 | 0 | 0 | 0 | 0 | 0 |
| 10 | 0 | 0 | 0 | 0 | 0 | 0 | 0 |
| | $\Sigma(C(t)dt) = 25$ | | | $t_m = 4.64\,\text{min}$ | | $\sigma^2 = 1.8304\,\text{min}^2$ | |

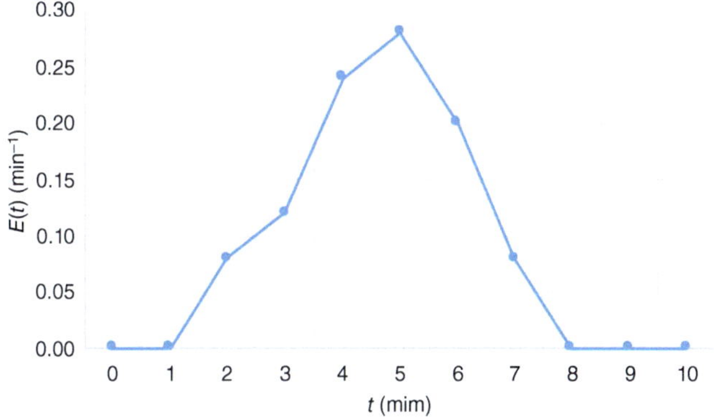

In the ideal reactor:

$$k = \frac{X_A}{\bar{t}(1 - X_A)} = \frac{0.9}{4.64 \cdot (1 - 0.9)} = 1.94 \, \text{min}^{-1}$$

Using the TIS model: $n_t = 4.64^2/1.8304 = 11.76$ tanks, and $t_i = 4.64/11.76 = 0.394$ min

$$X_A = 1 - \frac{1}{(1 + \bar{t_i} \cdot k)^{n_t}} = 0.9987$$

We can also use the segregation model and the dispersion model to obtain the same conversion (as it is a first-order reaction).

**Problem 2.23** A novel reactor has been installed to streamline the production of B via a first-order irreversible reaction, $2A \rightarrow B$. Suspicions arise regarding the ideal behavior of the reactor, prompting an investigation into its non-ideal characteristics using a pulse marker. The concentrations of the marker observed at the reactor's exit, following its introduction at $t = 0$ at the reactor's inlet, are outlined below:

| Time (min)      | 10 | 20 | 30 | 40 | 50 | 60 | 70 | 80 |
| --------------- | -- | -- | -- | -- | -- | -- | -- | -- |
| Marker (mol/l)  | 0  | 3  | 5  | 5  | 4  | 2  | 1  | 0  |

Drawing upon these measurements, we can extrapolate the conversion anticipated in the new reactor. It is pertinent to note that in an ideal CSTR operating at the designated space time for the new reactor, an 80% conversion rate is typically achieved.

**Solution to Problem 2.23**

For details refer the Wiley website at http://www.wiley-vch.de/ISBN9783527354115

Similarly to problem 2.22, we can estimate the mean residence time:

| $t$ (min) | $C$ (mol/l) | $C \cdot dt$ (mol $\cdot$ min/l) | $E(t)$ | $t \cdot E(t)$ | $t \cdot E(t) \cdot dt$ |
| --- | --- | --- | --- | --- | --- |
| 10 | 0 |    | 0.000 | 0.00 |       |
| 20 | 3 | 15 | 0.015 | 0.30 | 1.50  |
| 30 | 5 | 40 | 0.025 | 0.75 | 5.25  |
| 40 | 5 | 50 | 0.025 | 1.00 | 8.75  |
| 50 | 4 | 45 | 0.020 | 1.00 | 10.00 |
| 60 | 2 | 30 | 0.010 | 0.60 | 8.00  |
| 70 | 1 | 15 | 0.005 | 0.35 | 4.75  |
| 80 | 0 | 5  | 0.000 | 0.00 | 1.75  |

$$t_m = 40 \, \text{min}$$

The kinetic constant is:

$$k = \frac{X_A}{\bar{t}(1 - X_A)} = \frac{0.8}{40 \cdot (1 - 0.8)} = 0.1 \, \text{min}^{-1}$$

Using the segregation model with:

$$X_A(t) = 1 - \exp(-kt)$$

We can easily obtain: $\overline{X_A} = 0.961$.

**Problem 2.24**   A continuous process is used to convert pure A (flow rate of 12 m³/min, initial concentration $C_{A0} = 2$ kmol/m³) into B using two parallel ideal PFRs. The ongoing reaction involves a second-order irreversible liquid-phase process: A → B, with a rate constant of $k = 0.5$ m³/kmol · min.
   The reactor system is depicted in the subsequent diagram:

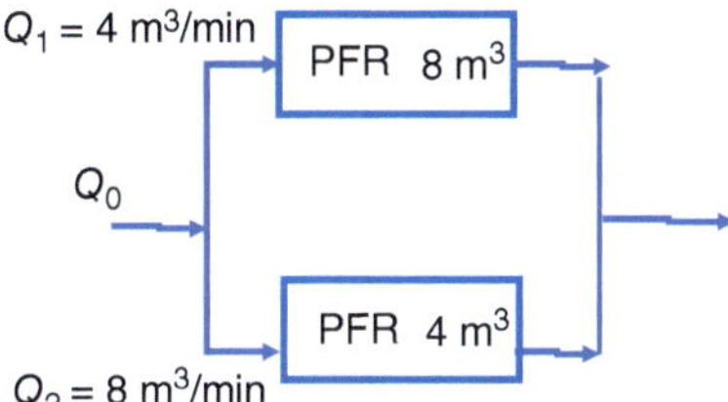


(a) Compute the system's conversion.
(b) Draw the age distribution function, $E(t)$, against time, $t$, for the system, clearly marking the scale divisions and units of the axes.
(c) Illustrate the cumulative age distribution function, $F(t)$, plotted against time, $t$, for the system, indicating the divisions and units of the axes clearly.

**Solution to Problem 2.24**
(a) In that system:

$$\overline{t_1} = 8/4 = 2 \, \text{min}$$
$$\overline{t_2} = 4/8 = 0.5 \, \text{min}$$

For the second-order reaction in one PFR:

$$X_A = \frac{kC_{A0}\overline{t}}{1 + kC_{A0}\overline{t}}$$

Using the data:

$$X_{A1} = 0.67$$
$$X_{A2} = 0.33$$

And:

$$X_A = \frac{Q_1 X_{A1} + Q_2 X_{A2}}{Q_1 + Q_2} = 0.443$$

(b) and (c) The required figures would be:

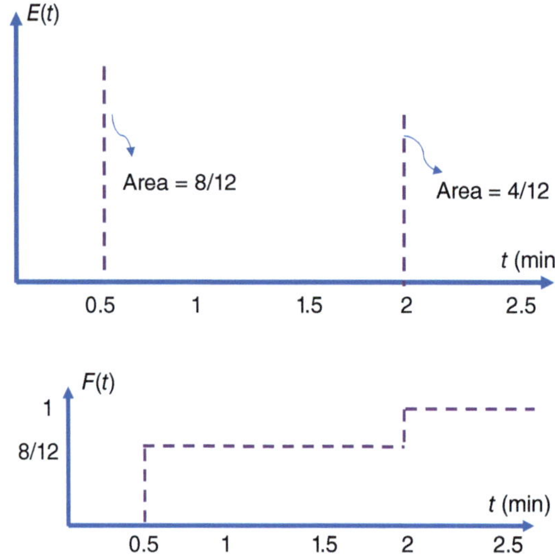

The $E(t)$ is a combination of two delta Dirac functions, centered at the residence times of each reactor. One of the deltas would have a higher height, with a total area of 8/12 (ratio of flow rate passing through the reactor versus total flow rate), whereas the other is 4/12. The $F(t)$ accounts for 8/12 of the total and the rest exits at 2 min.

### Problem 2.25

(a) Sketch and formulate the $E(t)$ versus $t$ curve for a short-circuited ideal CSTR and a short-circuited continuous ideal PFR, respectively.
(b) Calculate the concentration at the exit of the short-circuited ideal for a first-order irreversible reaction, CSTR on the basis of $E(t)$.

### Solution to Problem 2.25

(a) Schematically, we have:

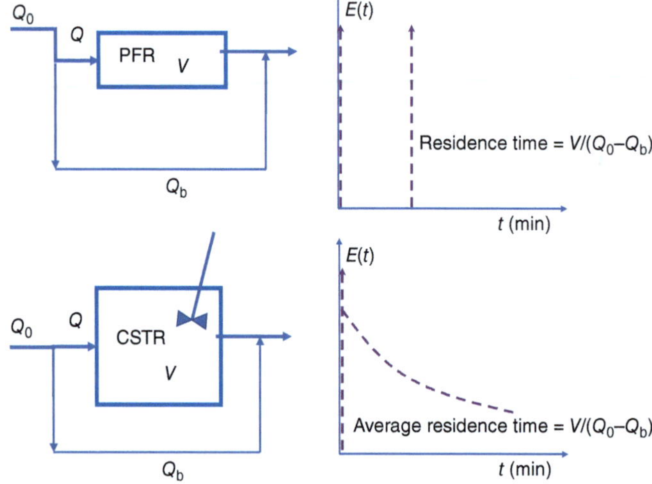

(b) For the CSTR:

$$E(t) = \frac{1}{\bar{t}} \cdot \exp\left(-\frac{t}{\bar{t}}\right) = \frac{Q_0 - Q_b}{V} \cdot \exp\left(-t \cdot \frac{Q_0 - Q_b}{V}\right)$$

Using the segregation model:

$$\begin{aligned}
\overline{X_{A,\text{CSTR}}} &= \int_0^\infty X_A(t)E(t)\,dt \\
&= \int_0^\infty (1 - \exp(-kt))\frac{Q_0 - Q_b}{V} \cdot \exp\left(-t \cdot \frac{Q_0 - Q_b}{V}\right) dt \\
&= \frac{k\frac{Q_0 - Q_b}{V}}{1 + k\frac{Q_0 - Q_b}{V}}
\end{aligned}$$

We can also calculate the final conversion, bearing in mind that the bypass has a zero conversion:

$$\begin{aligned}
X_{A,\text{final}} &= \frac{\overline{X_{A,\text{CSTR}}} \cdot (Q_0 - Q_b) + 0 \cdot Q_b}{Q_0} \\
&= \frac{k\frac{Q_0 - Q_b}{V}}{1 + k\frac{Q_0 - Q_b}{V}} \frac{(Q_0 - Q_b)}{Q_0} = \frac{k(Q_0 - Q_b)^2}{V + k(Q_0 - Q_b)} \cdot \frac{1}{Q_0}
\end{aligned}$$

**Problem 2.26**  In an ideal CSTR equipped with three blades, B is generated from A through a first-order irreversible liquid-phase reaction A $\rightarrow$ B (with a rate constant of $k = 9\,\text{min}^{-1}$). The CSTR possesses a volume of $10\,\text{m}^3$, and the feed rate is $5\,\text{m}^3/\text{min}$, with $C_A = 1\,\text{kmol/m}^3$.

During the CSTR's operation, the middle blades of the stirrer break off. Experimental findings reveal that the tank no longer behaves as an ideal CSTR but can be treated as two ideal CSTRs in series, each with a volume of $5\,\text{m}^3$.

(a) Is the stirrer to be repaired? Provide reasoning for your response.
(b) Establish an expression for $E(t)$ under the new circumstances.
(c) Plot $E(t)$ versus $t$ for the revised scenario.
(d) Compute the conversion achieved in the reactor under the revised conditions. Perform this calculation based on $E(t)$.

**Solution to Problem 2.26**

For details refer the Wiley website at http://www.wiley-vch.de/ISBN9783527354115

(a).- (d)

Let us try to calculate the conversion in the two systems (new and broken). In the new reactor:

$$\bar{t} = 10/5 = 2\,\text{min}$$

And, in the CSTR with first-order kinetics:

$$X_A = \frac{k\bar{t}}{1 + k\bar{t}} = \left(\frac{0.9 \cdot 2}{1 + 0.9 \cdot 2}\right) = 0.642$$

In the broken reactor, two CSTR in series of average residence time:

$$\overline{t_1} = \overline{t_2} = 5/5 = 1 \text{ min}$$

So we have:

$$X_A = 1 - \frac{1}{(1 + \overline{t_i} \cdot k)^{n_t}} = 1 - \frac{1}{(1 + 1 \cdot 0.9)^2} = 0.722$$

As we see, the conversion is higher in the broken reactor, so surely it will not be repaired.

In the TIS model:

$$E(t) = \frac{t^{n_t-1}}{(n_t - 1)!\,\overline{t_i}^{n_t}} \cdot \exp\left(-\frac{t}{\overline{t_i}}\right) = \frac{t}{1 \cdot 1^2} \cdot \exp\left(-\frac{t}{1}\right) = t \cdot \exp(-t)$$

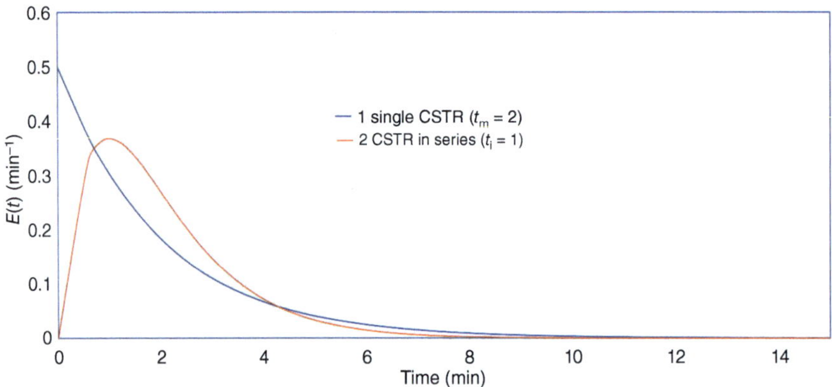


Using the $E(t)$ data:

$$\overline{X_A} = \int_0^\infty X_A(t)E(t)\mathrm{d}t = \int_0^\infty (1 - \exp(-kt))\, t \cdot \exp(-t)\mathrm{d}t$$

With $k = 9 \text{ min}^{-1}$, the final value of average conversion is 0.99.

**Problem 2.27**  In a tubular reactor, a hydrocarbon fraction undergoes conversion through a first-order reaction with a rate constant ($k$) of $1.06 \text{ s}^{-1}$. The feed flow rate is 15 l/s, and the properties of the feed are a density ($\rho$) of $800 \text{ kg/m}^3$ and a viscosity ($\eta$) of $5.0 \cdot 10^{-4} \text{ N} \cdot \text{s/m}^2$. The reactor itself has a non-ideal behavior and is characterized by the axial dispersion model.

The reactor measures 1 m in length and possesses an internal diameter of 30 cm. Our task involves determining the conversion in the reactor through the application of the dispersion coefficient.

**Solution to Problem 2.27**

For details refer the Wiley website at http://www.wiley-vch.de/ISBN9783527354115

Using the data in the text, it is easy to find that:

$$V = 1 \cdot \pi \cdot \left(\frac{0.3}{2}\right)^2 = 0.071 \text{ m}^3$$

$$\bar{t} = \frac{V}{Q} = 4.712 \text{ s}$$

$$\text{Da} = k \cdot \bar{t} = 5$$

$$\text{Section} = 0.0707 \text{ m}^2$$

$$v = \frac{Q}{\text{Section}} = 0.212 \frac{\text{m}}{\text{s}}$$

$$\text{Re} = \frac{\rho \cdot v \cdot L}{\mu} = 1.02 \cdot 10^5$$

The value of Re can be used to estimate the dispersion number $\text{Bo} = D_e/(u \cdot L)$ by using some correlation. At this so turbulent regime (Levenspiel, O. (1999). *Chemical Reaction Engineering*. 3e. John Wiley & Sons, New York, 54. http://dx.doi.org/10.1021/ie990488g), getting a value of $\text{Bo} = 0.2$.

Using the equation:

$$1 - X_A = \frac{4a \cdot \exp\left(\frac{1}{2\text{Bo}}\right)}{(1+a)^2 \cdot \exp\left(\frac{a}{2\text{Bo}}\right) - (1-a)^2 \cdot \exp\left(-\frac{a}{2\text{Bo}}\right)}$$

$$a = [1 + 4(\bar{t} \cdot k) \cdot \text{Bo}]^{1/2}$$

We finally get a conversion of 0.9610.


**Problem 2.28**  In a practical tubular reactor with a volume of $6 \text{ m}^3$, a stream of pure component A ($Q_0 = 3 \text{ m}^3/\text{s}$) undergoes conversion to produce product B via a first-order reaction with a rate constant ($k$) of $2.5 \text{ s}^{-1}$. The real tubular reactor's behavior can be modeled by a series of four ideal CSTRs, all possessing the same volume.

(a) Illustrate (in a single figure) the $E(t)$ versus $t$ curves for an ideal CSTR, a continuous ideal PFR, and the real reactor.
(b) Utilizing the "four CSTRs in series" model, compute the conversion in the real reactor.
(c) Alternatively, the behavior of the real reactor can be described by the dispersion model. Based on the conversion calculated in part (b), determine the dispersion coefficient and the anticipated conversion, considering that the reactor's cross section is $1 \text{ m}^2$.

**Solution to Problem 2.28**

For details refer the Wiley website at http://www.wiley-vch.de/ISBN9783527354115

(a) In this situation, we have:

$$\bar{t} = 2 \text{ s} \quad n_t = 4 \quad \bar{t_i} = 0.5 \text{ s}$$

With the help of a spreadsheet, we can plot the corresponding $E(t)$:

| | | |
|---|---|---|
| PFR ideal | $E(t) = \delta(t - \bar{t})$ | $\bar{t} = 2$ |
| CSTR ideal | $E(t) = \dfrac{1}{\bar{t}} \cdot \exp(-t/\bar{t})$ | $\bar{t} = 2$ |
| 4 CSTR in series | $E(t) = \dfrac{t^3}{6\bar{t}_i^4} \cdot \exp\left(-\dfrac{t}{\bar{t}_i}\right)$ | $\bar{t}_i = 0.5\ \text{s}$ |

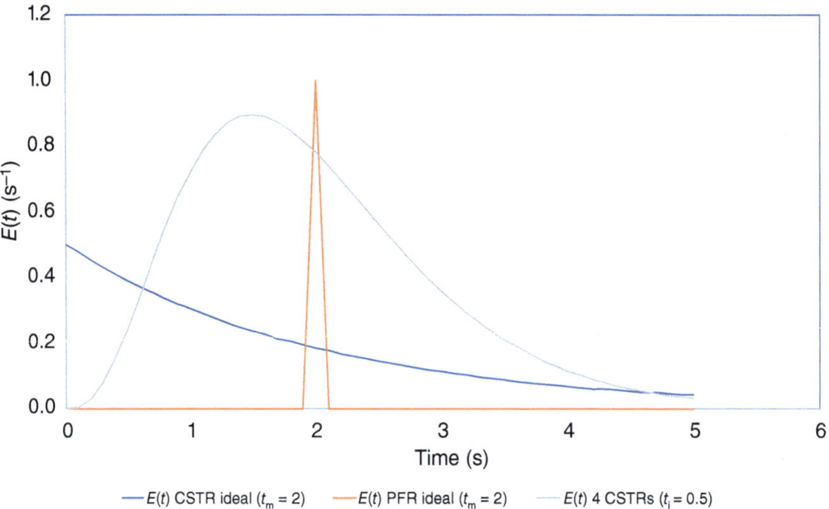


(b) For calculating the average conversion in this system:

$$X_A = 1 - \frac{1}{(1 + \bar{t}_i \cdot k)^{n_t}} = 1 - \frac{1}{(1 + 0.5 \cdot 2.5)^4} = 0.961$$

(c) First, we should fit the curve obtained with the TIS model ($n_t = 4$) to the dispersion model. In the TIS curve, we calculate $t_m$ and $\sigma^2$:

$$t_m = n_t \cdot \bar{t}_i = 2\ \text{s}$$

$$\sigma^2 = \frac{t_m^2}{n_t} = \frac{2^2}{4} = 1\ \text{s}^2$$

And then, with the dispersion model (closed–closed recipient):

$$\left(\frac{\sigma}{t_m}\right)^2 = 2 \cdot \text{Bo} - 2 \cdot \text{Bo}^2 \cdot \left[1 - \exp\left(-\frac{1}{\text{Bo}}\right)\right] = \left(\frac{1}{2}\right)^2$$

Solving, we found Bo $= 0.167$. Now, the conversion with the dispersion model is calculated:

$$1 - X_A = \frac{4a \cdot \exp\left(\frac{1}{2\text{Bo}}\right)}{(1 + a)^2 \cdot \exp\left(\frac{a}{2\text{Bo}}\right) - (1 - a)^2 \cdot \exp\left(-\frac{a}{2\text{Bo}}\right)}$$

$$a = [1 + 4(\bar{t} \cdot k) \cdot \text{Bo}]^{1/2}$$

We finally get a conversion of 0.965.

**Problem 2.29**   Identify the correct answers from the options below. Additional comments are unnecessary.

(a) The following $C$–$t$ plot is obtained in a batch reactor of constant volume. Ideal reaction A → P is being produced. With this behavior, the reaction order is:

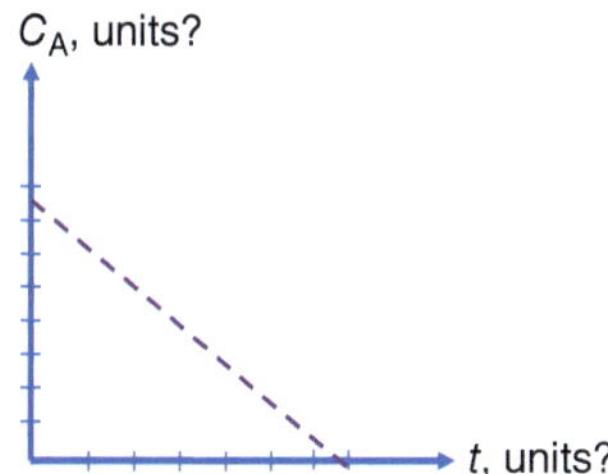


(a1) $n = 0$; (a2) $n = 1$; (a3) $n = 2$.

(b) With the conversion being equal regardless of the reactor type, the total selectivity for R in the following system is:

$$A \rightarrow R \quad (-r_A) = k_1 C_A^1$$
$$A \rightarrow S \quad (-r_A) = k_2 C_A^2$$

(b1)  Higher in a plug flow reactor than in an ideal tank reactor.
(b2)  Lower in a plug flow reactor than in an ideal tank reactor.
(b3)  Equal in an ideal plug flow and an ideal stirred tank reactor.

(c) In a system where reaction equations and kinetics are unknown, assuming that volumes of CSTR and PFR are equal, treat the conversion of A into P and Q:
   (c1)  With both reactors in similar conditions, the production of P and Q in the plug flow is larger than in the tank reactor.
   (c2)  The former assertion is only true when all reactions have zero-order kinetics.
   (c3)  Only under specific reaction conditions and kinetics the conversion can be compared.

(d) A system consisting of two reactors connected in series will be used for the production of P and Q from reactant A. Kinetics and reaction equations are unknown. We can use plug flow reactors P1 and P2 and the stirred tanks T1 and T2, all of them having the same volume.
   (d1)  It is not possible to know if a better conversion will be obtained by combinations P1 + T1 (plug flow followed by tank reactor) or T1 + P1.
   (d2)  Although ignoring kinetics, conversion in the system P1 + P2 will be higher than T1 + T2.
   (d3)  Although ignoring kinetics, conversion in the system P1 + T1 will be higher than T1 + T2.

(e) The product P is formed from reactant A in a first-order irreversible reaction. At the exit of the reactor, part of the product is recirculated as is, i.e., with no separation, to the inlet of the reactor. The production of P, comparing the system with and without the recycling, is:

(e1) Higher.

(e2) Lower.

(e3) Equal.

(f) A tubular reactor will be used as a fixed bed with porous catalyst K. The catalyzed reaction is first-order and irreversible, producing P from reactant A. In the system, positive activation energy $E_a$ was measured, and the effective diffusion coefficient $(D_e)$ does not depend on temperature. We will compare two experiments where only the radius of the catalyst and temperature are changed.

Exp. (1) Radius equals 2 mm and temperature is 100 °C.

Exp. (2) Radius equals 4 mm and temperature is 80 °C.

(f1) The minor radius and higher temperature of experiment 1 make the conversion higher, regardless of Ea.

(f2) The conversion of A in experiment 2 is higher.

(f3) The value of the activation energy can influence whether experiment 1 or 2 gives the higher conversion.

(g) The equation: $-r_A = -(dC_A/dt)$

(g1) Defines the rate of A's conversion.

(g2) Represents the molar balance of species A for an ideal batch reactor with a constant volume.

(g3) Remains valid until the system reaches steady state.

(h) For the liquid-phase reaction $A + 2B \rightarrow P$, the molar balance is:

(h1) $(C_{A0} - C_A) = (C_{B0} - C_B)$

(h2) $(C_{A0} - C_A) = 2(C_{B0} - C_B)$

(h3) $2(C_{A0} - C_A) = (C_{B0} - C_B)$

(i) Two continuous ideal PFRs, with volumes V1 and V2, respectively, where V1 > V2, are connected in series. These reactors undergo a second-order irreversible reaction.

(i1) The highest overall conversion is achieved when V1 precedes V2.

(i2) The highest overall conversion is achieved when V2 precedes V1.

(i3) The highest overall conversion is independent of the sequence of reactors.

**Problem 2.30**  Our objective is to convert a stream of pure component A, with an initial concentration $(C_{A0})$ of 5 kmol/m³ and a volumetric flow rate $(Q_0)$ of 0.5 m³/min, into products P and Q within an unidentified reactor type. The conversion process follows a first-order irreversible reaction:

$$A \rightarrow P + Q$$

with kinetics described by:

$$-r_A = 0.2C_A \text{kmol/m}^3 \text{ min)}$$

The following figure provides the $E(t)$ curve obtained from residence time distribution measurements.

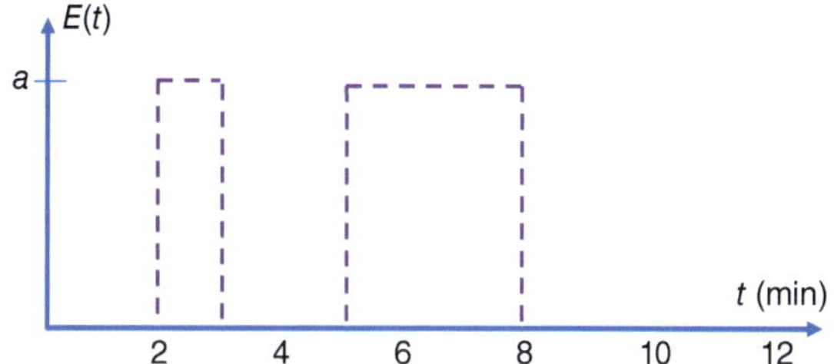


(a) Initially, our task involves ascertaining the volume of the unidentified reactor.
(b) Prior to commencing the intended process, a sudden alteration in tracer concentration is introduced into the reactor feed. What will the tracer concentration profile at the reactor outlet look like under these circumstances?
(c) Utilizing the residence time distribution curve, we can compute the conversion at the reactor outlet for the aforementioned first-order process.
(d) Would an ideal PFR with identical volume and operating conditions result in a lower, identical, or higher conversion?

**Solution to Problem 2.30**

(a) First of all, let us calculate the value of $E(t) = a$ not included in the graph. We know that the total area of an $E(t)$ curve should be unity, so from the plot:

$$1 = a \cdot (3 - 2) + a \cdot (8 - 5) = 4a \rightarrow a = 0.25$$

Let us consider two different compartments. In this sense, the area under the first square is $A_1 = 0.25$ and $A_2 = 0.75$. In a system with these two compartments in parallel, the area would represent the ratio of flow rate flowing through each of the compartments, i.e., $Q_1 = 0.25Q_0$ and $Q_2 = 0.75Q_0$. As we know that $Q_0 = 0.5\,\mathrm{m^3/min}$, we find: $Q_1 = 0.125$ and $Q_2 = 0.375\,\mathrm{m^3/min}$.

From the data in the plot, we can estimate the average residence time in both systems and in the complete reactor:

$$t_m = \int_0^\infty t \cdot E(t)\mathrm{d}t = \int_2^3 t \cdot 0.25\mathrm{d}t + \int_5^8 t \cdot 0.25\mathrm{d}t = 0.625 + 4.875 = 5.5\,\mathrm{min}$$

Using the relationship:

$$t_m = \bar{t} = \frac{V}{Q_0} \rightarrow V = 5.5 \cdot 0.5 = 2.75\,\mathrm{m^3}$$

(b) In the system, a step tracer run would give the following:

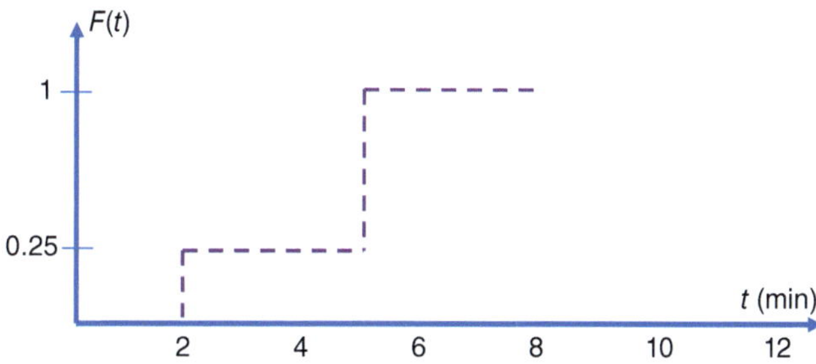

(c) As the kinetics of the reaction is first-order, we can use, for example, the complete segregation model.

$$\overline{X_A} = \int_0^\infty X_A(t)E(t)dt$$

$$= \int_0^\infty (1 - \exp(-kt))E(t)dt$$

$$\overline{X_A} = \int_2^3 (1 - \exp(-0.2t)) \cdot 0.25dt + \int_5^8 (1 - \exp(-0.2t)) \cdot 0.25dt$$

$$= 0.0981 + 0.5425 = 0.6406$$

(d) A PFR of the same volume and flow rate would give:

$$X_A = 1 - \exp(-k\bar{t}) = 1 - \exp(-0.2 \cdot 5.5) = 0.667$$

**Problem 2.31** In a CSTR with a volume $(V)$ of $3\,\text{m}^3$, a stream of component A is reacted to form some products through an irreversible first-order reaction. The initial conditions are a volumetric flow rate $(Q_0)$ of $1.5\,\text{m}^3/\text{min}$ and an initial concentration $(C_{A0})$ of $5\,\text{kmol/m}^3$. The constant of the rate equation for the process is $1.5\,\text{min}^{-1}$.

(a) To begin, let us calculate the concentration $C_{A\text{out}}$ assuming an ideal CSTR. Upon examining the flow profile within the glass reactor, it becomes evident that the reactor's configuration can also be likened to that of two parallel PFRs, each possessing the volumes as indicated (see figure).

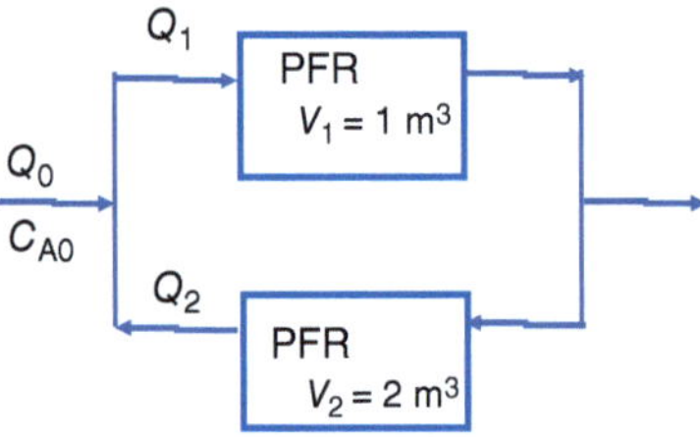


(b) Assuming both PFRs are ideal, we need to determine the ratio $Q_1/Q_0$ in order to achieve the same concentration $C_{A\text{out}}$ as that assumed for the ideal CSTR.

(c) Now, consider the scenario of a reaction with an order different from unity. Please provide an explanation of your answer(s) regarding the ratio $Q_1/Q_0$ in this case.

**Solution to Problem 2.31**

(a) In a CSTR with a first-order reaction:

$$C_A = \frac{C_{A0}}{1 + k\bar{t}} = \frac{5}{1 + 1.5 \cdot \frac{3}{1.5}} = 1.25\ \frac{\text{kmol}}{\text{m}^3}$$

This is a conversion of $X_A = 1 - \frac{C_A}{C_{A0}} = 0.75$

(b) With two PFRs in parallel, we have:
In the first one:

$$X_{A1} = 1 - \exp(-k\bar{t_1}) = 1 - \exp\left(-k \cdot \frac{V_1}{Q_1}\right) = 1 - \exp\left(-1.5 \cdot \frac{1}{Q_1}\right)$$

And, in the second one:

$$X_{A2} = 1 - \exp\left(-1.5 \cdot \frac{2}{Q_2}\right)$$

In the mixing point:

$$X_{A,\text{final}} = \frac{Q_1 X_{A1} + Q_2 X_{A2}}{Q_0}$$

$$= \frac{Q_1\left(1 - \exp\left(-1.5 \cdot \frac{1}{Q_1}\right)\right) + (Q_0 - Q_1)\left(1 - \exp\left(-1.5 \cdot \frac{2}{(Q_0 - Q_1)}\right)\right)}{Q_0}$$

Bearing in mind that $X_{A,\text{final}} = 0.75$ and that $Q_0 = 1.5\,\text{m}^3/\text{min}$

$$\frac{Q_1}{1.5}\left(1 - \exp\left(-\frac{1.5}{Q_1}\right)\right) + \left(1 - \frac{Q_1}{1.5}\right)\left(1 - \exp\left(-\frac{3}{(1.5 - Q_1)}\right)\right) = 0.75$$

We can solve and find that $Q_1 = 1.05\,\text{m}^3/\text{min}$, so $Q_1/Q_0 = 0.7$.

(c) For an order different from unity, we have in a PFR:

$$Q dC_A = -kC_A^n dV$$

$$\int_{C_{A0}}^{C_A} \frac{dC_A}{C_A^n} = \int_{C_{A0}}^{C_A} C_A^{-n} dC_A = -k \int_0^V \frac{dV}{Q}$$

$$\left. \frac{1}{1-n} C_A^{1-n} \right]_{C_{A0}}^{C_A} = -\frac{kV}{Q}$$

$$\frac{1}{1-n}\left(C_A^{1-n} - C_{A0}^{1-n}\right) = -\frac{kV}{Q} = -k\bar{t}$$

$$C_A = \left(k\bar{t}(n-1) + C_{A0}^{1-n}\right)^{\frac{1}{1-n}}$$

For the final concentration, we will have:

$$C_{A,\text{final}} = \frac{Q_1 C_{A1} + Q_2 C_{A2}}{Q_0}$$

$$= \frac{Q_1\left(k\bar{t_1}(n-1) + C_{A0}^{1-n}\right)^{\frac{1}{1-n}} + Q_2\left(k\bar{t_2}(n-1) + C_{A0}^{1-n}\right)^{\frac{1}{1-n}}}{Q_0}$$

We can do a similar study to that of previous question.


**Problem 2.32** According to our calculations, a PFR would obtain 99.99% conversion of a reagent in aqueous solution. However, the DTR shown in the figure was obtained. If $C_{A0} = 1000\,\text{mol/m}^3$, calculate the concentration that can be obtained in the reactor if the reaction is first-order.

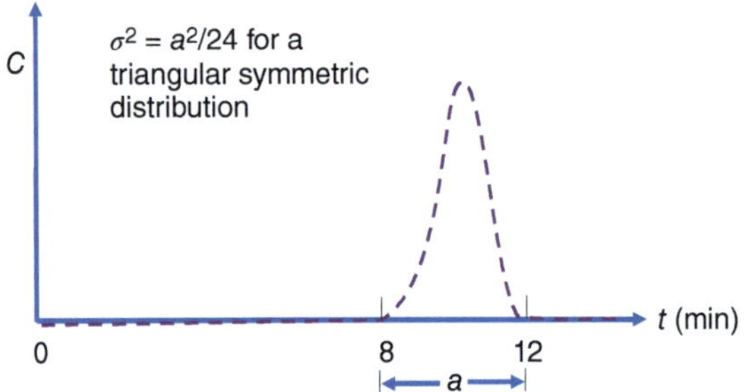


**Solution to Problem 2.32**

In a PFR, we obtain a 0.9999 conversion for the reactant. In the DTR, we can check that the residence time for the vessel is $t_m = 10\,\text{min} = \bar{t}$ and then:

$$X_A = 1 - \exp(-k\bar{t})$$

$$0.9999 = 1 - \exp(-k \cdot 10)$$

$$k = 0.921 \text{ min}^{-1}$$

In a dispersed flow reactor, we know that:

$$\left(\frac{\sigma}{t_m}\right)^2 = 2 \cdot \text{Bo} - 2 \cdot \text{Bo}^2 \cdot \left[1 - \exp\left(-\frac{1}{\text{Bo}}\right)\right]$$

And:

$$\sigma^2 = \frac{a^2}{24} = \frac{16}{24}$$

$$\left(\frac{\sigma}{t_m}\right)^2 = 0.000\,666\,7$$

By trial and error: $\text{Bo} = 0.017\,22$

And for a first-order reaction in the dispersed flow:

$$1 - X_A = \frac{4a \cdot \exp\left(\frac{1}{2\text{Bo}}\right)}{(1 + a)^2 \cdot \exp\left(\frac{a}{2\text{Bo}}\right) - (1 - a)^2 \cdot \exp\left(-\frac{a}{2\text{Bo}}\right)}$$

$$a = [1 + 4(\bar{t} \cdot k) \cdot \text{Bo}]^{1/2}$$

$$a = [1 + 4(10 \cdot 0.921) \cdot 0.017\,22]^{1/2} = 1.27$$

$$1 - X_A = \ldots = 0.0004$$

So, the expected conversion is $X_A = 0.996$.

If we prefer the TIS model (not very much appropriate in this case), we can do:

$$n_t = \frac{t_m^2}{\sigma^2} = 150 \text{ tanks}$$

$$\bar{t_i} = \frac{\bar{t}}{n_t} = \frac{1}{15} \text{ min}$$

And:

$$X_A = 1 - \frac{1}{(1 + \bar{t_i} \cdot k)^{n_t}} = 1 - \frac{1}{\left(1 + \frac{1}{15} \cdot 0.921\right)^{150}} = 0.9998$$

**Problem 2.33**   The second-order aqueous phase reaction $A + B \rightarrow R + S$ is carried out in a large tank reactor ($V = 3\ m^3$) and with a feed mixture ($C_{A0} = C_{B0}$); the conversion of the reactants is 90%. Unfortunately, the agitation in the tank is quite defective, and the tracer tests indicate that the reactor follows a model like the one schematized in the figure, with $\alpha = 0.3$ and $R = 0.5$ ($V_d$ is dead volume and $V_m$ well-mixed volume):

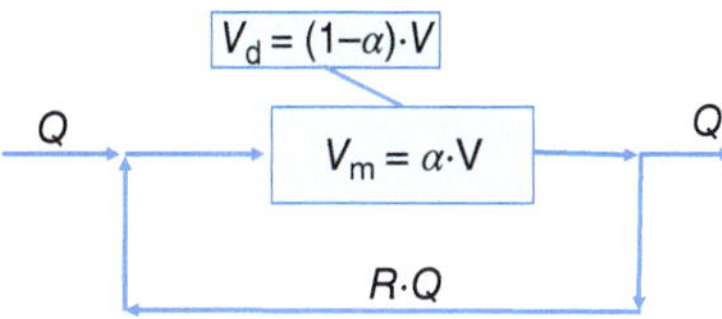


(a)  How will the RTD of this reactor be qualitatively?
(b)  Calculate the size of the stirred tank reactor that will give the same conversion as the real reactor.

**Solution to Problem 2.33**

For details refer the Wiley website at http://www.wiley-vch.de/ISBN9783527354115

(a)  In this system, the residence time compared to that expected is:

$$\overline{t_{actual}} = \frac{V_m}{Q_{reactor}} = \frac{\alpha V}{(1 + R)Q} = \frac{\alpha}{(1 + R)}\overline{t_{expected}}$$

As we have $\alpha = 0.3$ and $R = 0.5$, $\overline{t_{actual}} = 0.2\ \overline{t_{expected}}$, i.e., the residence time is much smaller than that expected.

Recirculation only promotes the mixing of different elements of the fluid that are already well mixed in the CSTR, so the $E(t)$ will be very similar to that of a CSTR, but changing the average residence time:

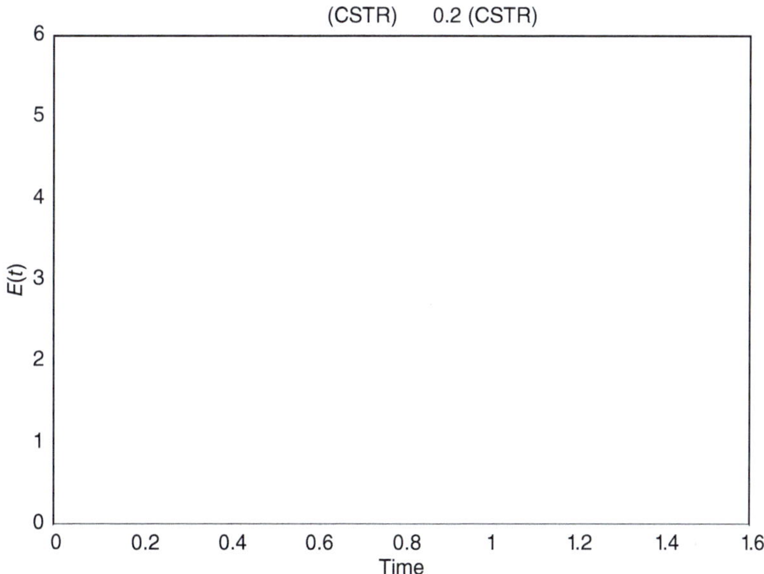

(b) Calculate the size of the stirred tank reactor that will give the same performance as the real reactor.

In the real reactor, the mass balance is:

$$E + G = A + S$$

$$C_{A0}(1 + R)Q + r_A V_m = C_{A0}(1 - X_A)(1 + R)Q$$

$$-r_A = kC_A C_B = k(C_{A0}(1 - X_A))(C_{B0}(1 - X_A)) = kC_{A0}^2(1 - X_A)^2$$

$$C_{A0}(1 + R)Q - kC_{A0}^2(1 - X_A)^2 V_m = C_{A0}(1 - X_A)(1 + R)Q$$

$$QX_A(1 + R) = kC_{A0}(1 - X_A)^2 V_m$$

From data: $V_m = 0.3 \cdot 3 = 0.9\,\text{m}^3$, $R = 0.5$, $X_A = 0.9$, and we get:

$$\frac{kC_{A0}}{Q} = 150$$

On the other hand, in the ideal reactor:

$$C_{A0}Q + r_A V = C_{A0}(1 - X_A)Q$$

$$Q - kC_{A0}(1 - X_A)^2 V = (1 - X_A)Q$$

$$1 - \frac{kC_{A0}}{Q}(1 - X_A)^2 V = (1 - X_A)$$

With $kC_{A0}/Q = 150$ and $X_A = 0.9$, the necessary volume is $0.61\,\text{m}^3$.

**Problem 2.34**   In a tubular reactor, a first-order reaction is carried out, in which $k = 2\,\text{s}^{-1}$. For a feed of $2\,\text{l/s}$, a conversion of 90% is expected to be obtained, but the reactor does not work quite well, and a conversion of 75% is obtained. After a characterization experiment, it is concluded that the reactor can be modeled as follows:

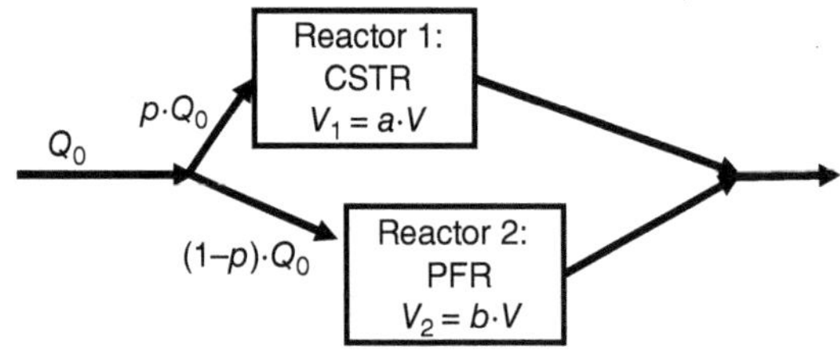


The fraction "$a$" is estimated to be 0.5. Calculate the fraction of flow that would pass through each branch of the system.

**Solution to Problem 2.34**

In the ideal reactor:

$$C_A = C_{A0} \cdot \exp(-k\bar{t})$$

And equally:

$$X_A = 1 - \exp(-k\bar{t})$$

we obtain a conversion of 0.9, so:

$$0.9 = 1 - \exp\left(-2\frac{V}{2}\right)$$

And the volume is: $V = 2.3\,l$, and so $V_1 = V_2 = 1.15\,l$.

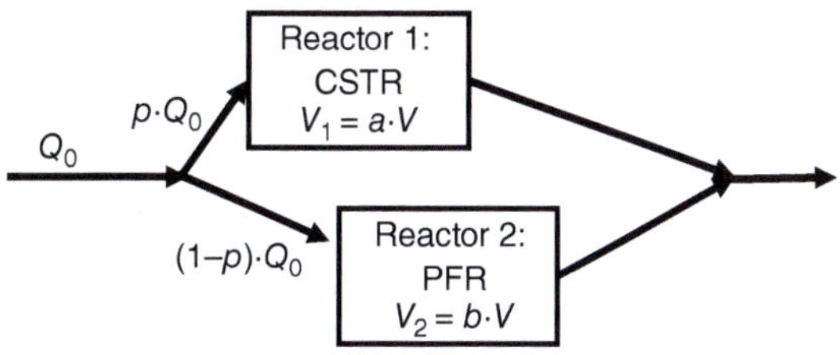


In the system, we have a part of the flow going through the CSTR and the other through the PFR. In the Reactor 1, a mass balance is:

$$pQ_0 C_{A0} + r_A \cdot V_1 = pQ_0 C_{A1}$$

$$pQ_0 C_{A0} - k \cdot C_{A1} \cdot V_1 = pQ_0 C_{A1}$$

$$C_{A1} = \frac{pQ_0 C_{A0}}{pQ_0 + kV_1} = \frac{1}{1 + k\frac{V_1}{pQ_0}} C_{A0}$$

In Reactor 2:

$$C_{A2} = C_{A0} \cdot \exp\left(-\frac{kV_2}{(1-p)Q_0}\right)$$

And proportionally, mixing both flows:

$$C_{A\text{final}} = \frac{pQ_0 C_{A1} + (1-p)Q_0 C_{A2}}{Q_0} = pC_{A1} + (1-p)C_{A2}$$

$$= p\frac{1}{1 + k\frac{V_1}{pQ_0}} C_{A0} + (1-p)C_{A0} \cdot \exp\left(-\frac{kV_2}{(1-p)Q_0}\right)$$

With the data:

$$0.1 C_{A0} = p\frac{1}{1 + \frac{1.15}{2p}} C_{A0} + (1-p)C_{A0} \cdot \exp\left(-\frac{2 \cdot 1.15}{(1-p)2}\right)$$

The only unknown is the ratio "$p$" resulting: $p = 0.65$.

**Problem 2.35**  The second-order aqueous reaction $A + B \rightarrow R + S$ is carried out in a large tank reactor ($V = 30\,m^3$) and with a feed mixture ($C_{A0} = 2C_{B0}$); the conversion of the reactants is 70%. Unfortunately, the agitation in the tank is quite defective, and the tracer tests indicate that the reactor follows a model like the one schematized in the figure. Calculate the size of the stirred tank reactor that will give the same performance as the real reactor.

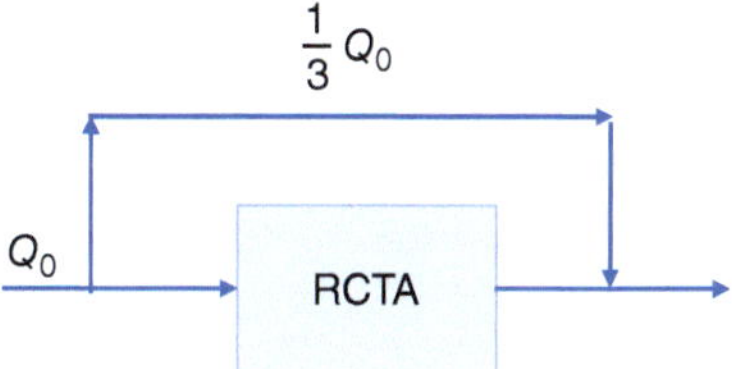

**Solution to Problem 2.35**

The reaction rate is:

$$-r_A = k \cdot C_A \cdot C_B$$

In any moment, we have that:

$$C_A = C_{A0} \cdot (1 - X_A)$$

$$C_B = C_{B0} + C_{A0} \cdot \left(\frac{-1}{-(-1)}\right) X_A = C_{B0} - C_{A0} \cdot X_A = \frac{C_{A0}}{2} - C_{A0} \cdot X_A$$

$$= C_{A0} \cdot \left(\frac{1}{2} - X_A\right)$$

In a CSTR with flow rate $Q$ and $X_A$ conversion at the exit:

$$Q \cdot C_{A0} + r_A \cdot V = Q \cdot C_A$$

In this case:

$$Q \cdot C_{A0} - k \cdot C_{A0}(1 - X_A) \cdot C_{A0} \cdot \left(\frac{1}{2} - X_A\right) \cdot V = Q \cdot C_{A0} \cdot (1 - X_A)$$

Dividing by $(Q \cdot C_{A0})$:

$$1 - k \cdot (1 - X_A) \cdot C_{A0} \cdot \left(\frac{1}{2} - X_A\right) \cdot \frac{V}{Q} = (1 - X_A)$$

From this equation:

$$k \cdot C_{A0} \cdot \frac{V}{Q} = \frac{X_A}{(1 - X_A) \cdot \left(\frac{1}{2} - X_A\right)}$$

Note that in this reactor, the conversion cannot be greater than 0.5 since we would lack reagent "B." Now, in the REAL reactor, with recirculation, the flow rate inside the reactor is $2/3 \cdot Q_0$, and the conversion at the outlet is:

$$X'_A \cdot Q_0 = \left(\frac{2}{3} \cdot Q_0\right) \cdot X_A + \left(\frac{Q_0}{3}\right) \cdot 0$$

So, taking into account that $X'_A = 0.7$, we have:

$$0.7 = \left(\frac{2}{3}\right) \cdot X_A$$

And we obtain $X_A = 1.05$, which indicates that either the recirculation flow or the conversion is incorrect. The configuration that gives us the problem is not possible. Assuming a conversion of $X'_A = 0.3$, we would have $X_A = 0.45$, and:

$$k \cdot C_{A0} \cdot \frac{V}{Q} = \frac{X_A}{(1 - X_A) \cdot \left(\frac{1}{2} - X_A\right)} = 16.36$$

As in the reactor, the flow is 2/3 of the total, and the volume is 30 m³:

$$k \cdot C_{A0} \cdot \frac{30}{\frac{2}{3} \cdot Q_0} = 16.36$$

And so:

$$\frac{k \cdot C_{A0}}{Q_0} = 0.363$$

In the IDEAL reactor, it is also true that:

$$k \cdot C_{A0} \cdot \frac{V_{CSTR}}{Q_{CSTR}} = \frac{X_A}{(1 - X_A) \cdot \left(\frac{1}{2} - X_A\right)}$$

But now:

$$k \cdot C_{A0} \cdot \frac{V_{CSTR}}{Q_0} = \frac{0.3}{(1 - 0.3) \cdot \left(\frac{1}{2} - 0.3\right)} = 2.14$$

And:

$$0.363 \cdot V_{CSTR} = 2.14$$

Finally, oh surprise:

$$V_{CSTR} = 5.89 \text{ m}^3$$

which is logically much smaller than the REAL reactor.

**Problem 2.36** Researchers undertook experiments to investigate the residence time distribution curves of a twin-screw extruder employed in the gelatinization process of cornmeal. In this process, solid cornmeal undergoes heating to the reaction temperature and is extruded as a gel. The following data illustrate the effluent concentration of a dye subsequent to a pulse injection under operational parameters of 250 rpm, a mass flow rate of 25 kg/h, and a temperature of 150 °C (concentration in % weight):

| *t* (s) | *C(t)* |
| --- | --- |
| 0.0 | 0 |
| 42.0 | 0 |
| 44.0 | 0.023 |
| 46.0 | 0.038 |
| 48.0 | 0.05 |
| 50.0 | 0.054 |
| 52.0 | 0.055 |
| 54.0 | 0.052 |
| 56.0 | 0.046 |
| 58.0 | 0.042 |
| 60.0 | 0.034 |
| 62.0 | 0.028 |
| 64.0 | 0.022 |

| *t* (s) | *C*(*t*) |
| --- | --- |
| 66.0 | 0.018 |
| 68.0 | 0.012 |
| 70.0 | 0.006 |
| 72.0 | 0.003 |
| 74.0 | 0.002 |
| 76.0 | 0.001 |
| 78.0 | 0.001 |
| 80.0 | 0.0005 |
| 82.0 | 0 |

(a) Utilize the provided dataset to create a plot illustrating the cumulative residence time distribution function, F(t), and determine the mean residence time.

(b) The gelatinization process of cornmeal can be modeled as a first-order reaction with a rate constant given by:

$k = 1.28 \times 10^{11} \exp(-10\,367/T) \ \text{min}^{-1}$ with $T$ in K

Hence, at 150 °C, $k = 2.90 \ \text{min}^{-1}$

(b1) Calculate the conversion predicted using the segregated flow model when the extruder operates isothermally at 150 °C.

(b2) Determine the conversion predicted using the TIS model.

(b3) Find the conversion that would be achieved at 150 °C in a PFR with a mean residence time equal to that in the extruder.

(c) Provide insights on the obtained results. The experimentally observed conversion is approximately 70%. Discuss potential reasons why the observed conversion might be notably lower than the conversion predicted by the various models. Additionally, assuming an uncertainty of ±5 °C in the operating temperature, compute the corresponding uncertainty in the conversion predicted using the PFR model.

**Solution to Problem 2.36**

(a) Using the same procedure as in previous problems:

| *t* (s) | *C*(*t*) | *C*(*t*) · Δ*t* | *E*(*t*) | *F*(*t*) | *t* · *E*(*t*) · Δ*t* | $(t - t_m)^2 \cdot E(t)$ | $(t - t_m)^2 \cdot E(t) \cdot \Delta t$ |
| --- | --- | --- | --- | --- | --- | --- | --- |
| 0.0 | 0 | 0 | 0 | 0 | 0 | 0 | 0 |
| 42.0 | 0 | $2.30 \cdot 10^{-2}$ | 0.000 | 0.000 | 0.000 | 0.000 | 0.000 |
| 44.0 | 0.023 | $6.10 \cdot 10^{-2}$ | 0.024 | 0.048 | 1.051 | 2.648 | 2.648 |
| 46.0 | 0.038 | $8.80 \cdot 10^{-2}$ | 0.039 | 0.127 | 2.866 | 2.871 | 5.519 |
| 48.0 | 0.05 | $1.04 \cdot 10^{-1}$ | 0.052 | 0.231 | 4.307 | 2.214 | 5.085 |
| 50.0 | 0.054 | $1.09 \cdot 10^{-1}$ | 0.056 | 0.343 | 5.296 | 1.150 | 3.364 |
| 52.0 | 0.055 | $1.07 \cdot 10^{-1}$ | 0.057 | 0.457 | 5.774 | 0.365 | 1.516 |

| $t$ (s) | $C(t)$ | $C(t) \cdot \Delta t$ | $E(t)$ | $F(t)$ | $t \cdot E(t) \cdot \Delta t$ | $(t - t_m)^2 \cdot E(t)$ | $(t - t_m)^2 \cdot E(t) \cdot \Delta t$ |
|---|---|---|---|---|---|---|---|
| 54.0 | 0.052 | $9.80 \cdot 10^{-2}$ | 0.054 | 0.565 | 5.886 | 0.015 | 0.381 |
| 56.0 | 0.046 | $8.80 \cdot 10^{-2}$ | 0.048 | 0.660 | 5.591 | 0.103 | 0.118 |
| 58.0 | 0.042 | $7.60 \cdot 10^{-2}$ | 0.044 | 0.748 | 5.205 | 0.525 | 0.629 |
| 60.0 | 0.034 | $6.20 \cdot 10^{-2}$ | 0.035 | 0.818 | 4.648 | 1.057 | 1.582 |
| 62.0 | 0.028 | $5.00 \cdot 10^{-2}$ | 0.029 | 0.876 | 3.921 | 1.623 | 2.679 |
| 64.0 | 0.022 | $4.00 \cdot 10^{-2}$ | 0.023 | 0.922 | 3.265 | 2.049 | 3.672 |
| 66.0 | 0.018 | $3.00 \cdot 10^{-2}$ | 0.019 | 0.960 | 2.696 | 2.459 | 4.508 |
| 68.0 | 0.012 | $1.80 \cdot 10^{-2}$ | 0.012 | 0.984 | 2.081 | 2.261 | 4.720 |
| 70.0 | 0.006 | $9.00 \cdot 10^{-3}$ | 0.006 | 0.997 | 1.283 | 1.491 | 3.752 |
| 72.0 | 0.003 | $5.00 \cdot 10^{-3}$ | 0.003 | 1.003 | 0.660 | 0.951 | 2.442 |
| 74.0 | 0.002 | $3.00 \cdot 10^{-3}$ | 0.002 | 1.007 | 0.378 | 0.787 | 1.738 |
| 76.0 | 0.001 | $2.00 \cdot 10^{-3}$ | 0.001 | 1.009 | 0.233 | 0.479 | 1.266 |
| 78.0 | 0.001 | $1.50 \cdot 10^{-3}$ | 0.001 | 1.011 | 0.160 | 0.572 | 1.051 |
| 80.0 | 0.0005 | $5.00 \cdot 10^{-4}$ | 0.001 | 1.012 | 0.123 | 0.337 | 0.909 |
| 82.0 | 0 | 0 | | | | | |

$$\Sigma(C(t)\mathrm{d}t) = 0.963 \qquad t_m = 54.53\,\text{s} \qquad \sigma^2 = 42.615\,\text{s}^2$$

Graphically:

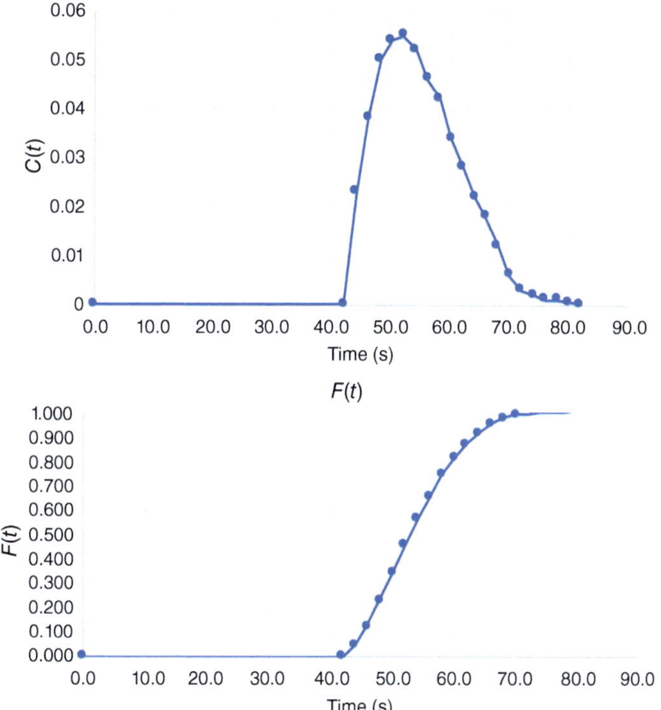

(b) Using the segregation model:

$$\overline{X_A} = \int_0^\infty X_A(t)E(t)\,dt$$

We can use the spreadsheet to calculate the conversion. $X_A(t)$ is given by the conversion obtained in a batch reactor that in the case of a first-order reaction is:

$$X_A(t) = 1 - \exp(-k \cdot t)$$

With $k = 2.9\,\text{min}^{-1}$ we have:

| $t$ (s) | $X_A(t)$ | $E(t) \cdot X_A(t) \cdot \Delta t$ |
|---------|----------|-------------------------------------|
| 0.0 | 0.000 | 0.000 |
| 42.0 | 0.838 | 0.020 |
| 44.0 | 0.851 | 0.054 |
| 46.0 | 0.864 | 0.079 |
| 48.0 | 0.875 | 0.095 |
| 50.0 | 0.885 | 0.100 |
| 52.0 | 0.895 | 0.099 |
| 54.0 | 0.904 | 0.092 |
| 56.0 | 0.912 | 0.083 |
| 58.0 | 0.919 | 0.073 |
| 60.0 | 0.926 | 0.060 |
| 62.0 | 0.932 | 0.048 |
| 64.0 | 0.938 | 0.039 |
| 66.0 | 0.943 | 0.029 |
| 68.0 | 0.947 | 0.018 |
| 70.0 | 0.952 | 0.009 |
| 72.0 | 0.956 | 0.005 |
| 74.0 | 0.960 | 0.003 |
| 76.0 | 0.963 | 0.002 |
| 78.0 | 0.966 | 0.002 |
| 80.0 | 0.969 | 0.001 |
| 82.0 | | |
| | $X_A$ mean = 0.903 | |

Second column is calculated using the previous equation, and the last one by multiplying $X(t)$ by $E(t)\Delta t$. Finally, summing all values in the last column:

$$\overline{X_A} = \sum_{\text{all points}} X_A(t)E(t)\Delta t = 0.903$$

In a PFR under the same conditions:

$$\overline{X_A} = 1 - \exp(-k\bar{t}) = 0.906$$

Using the TIS model:

$$n_t = \frac{t_m^2}{\sigma^2} = \frac{54.53^2}{42.61} = 69.0 \text{ tanks}$$

So the reactor is similar to a lot of CSTRs connected in series.

Note that the segregation model is hardly applicable to the situation, as the RTD indicated that the mixing is quite important.

(c) Considering a temperature of 145 °C, the repetition of the former calculations gives:

$$k = 2.16 \text{ min}^{-1}$$

$$\overline{X_A} \text{ (segregation model)} = 0.855$$

$$X_A \text{ (PFR)} = 0.860$$

Now, a temperature of 155 °C gives:

$$k = 3.87 \text{ min}^{-1}$$

$$\overline{X_A} \text{ (segregation model)} = 0.971$$

$$X_A \text{ (PFR)} = 0.97$$

As we can see, the conversion is very unsensitive both to the temperature and the RTD being held in different situations. This can be the reason why the experimental conversion is quite low compared to that expected.

**Problem 2.37**  In the river scenario outlined in Problem 1.20, an organic species A poses a potential pollutant found in the waste stream discharged by a factory into the upper end of the reach. The average concentration of species A at the upper end of the reach measures 100 mg/l. Species A undergoes a first-order reaction in solution, converting into a harmless product. The effective rate constant at the stream temperature is 0.01 min$^{-1}$. Employing the segregated flow model, ascertain the fraction of the initial concentration of species A present at the downstream end of the reach.

**Solution to Problem 2.37**

For details refer the Wiley website at http://www.wiley-vch.de/ISBN9783527354115

In Problem 1.20 we found that:

$$t_m = 143.44 \text{ min}$$
$$\sigma^2 = 249.42 \text{ min}^2$$

On the other hand, for a first-order reaction:

$$X_A = 1 - \exp(-kt)$$

Using the segregated flow model:

$$\overline{X_A} = \int_0^\infty X_A(t)E(t)\mathrm{d}t = \int_0^\infty (1 - \exp(-kt))E(t)\mathrm{d}t$$

$$= \int_0^\infty (1 - \exp(-0.01t))E(t)\,dt$$

$$= \sum_{\text{all points}} (1 - \exp(-0.01t))E(t)\Delta t$$

In the spreadsheet, we can do:

| $t$ (min) | $E(t)$ | $X_A(t)$ | $E(t) \cdot X_A(t) \cdot \Delta t$ |
|---|---|---|---|
| 0.0 | 0.000 | 0.000 | 0.000 |
| 60.0 | 0.000 | 0.451 | 0.000 |
| 75.0 | 0.000 | 0.528 | 0.000 |
| 90.0 | 0.000 | 0.593 | 0.004 |
| 105.0 | 0.001 | 0.650 | 0.033 |
| 120.0 | 0.006 | 0.699 | 0.100 |
| 130.0 | 0.023 | 0.727 | 0.046 |
| 132.5 | 0.028 | 0.734 | 0.031 |
| 134.0 | 0.028 | 0.738 | 0.081 |
| 138.0 | 0.027 | 0.748 | 0.081 |
| 142.5 | 0.021 | 0.759 | 0.116 |
| 150.0 | 0.019 | 0.777 | 0.158 |
| 165.0 | 0.008 | 0.808 | 0.072 |
| 180.0 | 0.004 | 0.835 | 0.026 |
| 195.0 | 0.000 | 0.858 | 0.000 |
| 210.0 | 0.000 | 0.878 | 0.000 |
| 300.0 | 0.000 | 0.950 | 0.000 |

$$X_A \text{ mean} = \text{sum of all increments} = 0.748$$

so:

$$C_A = C_{A0}(1 - X_A) = 100\,(1 - 0.748) = 25.2\,\frac{\text{mg}}{\text{l}}$$

**Problem 2.38**  In a CSTR, the reaction between organic oxide and calcium carbonate (1 mol/l at the inlet) takes place:

$$R_2O + CaCO_3 \rightarrow R_2CO_3 + CaO$$

It is known that the reaction is second-order or first-order with respect to each reactant. The rate constant is $k = 25\,\text{l/mol} \cdot \text{min}$, and the average residence time is 10 min.

(a) What concentration of $R_2O$ in the reactor causes the $CaCO_3$ conversion to be 0.9?

(b) If there are no excess reactants, what is the concentration of $R_2O$ at the inlet?

(c) If the reactor were PFR, under the same input conditions, what conversion would be obtained?

**Solution to Problem 2.38**

Calling "A" to the carbonate and "B" to the oxide, we have that:

$$-r_A = kC_A C_B$$

$$k = 25\,\frac{\text{mol}}{\text{l}\cdot\text{min}}$$

$$\bar{t} = 10\,\text{min}$$

As the fluid is not compressible, with a $1:1$ stoichiometry, the following is fulfilled in any case:

$$C_A = C_{A0}(1 - X_A)$$

$$C_B = C_{B0} - C_{A0}X_A$$

(a) In the CSTR, a balance of "A" gives:

$$n_{A0} + r_A V = n_A$$

$$C_{A0} + r_A \bar{t} = C_A$$

$$C_{A0} - kC_A C_B \bar{t} = C_A$$

With the data $X_A = 0.9$, we have that $C_A = 0.1 \cdot C_{A0}$, and:

$$C_{A0} - k(0.1C_{A0})C_B\bar{t} = 0.1C_{A0}$$

The value of $C_{A0}$ is known (1 mol/l), giving: $C_B = 0.036$ mol/l.

(b) We can write:

$$C_B = C_{B0} - C_{A0}X_A$$

$$0.036 = C_{B0} - 1 \cdot 0.9$$

So:

$$C_{B0} = 0.936\,\frac{\text{mol}}{\text{l}}$$

(c) In a PFR, a mass balance of "A" gives:

$$dn_A = r_A \cdot dV$$

$$C_{A0} \cdot Q \cdot dX_A = -k(C_{A0}(1 - X_A))(C_{B0} - C_{A0}X_A)dV$$

That, rearranging:

$$\int_{C_{A0}}^{C_A} \frac{dC_A}{C_{B0} - C_{A0} + C_A} = -k\bar{t}$$

Solving the integral and using the limits:

$$C_A = C_{B0} \cdot \exp(-k\bar{t}) + C_{A0} - C_{B0}$$

$$C_A = 0.065\,\frac{\text{mol}}{\text{l}}$$

$$X_A = 1 - \frac{C_A}{C_{A0}} = 0.935$$

**Problem 2.39**   In a batch reactor, the dehydration of organic alcohols occurs at room temperature, following the typical equation:

$$CH_3 - CH_2 - OH \rightarrow CH_2 = CH_2 + H_2O$$

For a reaction time of 30 min, the conversion is 0.7.

(a) What is the rate constant?
(b) By increasing the temperature by $10\,°C$, the rate increases by a factor of 100. What activation energy does the reaction have?

**Solution to Problem 2.39**

(a) Let us use the following nomenclature:

$$A \rightarrow B + H_2O$$

We know the following data:

| $t$ (min) | $X_A$ |
|---|---|
| 30 | 0.7 |

In the batch reactor:

$$r_A = \frac{dC_A}{dt}$$

$$-kC_A = \frac{dC_A}{dt}$$

$$-kC_{A0}(1 - X_A) = -C_{A0}\frac{dX_A}{dt}$$

$$k \int_0^t dt = \int_0^{X_A} \frac{dX_A}{(1 - X_A)}$$

And integrating:

$$-\ln(1 - X_A)\big]_0^{X_A} = kt$$

$$\ln(1 - X_A) = -kt$$

Using the data given, $k$ can be calculated:

| $t$ (min) | $X_A$ | $\ln(1 - X_A)$ |
|---|---|---|
| 30 | 0.7 | −1.2059 |

Obtaining: $k = 0.233\ \text{min}^{-1}$

(b) Using the Arrhenius law:

$$k = k_0 \exp\left(-\frac{E}{R_g T}\right)$$

At two different temperatures:

$$k_{T1} = k_{T_2} \exp\left(-\frac{E}{R_g}\left(\frac{1}{T_1} - \frac{1}{T_2}\right)\right)$$

And also:

$$\frac{E}{R_g} = \frac{\ln\left(\frac{k_{T2}}{k_{T1}}\right)}{\frac{1}{T_1} - \frac{1}{T_2}}$$

So:

$$\frac{E}{R_g} = \frac{\ln 100}{\frac{1}{298} - \frac{1}{308}} = 422\,68\ \text{K}$$

**Problem 2.40**   Using data of RTD given in Problem 1.19, suppose that a catalyst is present in a dilute solution of species A (100 pmol/ml) in the liquid carrier. Additionally, assume that this solution does not have a measurable impact on flow conditions or residence times of fluid elements in the column. If the reaction taking place is:

$$A \rightarrow 2B$$

with a rate constant ($k$) of 0.12 min$^{-1}$, use the segregated flow model to determine the average effluent concentration of species B.

**Solution to Problem 2.40**

For details refer the Wiley website at http://www.wiley-vch.de/ISBN9783527354115

In the spreadsheet:

| $t$ (min) | $C_A$ (pmol/ml) | $E(t)$ | $X_A(t)$ | $E(t) \cdot X_A(t) \cdot \Delta t$ |
|---|---|---|---|---|
| 0.0 | 0 | 0 | 0 | 0 |
| 0.5 | 0 | 0 | 0.058 | 0 |
| 1.0 | 0 | 0 | 0.113 | 0 |
| 1.5 | 0 | 0 | 0.165 | 0 |
| 2.0 | 0 | 0 | 0.213 | 0 |
| 2.5 | 0 | 0 | 0.259 | 0 |
| 3.0 | 0 | 0 | 0.302 | 0 |
| ... | | | | |
| 12.5 | $1.39 \cdot 10^{-6}$ | 0.001 | 0.777 | 0.000 |
| 13.0 | $5.11 \cdot 10^{-7}$ | 0.000 | 0.790 | 0.000 |
| 13.5 | $1.88 \cdot 10^{-7}$ | 0.000 | 0.802 | 0.000 |
| 14.0 | $6.91 \cdot 10^{-8}$ | 0.000 | 0.814 | 0.000 |
| 14.5 | $2.54 \cdot 10^{-8}$ | | | |
| | | | | $X_A$ mean $= 0.661$ |

Using the segregation model:

$$\overline{X_A} = \int_0^\infty X_A(t)E(t)dt$$

We can use the spreadsheet to calculate the conversion. $X_A(t)$ is given by the conversion obtained in a batch reactor that in the case of a first-order reaction is:

$$X_A(t) = 1 - \exp(-k \cdot t)$$

With $k = 0.12\,\mathrm{min}^{-1}$, we have $\overline{X_A} = 0.661$
We know then that:

$$C_B = C_{B0} + C_{A0} \cdot X_A \cdot 2 = 0 + 100 \cdot 0.661 \cdot 2 = 132.2\,\frac{\mathrm{pmol}}{\mathrm{l}}$$

**Problem 2.41**   In a tubular reactor with a section of $0.1\,\mathrm{m}^2$, it is desired to obtain two products with different conversions of a reactant A that follows the kinetics $-r_A = kC_A$, with an estimated $k$ value of $0.12\,\mathrm{s}^{-1}$. The reagent flow rate is $0.1\,\mathrm{m}^3/\mathrm{s}$, and conversion products of 0.5 and 0.9 are desired. Estimate the distance at which the first product must be collected and the total length of the reactor, if:

(a) The reactor is ideal.
(b) The reactor behaves following the dispersed flow model, in a closed container, and is characterized by a pulse tracer obtaining $t_m = 2\,\mathrm{s}$ and $\sigma^2 = 3\,\mathrm{s}^2$.

**Solution to Problem 2.41**

For details refer the Wiley website at http://www.wiley-vch.de/ISBN9783527354115

(a) In the PFR: $X_A = 1 - \exp(-k \cdot \bar{t})$
  For $X_A = 0.5 \to \bar{t_1} = 5.77\,\mathrm{s} \to V_1 = \bar{t_1} \cdot Q = 0.57\,\mathrm{m}^3 \to L_1 = \frac{V_1}{S} = 5.7\,\mathrm{m}$
  For $X_A = 0.9 \to \bar{t_2} = 19.18\,\mathrm{s} \to V_2 = \bar{t_2} \cdot Q = 1.92\,\mathrm{m}^3 \to L_2 = \frac{V_2}{S} = 19.2\,\mathrm{m}$
(b) If the system were not ideal, following dispersed flow closed–closed container, we have that:

$$\left(\frac{\sigma}{t_m}\right)^2 = 2 \cdot \mathrm{Bo} - 2 \cdot \mathrm{Bo}^2 \cdot \left[1 - \exp\left(-\frac{1}{\mathrm{Bo}}\right)\right]$$

So, using the data given in the pulse test, $t_m = 2\,\mathrm{s}$ and $\sigma^2 = 3\,\mathrm{s}^2$, we found that $\mathrm{Bo} = 1.02$. This is a dimensionless number, valid for the specific reactor being tested in any condition. Now, we know the kinetics and the conversion as follows:

$$1 - X_A = \frac{4a \cdot \exp\left(\frac{1}{2}\right)}{(1+a)^2 \cdot \exp\left(\frac{a}{2}\right) - (1-a)^2 \cdot \exp\left(-\frac{a}{2}\right)}$$

$$a = [1 + 4 \cdot 0.12 \cdot \bar{t}]^{1/2}$$

We should solve by iterations or use a graphical tool. In both cases, residence times for the two needed conversion values should be calculated.

Using optimization method, in the spreadsheet we will assume a value of residence time for the first value of conversion. Then we will calculate the conversion obtained by using previous equation, and a comparison of both values is done, by defining an objective function:

$$(\text{Objective function}) = \sum (X_{A_{\text{calculated}}} - X_{A_{\text{desired}}})^2$$

For optimizing the value of residence time, we can do a trial and error procedure, observing the values of the objective function. Also, we can use the "Solver" tool in Excel (and similar spreadsheet software) that allows us to define a function to minimize by changing one or more values of assumed parameters.

In any way we can check that:

$$\text{For } X_A = 0.5 \rightarrow \bar{t_1} = 7.42\,\text{s} \rightarrow V_1 = \bar{t_1} \cdot Q = 0.742\,\text{m}^3 \rightarrow L_1 = \frac{V_1}{S} = 7.42\,\text{m}$$

$$\text{For } X_A = 0.9 \rightarrow \bar{t_2} = 41.15\,\text{s} \rightarrow L_2 = 41.15\,\text{m}$$

# 3

# Transfer Function in Chemical Reactor Design

## Summary of the Equations and Concepts

### Transfer Function

A transfer function can be determined by the expression:

$$H(s) = \frac{\text{Response to stimulus}}{\text{Forcing function}} = \frac{Y(s)}{X(s)}$$

### Laplace Transform of Some Functions

| $f(t)$ | $F(s) = L\{f(t)\}$ |
|---|---|
| Pulse function, Dirac delta | $1$ |
| $c$ | $\dfrac{c}{s}$ |
| $U(t)$, unitary step | $\dfrac{1}{s}$ |
| $f(t) = \begin{cases} f(t-a)\, t \geq a \\ 0 \quad t < a \end{cases}$ | $e^{-as} f(s)$ |
| $t^n$ | $\dfrac{n!}{s^{n+1}}$ |
| $e^{at}$ | $\dfrac{1}{s-a} \quad s > a$ |
| $t^n \cdot e^{at}$ | $\dfrac{n!}{(s-a)^{n+1}} \quad s > a$ |
| $\sin(at)$ | $\dfrac{a}{s^2 + a^2}$ |
| $\cos(at)$ | $\dfrac{s}{s^2 + a^2}$ |
| $\sinh(at)$ | $\dfrac{a}{s^2 - a^2}$ |
| $\cosh(at)$ | $\dfrac{s}{s^2 - a^2}$ |
| $e^{-at}\cos(wt)$ | $\dfrac{s+a}{(s+a)^2 + w^2}$ |
| $e^{-at}\sin(wt)$ | $\dfrac{w}{(s+a)^2 + w^2}$ |
| $t^{-1/2}$ | $\sqrt{\dfrac{\pi}{s}}$ |

## Transfer Function in Ideal Reactors

### CSTR

$$E(s) = L\left[\frac{1}{\bar{t}}\exp\left(-\frac{t}{\bar{t}}\right)\right] = \frac{1}{1 + \bar{t}s}$$

### PFR

$$E(s) = L(\delta(t - \bar{t})) = \exp(-\bar{t}s)$$

For more details, please consult Conesa (2019).

**Problem 3.1**   In a reactive system, the model of two continuous stirred tank reactor (CSTR) reactors with exchange can be applied, which is shown in the figure. Derive the transfer function of this model.

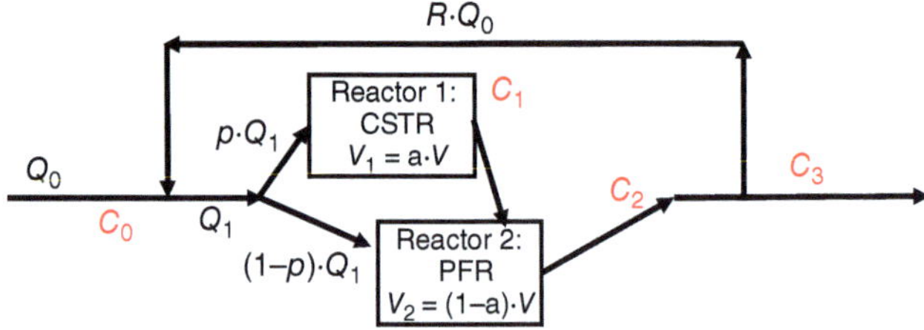


**Solution to Problem 3.1**

At the recirculation point:

$$Q_0 + RQ_0 = Q_1$$

The residence time for Reactor 1 is:

$$\bar{t_1} = \frac{V_1}{Q_{\text{reactor1}}} = \frac{V_1}{pQ_1} = \frac{V_1}{p(1+R)Q_0} = \frac{aV}{p(1+R)Q_0} = \frac{a}{p(1+R)}\bar{t}$$

In a similar way:

$$\bar{t_2} = \frac{V_2}{Q_{\text{reactor2}}} = \frac{(1-a)}{(1-p)(1+R)}\bar{t}$$

Applying the transfer function definition:

$$E_1 = \frac{C_1}{C_X} \rightarrow C_1 = E_1 C_X$$

Being $C_X$ the tracer concentration after the recirculation, entering both Reactor 1 and Reactor 2. We can say that:

$$pQ_1C_1 + (1-p)Q_1C_X = E_2C_2Q_1$$

and then:

$$E_{\text{global}} = \frac{C_2}{C_0} = \frac{\dfrac{pQ_1C_1 + (1-p)Q_1C_X}{E_2Q_1}}{C_0}$$

On the other hand:

$$C_X Q_1 = C_2 Q_0 R + C_0 Q_0$$

$$C_X(1 + R) = C_2 R + C_0$$

We can finally arrive to:

$$E_{\text{global}} = \frac{pE_1 + (1 - p)}{E_2 + E_2 R - pE_1 R + pR - R}$$

**Problem 3.2**  In a non-ideal reactor, the model of three ideal exchange reactors shown in the figure (one plug flow reactor and two stirred tank reactors) can be applied

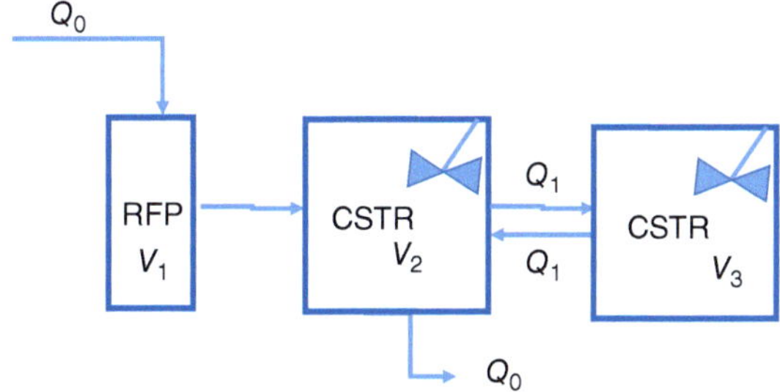


(a)  Derive the transfer function of this model. Use the relations:

$$\alpha_1 = \frac{V_1}{V_{\text{total}}}$$

$$\alpha_2 = \frac{V_2}{V_{\text{total}}}$$

$$\beta = \frac{Q_1}{Q_0}$$

(b)  If a signal of the form $C_{\text{in}} = 3t^2$ is fed to this system, what signal is expected at the output?

**Solution to Problem 3.2**
With the mentioned variables, we can do:

$$V_1 = \alpha_1 V_{\text{total}}$$

$$V_2 = \alpha_2 V_{\text{total}}$$

$$V_3 = (1 - \alpha_1 - \alpha_2)V_{\text{total}}$$

And so:

$$\overline{t_1} = \frac{V_1}{Q_{\text{reactor1}}} = \frac{\alpha_1 V}{Q_0} = \alpha_1 \overline{t}$$

$$\overline{t_2} = \frac{V_2}{Q_{\text{reactor2}}} = \frac{\alpha_2 V}{Q_0} = \alpha_2 \overline{t}$$

$$\overline{t_3} = \frac{V_3}{Q_{\text{reactor3}}} = \frac{(1 - \alpha_1 - \alpha_2)}{\beta}\overline{t}$$

(a) Using the following nomenclature, we have:

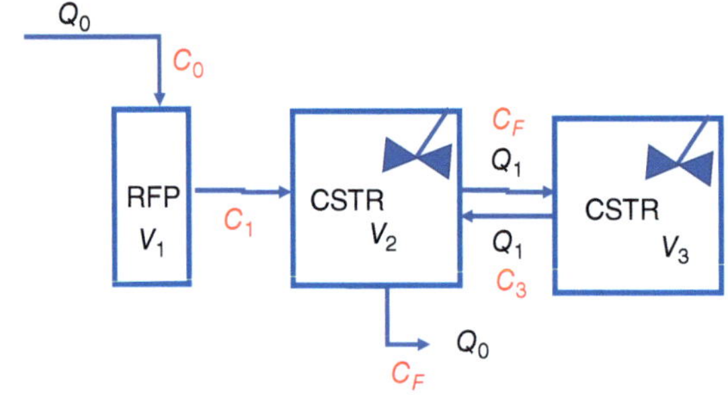


$$E_{\text{global}} = \frac{C_F}{C_0} \qquad E_1 = \frac{C_1}{C_0} \qquad E_3 = \frac{C_3}{C_F}$$

In Reactor 2, we can do:

$$(Q_0 C_1 + Q_1 C_3) \cdot E_2 = (Q_0 + Q_1)C_F$$

$$(C_1 + \beta C_3) \cdot E_2 = (1 + \beta)C_F$$

$$C_F = \frac{(C_1 + \beta C_3)}{(1 + \beta)} \cdot E_2$$

We can easily go to:

$$E_{\text{global}} = \frac{C_F}{C_0} = \frac{E_1 E_2}{1 + \beta - \beta E_3 E_2}$$

(b) $C_0(t) = 3t^2$, applying the Laplace transform:

$$C_0(s) = \frac{6}{s^3}$$

$$C_0(s) \cdot E_{\text{global}} = C_{\text{out}}(s)$$

$$C_{\text{out}}(s) = \frac{6}{s^3} \cdot \frac{E_1 E_2}{1 + \beta - \beta E_3 E_2} = \frac{6}{s^3} \cdot \frac{(\exp(-\overline{t_1}s)) \cdot \left(\frac{1}{1+\overline{t_2}s}\right)}{1 + \beta - \beta \left(\frac{1}{1+\overline{t_3}s}\right)\left(\frac{1}{1+\overline{t_2}s}\right)}$$

**Problem 3.3**   The curve representing the input to a CSTR has the equation:

$$\left\{ \begin{array}{ll} C(t) = 0.06t & 0 \leq t < 5 \\ C(t) = 0.06(10 - t) & 5 \leq t < 10 \end{array} \right\}$$

Calculate the outlet concentration over time if the mean residence time in the reactor is one second.

**Solution to Problem 3.3**

This problem can be solved in many ways. One of them is using the Laplace transform for calculating the response. The input is in the Laplace space:

$$\left\{ \begin{array}{ll} C(s) = 0.06/s^2 & 0 \leq t < 5 \\ C(s) = (0.6s - 0.06)/s^2 & 5 \leq t < 10 \end{array} \right\}$$

In the CSTR we have for a non-reacting species:

$$\bar{t}\frac{dC_{out}}{dt} + C_{out} = C_{in} \text{ with } C_{out} = 0 \text{ at } t = 0$$

In the interval $0 \leq t \leq 5$:

$$\bar{t}\frac{dC_{out}}{dt} + C_{out} = 0.06t$$

Doing the Laplace transform:

$$\bar{t} \cdot s \cdot C(s) + C(s) = \frac{0.06}{s^2}$$

Solving for $C(s)$:

$$C(s) = \frac{\frac{0.06}{s^2}}{1 + \bar{t} \cdot s}$$

And doing the inverse of the Laplace transform:

$$C_{out}(t) = 0.06\bar{t}\left[\frac{t}{\bar{t}} - 1 + \exp\left(-\frac{t}{\bar{t}}\right)\right]$$

This equation gives, with $\bar{t} = 1$, and at $t = 5 \rightarrow C_{out} = 0.11$
In the interval when $5 < t \leq 10$:

$$\bar{t}\frac{dC_{out}}{dt} + C_{out} = 0.06 \cdot (10 - t) \text{ with } C_{out} = 0.11 \text{ at } t = 5$$

Let's call $C_{out}(t = 5) = C_5$. Transforming:

$$\bar{t} \cdot s \cdot (C(s) - C_5) + C(s) = 0.06\left[\frac{10}{s} + \frac{1}{s^2}\right]$$

Solving for $C(s)$ and doing the inverse of the Laplace transform:

$$C_{\text{out}}(t) = 0.11 \cdot \exp\left(-\frac{t-5}{\bar{t}}\right) + 0.3 \cdot \left[1 - \exp\left(-\frac{t-5}{\bar{t}}\right)\right]$$
$$- 0.06 \cdot \bar{t}\left[\frac{t-5}{\bar{t}} - 1 + \exp\left(-\frac{t-5}{\bar{t}}\right)\right]$$

This gives $C_{\text{out}}(t = 10) = C_{10} = 0.122$.
In the interval $t > 10$:

$$\bar{t} \cdot s \cdot (C(s) - C_{10}) + C(s) = 0$$

$$C(s) = \frac{\bar{t} \cdot s \cdot C_{10}}{1 + \bar{t} \cdot s} = \frac{\bar{t} \cdot s \cdot 0.122}{1 + \bar{t} \cdot s}$$

And doing the inverse of the Laplace transform:

$$C_{\text{out}}(t) = 0.122\delta(t) - \frac{0.122\exp\left(-\frac{t}{\bar{t}}\right)}{\bar{t}}$$

**Problem 3.4**   Find the transfer function of the proposed network of reactors, given that each reactor has the transfer function $E(s) = 1/(s + a)$.

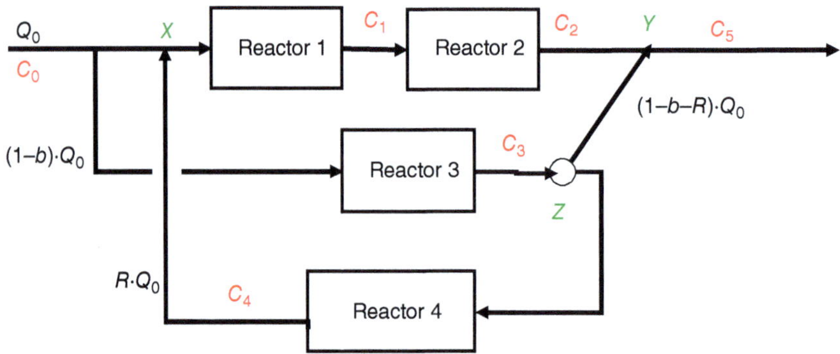


**Solution to Problem 3.4**
The global transfer function is:

$$E_{\text{global}} = \frac{C_5}{C_0}$$

By definition of transfer functions, we have that:

$$E_1 = \frac{C_1}{C_X} \qquad E_2 = \frac{C_2}{C_1} \qquad E_3 = \frac{C_3}{C_0} \qquad E_4 = \frac{C_4}{C_3}$$

where $C_X$ represents the tracer concentration after point $X$. At point $X$, we can do:

$$Q_0 C_0 + R Q_0 C_4 = (1 + R)Q_0 C_X$$

So:

$$C_X = \frac{C_0 + RC_4}{(1 + R)} = \frac{C_0 + RE_4C_3}{(1 + R)}$$

And, in point $Y$:

$$(1 + R)Q_0C_2 + (1 - bR)Q_0C_3 = Q_0C_5$$

With all these equations, we should operate until we obtain an expression just by defining the parameters of the system and the $E_{\text{global}}$. For example:

$$C_5 = (1 + R)C_2 + (1 - bR)C_3 = (1 + R)E_2C_1 + (1 - bR)E_3C_0$$
$$= (1 + R)E_2E_1C_X + (1 - bR)E_3C_0$$

$$C_5 = (1 + R)E_2E_1\frac{C_0 + RE_4C_3}{(1 + R)} + (1 - bR)E_3C_0$$
$$= E_2E_1(C_0 + RE_4E_3C_0) + (1 - bR)E_3C_0$$

So, finally:

$$E_{\text{global}} = E_2E_1(1 + RE_4E_3) + (1 - bR)E_3$$

**Problem 3.5**    There is a series of two reactors designed as RFP but with certain deficiencies in their behavior, so that they could be modeled with the following scheme:

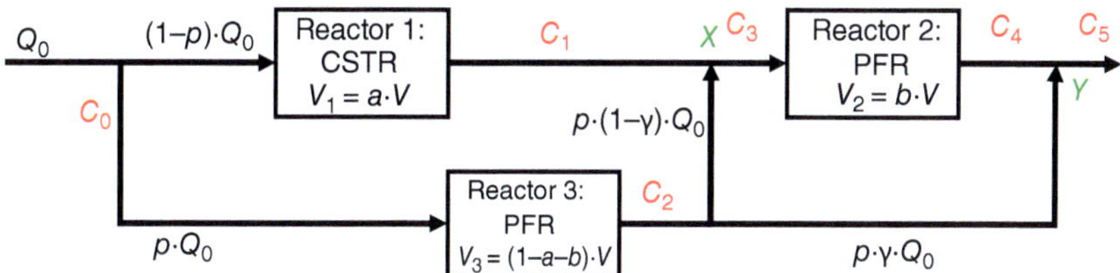


(a) Determine the values of the mean residence times in each of the three model reactors as a function of the total mean residence time, $\bar{t}$.
(b) Taking into account the transfer function in the CSTR and plug flow reactor (PFR), find the transfer function of the proposed reactor network.

**Solution to Problem 3.5**

(a) Using the nomenclature proposed in the figure, we can write:

$$\bar{t_1} = \frac{V_1}{Q_{\text{reactor1}}} = \frac{V_1}{(1-p)Q_0} = \frac{aV}{(1-p)Q_0} = \frac{a}{(1-p)}\bar{t}$$

$$\bar{t_2} = \frac{V_2}{Q_{\text{reactor2}}} = \frac{V_2}{(1-p)Q_0 + p(1-\gamma)Q_0} = \frac{bV}{(1-\gamma p)Q_0} = \frac{b}{(1-\gamma p)}\bar{t}$$

$$\bar{t_3} = \frac{V_3}{Q_{\text{reactor3}}} = \frac{V_3}{pQ_0} = \frac{(1-a-b)V}{pQ_0} = \frac{(1-a-b)}{p}\bar{t}$$

(b) The global residence time distribution (RTD) is given by:

$$E_{\text{global}} = \frac{C_5}{C_0}$$

Definitions:

$$E_1 = \frac{C_1}{C_0} \qquad E_2 = \frac{C_4}{C_3} \qquad E_3 = \frac{C_2}{C_0}$$

Balances, at point $X$:

$$(1 - p)Q_0C_1 + p(1 - \gamma)Q_0C_2 = (1 - p\gamma)Q_0C_3$$

So:

$$C_3 = \frac{(1 - p)C_1 + p(1 - \gamma)C_2}{(1 - p\gamma)} = \frac{(1 - p)E_1C_0 + p(1 - \gamma)E_3C_0}{(1 - p\gamma)}$$

Balances, at point $Y$:

$$(1 - p\gamma)C_4 + p\gamma C_2 = C_5$$

So:

$$C_5 = (1 - p\gamma)E_2C_3 + p\gamma E_3C_0 = (1 - p\gamma)E_2\frac{(1 - p)E_1C_0 + p(1 - \gamma)E_3C_0}{(1 - p\gamma)} + p\gamma E_3C_0$$

And, finally:

$$E_{\text{global}} = \frac{C_5}{C_0} = E_2((1 - p)E_1 + p(1 - \gamma)E_3) + p\gamma E_3$$

**Problem 3.6**  Determine, with the help of a diagram and the idea of a transfer function, the $E(t)$ of a combination of CSTR and RFP in the series in which there is a bypass. Remember that the transfer function in an isolated CSTR is $E(s) = 1/(1 + \bar{t}\,s)$ and that of RFP is $E(s) = \exp(-\bar{t}\,s)$.

**Solution to Problem 3.6**

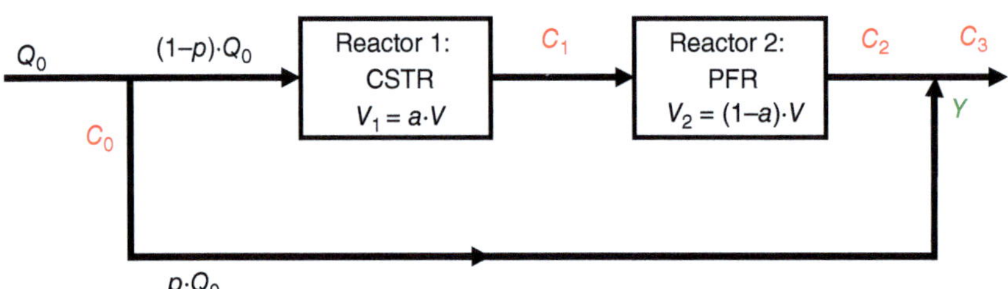

Using the nomenclature proposed in the figure, we can write:

$$\overline{t_1} = \frac{V_1}{Q_{\text{reactor1}}} = \frac{V_1}{(1-p)Q_0} = \frac{aV}{(1-p)Q_0} = \frac{a}{(1-p)}\overline{t}$$

$$\overline{t_2} = \frac{V_2}{Q_{\text{reactor2}}} = \frac{V_2}{(1-p)Q_0} = \frac{(1-a)V}{(1-p)Q_0} = \frac{(1-a)}{(1-p)}\overline{t}$$

The global RTD is given by:

$$E_{\text{global}} = \frac{C_3}{C_0}$$

Definitions:

$$E_1 = \frac{C_1}{C_0} \qquad E_2 = \frac{C_2}{C_1}$$

Balances, at point $Y$:

$$(1-p)\,Q_0 C_2 + pQ_0 C_0 = Q_0 C_3$$

So:

$$C_3 = (1-p)\,C_2 + pC_0$$

$$C_3 = (1-p)\,E_2 C_1 + pC_0 = (1-p)\,E_2 E_1 C_0 + pC_0$$

And, finally:

$$E_{\text{global}} = \frac{C_3}{C_0} = (1-p)\,E_2 E_1 + p$$

Using the transfer function for the ideal reactors:

$$E_{\text{global}} = (1-p)\exp(-\overline{t_2}s)\frac{1}{1+\overline{t_1}s} + p$$

$$E_{\text{global}} = (1-p)\exp\left(-\frac{(1-a)}{(1-p)}\overline{t}s\right)\frac{1}{1+\frac{(1-a)}{(1-p)}\overline{t}s} + p$$

**Problem 3.7**   A tracer concentration given by $C = (25t + 1)^2$ is fed into a CSTR for 1 second ($t$ in seconds, $C$ in mol/l). Make the necessary balances to calculate how the tracer concentration varies at the outlet between $t = 0$ and infinity: (a) in the d$t$–d$C$ space and (b) making the corresponding Laplace transform.

**Solution to Problem 3.7**

For details refer the Wiley website at http://www.wiley-vch.de/ISBN9783527354115

(a) In the d$t$–d$C$ space, things are as usual. A mass balance during the first second in the CSTR is:

$$\text{Input} + \text{Generation} = \text{Output} + \text{Accumulation}$$

$$Q_0 C_{in} + 0 = Q_0 C_{out} + \frac{dC_{out}}{dt} V$$

$$C_{in} = C_{out} + \frac{dC_{out}}{dt} \bar{t}$$

During the first second:

$$(25t + 1)^2 = C_{out} + \frac{dC_{out}}{dt} \bar{t}$$

$$\frac{(25t + 1)^2 - C_{out}}{\bar{t}} = \frac{dC_{out}}{dt}$$

and this cannot be separated for analytical integration. At most, we can come up with a numerical solution, for example, using the finite differences method:

$$\frac{C_{out}^{t+1} - C_{out}^t}{\Delta t} = \frac{(25t + 1)^2 - C_{out}^t}{\bar{t}}$$

$$C_{out}^{t+1} = \left[ \frac{(25t + 1)^2 - C_{out}^t}{\bar{t}} \right] \Delta t + C_{out}^t$$

(b) In the Laplace space:

$$C_{in} = C_{out} + \frac{dC_{out}}{dt} \bar{t}$$

$$C_{in}(s) = C_{out}(s) + s C_{out}(s) \bar{t}$$

$$C_{out}(s) = \frac{C_{in}(s)}{1 + \bar{t}s}$$

And we know that:

$$L[(25t + 1)^2] = 1250/s^3 + 50/s^2 + 1/s$$

And then:

$$C_{out}(s) = \frac{1250/s^3 + 50/s^2 + 1/s}{1 + \bar{t}s}$$

Doing the inverse:

$$C_{out}(t) = 1 - e^{-\frac{t}{\bar{t}}} - 50\bar{t} + 50 e^{-\frac{t}{\bar{t}}} \bar{t} + 1250\bar{t}^2 - 1250 e^{-\frac{t}{\bar{t}}} \bar{t}^2 + 50t - 1250\bar{t}t + 625t^2$$

This function is valid during the first second, as the $C_{in}$ is given by the function considered only during the first second.

We can compare both results by giving a value to $\bar{t}$. Assuming a value of one second, we can do in the spreadsheet the comparison, obtaining:

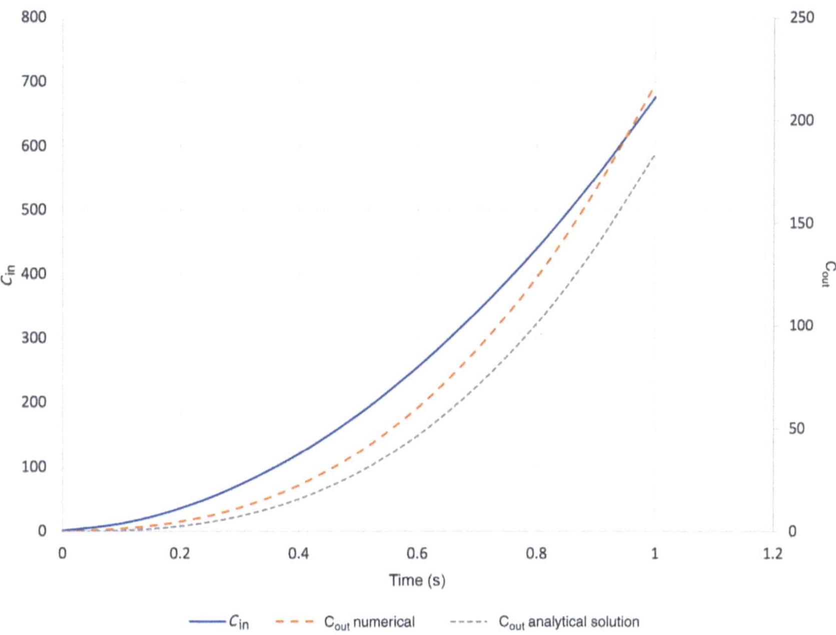


**Problem 3.8** There is a series of two reactors designed as CSTR but with certain deficiencies in their behavior, so that they could be modeled with the following scheme:

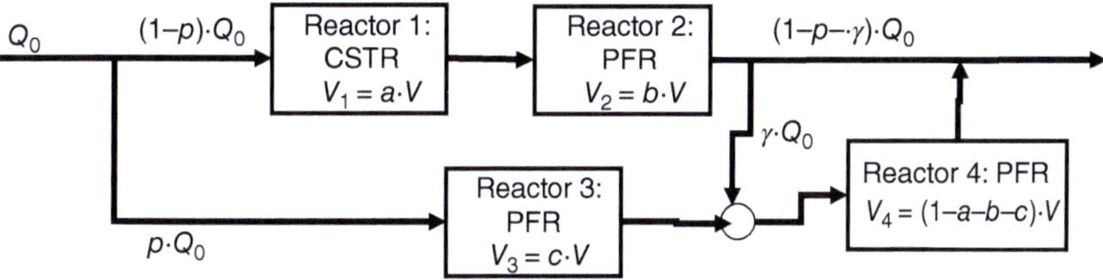


Taking into account the transfer function in the CSTR and RFP, find the transfer function of the proposed reactor network.

**Solution to Problem 3.8**

First, we are going to mark in the diagram the different currents:

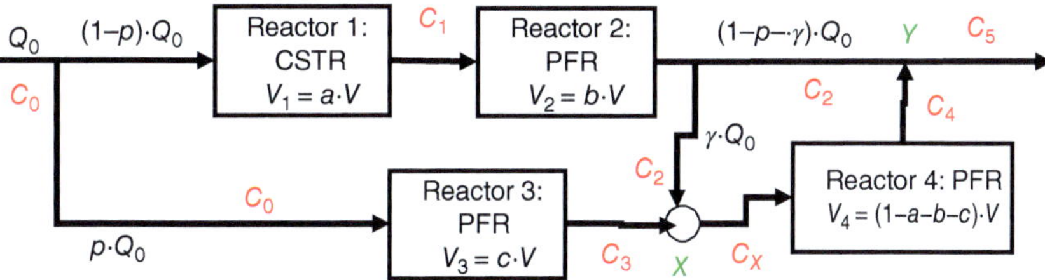

The global transfer function is:

$$E_{\text{global}} = \frac{C_5}{C_0}$$

By definition of transfer functions, we have:

$$E_1 = \frac{C_1}{C_0} \qquad E_2 = \frac{C_2}{C_1} \qquad E_3 = \frac{C_3}{C_0} \qquad E_4 = \frac{C_4}{C_X}$$

where $C_X$ represents the tracer concentration after point $X$. At point $X$, we can do:

$$pQ_0 C_3 + \gamma Q_0 C_2 = (p + \gamma)Q_0 C_X$$

So:

$$C_X = \frac{pC_3 + \gamma C_2}{(p + \gamma)} = \frac{pE_3 C_0 + \gamma E_2 C_1}{(p + \gamma)} = \frac{pE_3 C_0 + \gamma E_2 E_1 C_0}{(p + \gamma)}$$

And, at point $Y$:

$$(1 - p - \gamma)Q_0 C_2 + (p + \gamma)Q_0 C_4 = Q_0 C_5$$

And then:

$$(1 - p - \gamma)Q_0 E_1 E_2 C_0 + (p + \gamma)Q_0 E_4 C_X = Q_0 C_5$$

With all these equations, we should operate until we obtain an expression just with the parameters of the system and defining the $E_{\text{global}}$. For example:

$$C_5 = (1 - p - \gamma)E_1 E_2 C_0 + (p + \gamma)E_4 C_X$$

$$C_5 = (1 - p - \gamma)E_1 E_2 C_0 + (p + \gamma)E_4 \frac{p E_3 C_0 + \gamma E_2 E_1 C_0}{(p + \gamma)}$$

So, finally:

$$E_{\text{global}} = (1 - p - \gamma)E_1 E_2 + E_4(p E_3 + \gamma E_2 E_1)$$

**Problem 3.9** In a non-ideal reactor, the TWO ideal exchange reactor model shown in the figure (one plug flow reactor and one stirred tank reactor with dead volume) can be applied.

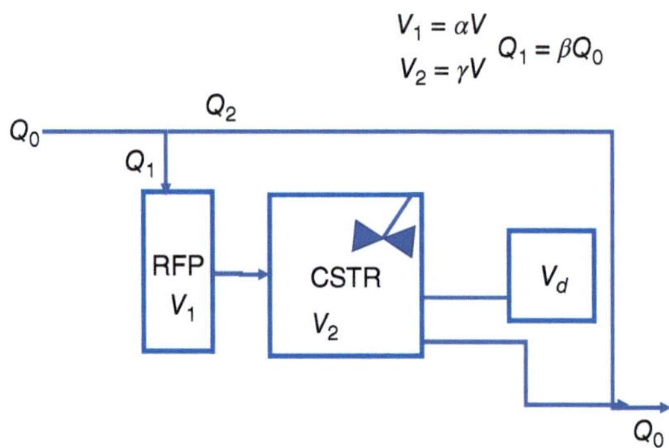

(a) Derive the transfer function of this model. Use the relationships indicated in the figure.

(b) If a signal of the form $C_{in} = 2 + 3t^2$ is fed to this system, what signal is expected at the exit?

**Solution to Problem 3.9**

(a) Let us use the following nomenclature:

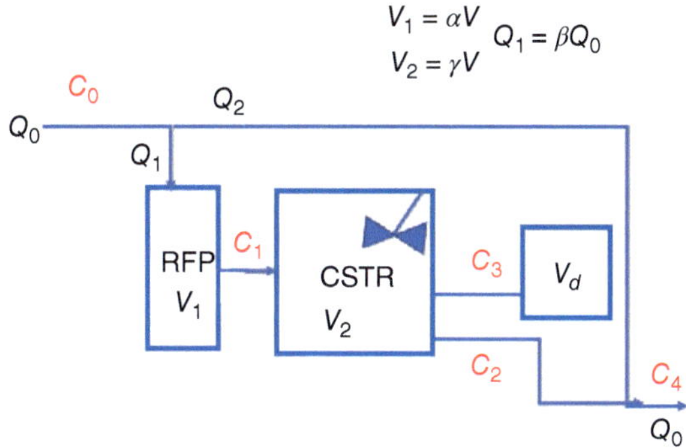


By definition, we have that:

$$E_1 = \frac{C_1}{C_0}$$

$$E_2 = \frac{C_2}{C_1}$$

So:

$$C_2 = E_2 C_1 = E_2 E_1 C_0$$

In the mixing point:

$$Q_1 C_2 + Q_2 C_0 = Q_0 C_4$$

$$\beta Q_0 C_2 + (1 - \beta) Q_0 C_0 = Q_0 C_4$$

So, the global transfer function is:

$$E_{global} = \frac{C_4}{C_0} = \beta E_1 E_2 + (1 - \beta)$$

Note that:

$$E_1 = \exp\left(-\frac{\alpha}{\beta} ts\right)$$

$$E_2 = \frac{1}{1 + \frac{1-\alpha}{\gamma} ts}$$

And:

$$E_{\text{global}} = \frac{\beta \exp\left(-\frac{\alpha}{\beta}\bar{t}s\right)}{1 + \frac{1-\alpha}{\gamma} = ts} + (1 - \beta)$$

The inverse Laplace transform of this transfer function is:

$$E_{\text{global}}(t) = \frac{\beta \exp\left(\frac{\frac{\alpha}{\beta}\bar{t}-t}{\frac{1-\alpha}{\gamma}\bar{t}}\right)}{\frac{1-\alpha}{\gamma} = t} U\left(t - \frac{\alpha}{\beta}\bar{t}\right) + (1 - \beta)\delta(t)$$

Where $U(t)$ is the unitary step function, and $\delta(t)$ is the Dirac delta function.

(b) If $C_{\text{in}} = 2 + 3t^2$, its Laplace transform is:

$$C_{\text{in}}(s) = \frac{2(e + s^2)}{s^3}$$

And the signal output is (in the Laplace space):

$$C_{\text{out}}(s) = E_{\text{global}} \cdot C_{\text{in}} = \left[\frac{\beta \exp\left(-\frac{\alpha}{\beta}\bar{t}s\right)}{1 + \frac{1-\alpha}{\gamma} = ts} + (1 - \beta)\right]\left(\frac{2(e + s^2)}{s^3}\right)$$

**Problem 3.10**  One non-ideal reactor is fed with a signal input following the function:

$$C_{\text{in}} = \exp\left(\frac{1 - 0.2t}{4}\right)$$

At the exit, the signal output is given by:

$$C_{\text{out}} = 2 * \exp(-2t)$$

What $E(t)$ is representing the behavior of the reactor?

**Solution to Problem 3.10**
For details refer the Wiley website at http://www.wiley-vch.de/ISBN9783527354115

First, let us take a look at the form of the input and output signals. In a spreadsheet, we can check:

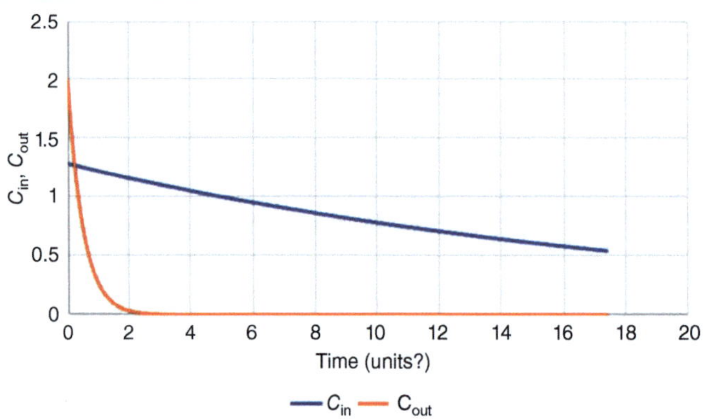

Using Laplace transforms:

$$C_{\text{in}}(s) = \frac{12.1825}{0.5 + s}$$

$$C_{\text{out}}(s) = \frac{2}{s + 2}$$

So:

$$E(s) = \frac{C_{\text{out}}(s)}{C_{\text{in}}(s)} = \frac{2}{s + 2} \cdot \frac{(0.5 + s)}{12.1825}$$

Doing the inverse transformation:

$$E(t) = 0.164\,17(\delta(t) - 1.5e^{-2t})$$

This $E(t)$ is probably due to the series combination of a PFR (first addition of the function) and a CSTR (second addition).

# Part II

# Convolution and Unsteady State in Chemical Reactors

# 4

# Convolution and Deconvolution of Signals in Chemical Reactor Engineering

## Summary of Equations and Methods

In a chemical reactor:

$$C_{\text{out}} = E \otimes C_{\text{in}}$$

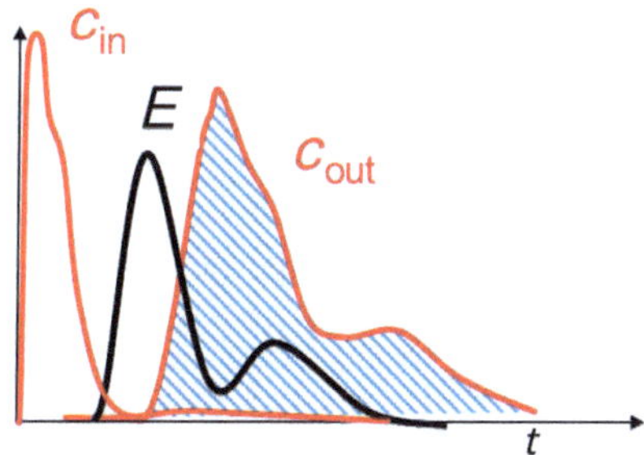


For functions of a discrete variable "$x$" (i.e. arrays of numbers), the definition of the convolution operation is:

$$C(u) = \sum_{u=-\infty}^{\infty} f(x) \cdot g(u - x)$$

It is not possible to calculate the convolution of two given functions if both are given at points with different $\Delta x$.

## Convolution

Assume that $f(x)$ is a vector of "$n$" values, i.e., $f(x) = [f_1, f_2, f_3, \ldots f_n]$ and $g(x)$ has "$m$" values so that $g(x) = [g_1, g_2, g_3, \ldots, g_m]$.

As an example, if $n = 5$ and $m = 4$, for the convolution operation, the desired result is:

$$C_1 = f_1 \cdot g_1$$

$$C_2 = f_2 \cdot g_1 + f_1 \cdot g_2$$

$$C_3 = f_3 \cdot g_1 + f_2 \cdot g_2 + f_1 \cdot g_3$$

$$C_4 = f_4 \cdot g_1 + f_3 \cdot g_2 + f_2 \cdot g_3 + f_1 \cdot g_4$$

*Problem Solving in Chemical Reactor Design*, First Edition. Juan A. Conesa.
© 2025 WILEY-VCH GmbH. Published 2025 by WILEY-VCH GmbH.

$$C_5 = f_5 \cdot g_1 + f_4 \cdot g_2 + f_3 \cdot g_3 + f_2 \cdot g_4$$

$$C_6 = \qquad\quad f_5 \cdot g_2 + f_4 \cdot g_3 + f_3 \cdot g_4$$

$$C_7 = \qquad\qquad\qquad\quad f_5 \cdot g_3 + f_4 \cdot g_4$$

$$C_8 = \qquad\qquad\qquad\qquad\qquad f_5 \cdot g_4$$

We can write this in matrix form as:

$$
\begin{pmatrix} C_1 \\ C_2 \\ C_3 \\ C_4 \\ C_5 \\ C_6 \\ C_7 \\ C_8 \end{pmatrix}
=
\begin{bmatrix}
f_1 & 0 & 0 & 0 \\
f_2 & f_1 & 0 & 0 \\
f_3 & f_2 & f_1 & 0 \\
f_4 & f_3 & f_2 & f_1 \\
f_5 & f_4 & f_3 & f_2 \\
0 & f_5 & f_4 & f_3 \\
0 & 0 & f_5 & f_4 \\
0 & 0 & 0 & f_5
\end{bmatrix}
\cdot
\begin{pmatrix} g_1 \\ g_2 \\ g_3 \\ g_4 \end{pmatrix}
$$

Or:

$$C = \mathbf{A} \cdot g$$

Being $\mathbf{A}$ the "matrix convolution," with values of $f(x)$ in the form of columns. For a general case, when vectors of $n$ and $m$ values are combined, the resulting matrix convolution is a $(n + m - 1) \times m$ matrix, and the resulting $C$ is a vector of $(n + m - 1)$ values:

$$C_{(n+m-1\times 1)} = [\mathbf{A}_{(n+m-1\times m)} \cdot g_{(m\times 1)}]$$

Let us call "$v$" the number of rows of the $C$ vector, i.e., $v = (n + m - 1)$. Let us apply these equations to the convolution of a residence time distribution (RTD) curve, $E$, and the signal input to a reactor, $C_{\text{in}}$. As mentioned before, we will need to convolute both signals, in such a way that:

$$C_{\text{out}} = E \otimes C_{\text{in}}$$

If $C_{\text{in}}$ has "$m$" values and $E$ has "$n$" values, the convolution is given by:

$$
\begin{pmatrix} C_{\text{out},1} \\ C_{\text{out},2} \\ C_{\text{out},3} \\ C_{\text{out},4} \\ \cdots \\ \cdots \\ \cdots \\ C_{\text{out},v} \end{pmatrix}
=
\begin{bmatrix}
E_1 & 0 & 0 & \cdots & 0 \\
E_2 & E_1 & 0 & \cdots & 0 \\
E_3 & E_2 & E_1 & \cdots & 0 \\
\cdots & \cdots & \cdots & \cdots & \cdots \\
E_n & E_{n-1} & \cdots & \cdots & \cdots \\
0 & E_n & E_{n-1} & \cdots & \cdots \\
 & & E_n & & \\
 & & \cdots & & \\
 & & \cdots & & \\
0 & 0 & 0 & 0 & 0 & E_n
\end{bmatrix}
\cdot
\begin{pmatrix} C_{\text{in},1} \\ C_{\text{in},2} \\ \cdots \\ C_{\text{in},m} \end{pmatrix}
$$

This is:

$$C_{\text{out}(v\times1)} = C_{\text{out}(n+m-1\times1)} = [\mathbf{A}_{(n+m-1\times m)} \cdot C_{\text{in}(m\times1)}]$$

To calculate the *time vector* associated with $C_{\text{out}}$, we would need the corresponding times associated with $C_{\text{in}}$ and $E$. In general, let us assume that the first and last elements of $C_{\text{in}}$ correspond to times $t_{0,C_{\text{in}}}$, and $t_{m,C_{\text{in}}}$, while the first and last elements of the $E$ vector are $t_{0,E}$, and $t_{n,E}$. The convolution can be done (as mentioned before), only if time increments in both vectors ($\Delta t$) are equal, i.e.,

$$\Delta t = \frac{t_{m,C_{\text{in}}} - t_{0,C_{\text{in}}}}{m-1} = \frac{t_{n,E} - t_{0,E}}{n-1}$$

$$t_{0,\text{convolution}} = t_{0,C_{\text{in}}} + t_{0,E} \quad \text{(initial time is the sum of the initial times)}$$

$$t_{v,\text{convolution}} = t_{m,C_{\text{in}}} + t_{n,E} \quad \text{(final time is the sum of the final times)}$$

$$\Delta t_{\text{convolution}} = \frac{t_{v,\text{convolution}} - t_{0,\text{convolution}}}{m+n-2}$$

Applying former equations, we can also see that the following is true:

$$\Delta t_{\text{convolution}} = \frac{(m-1)\cdot\Delta t + (n-1)\cdot\Delta t}{m+n-2} = \left(\frac{m+n-2}{m+n-2}\right)\cdot\Delta t = \Delta t$$

## Deconvolution

This is a very common problem because the calculation of $E$ from an impulse input signal ($C_{\text{in}}$ = Dirac delta function) or a step function is not always possible or easy. Assuming the dimension of vectors is "$m$" for $C_{\text{in}}$ and "$v$" for $C_{\text{out}}$ (as used above), we have that $C_{\text{out}} = E \otimes C_{\text{in}}$ and:

$$C_{\text{out}(v\times1)} = [\mathbf{A}_{(v\times m)} \cdot C_{\text{in}(m\times1)}]$$

Note that the dimensions of convolution matrix $\mathbf{A}$ are ($v \times m$), from that, we will calculate ($n = v - m + 1$) values of the vector $E$ curve. From this equation, it is not possible to directly calculate matrix $\mathbf{A}$. For the calculation of $n$ values of vector $E$, we have ($n + m - 1$) equations, so it is an overdetermined system of equations. This means that there are many solutions for the system, and we need to choose, in any way, the best solution. In this way, for the resolution of the deconvolution of two given signals ($C_{\text{in}}$ and a $C_{\text{out}}$), it is necessary to pose an optimization by least squared residuals of the components of matrix $\mathbf{A}$.

For calculating the $E$ curve, the easiest way to do the optimization is to minimize an objective function (O.F.) in the form of the sum of squared residuals between the $C_{\text{out}}$ curve (known) and the product $[\mathbf{A}\cdot C_{\text{in}}]$:

$$\text{O.F.} = [C_{\text{out}(v\times1)} - \mathbf{A}_{(v\times m)} \cdot C_{\text{in}(m\times1)}]^2$$

For that purpose, an initial value of the vector $E$ is taken, and the use of an optimization method is necessary.

For the calculation of the corresponding time vector for $E$ distribution (deconvoluted signal), we have:

$$t_{0,E} = t_{0,C_{out}} - t_{0,C_{in}} \quad \text{(initial time of the convolution is the sum of the initial times)}$$

$$t_{v,E} = t_{v,C_{out}} - t_{m,C_{in}} \quad \text{(final time of the convolution is the sum of the final times)}$$

$$\Delta t_{deconvolution} = \frac{t_{v,E} - t_{0,E}}{v - 1} = \Delta t$$

For more details, please consult Conesa (2019).

**Problem 4.1**  In an experimental reactor, a tracer is injected with the following distribution:

| $t$ (s) | 0 | 1 | 2 | 3 | 4 |
| --- | --- | --- | --- | --- | --- |
| $C$ (mol/l) | 0 | 8 | 4 | 6 | 0 |

In the reactor, the RTD is given by:

| $t$ (s) | 2 | 3 | 4 | 5 | 6 |
| --- | --- | --- | --- | --- | --- |
| $E$ | 0 | 0.05 | 0.50 | 0.35 | 0 |

Calculate the evolution of the output signal with time.

**Solution to Problem 4.1**

We must calculate both the concentration and the time vectors. The time vector is easy to calculate. We have $m = 5$ points and $n = 5$ points, so the resulting vector will have 9 points $(m + n - 1)$. The first value of time is $(0 + 2) = 2$ and the last one is $(6 + 4) = 10$. The corresponding $\Delta t_{convolution}$ is $(10 - 2)/(m + n - 2) = 8/8 = 1$, so:

$$t_{convolution} \ (s) = [2 \quad 3 \quad 4 \quad 5 \quad 6 \quad 7 \quad 8 \quad 9 \quad 10]$$

To calculate the concentration:

$$C_{convolution1} = C_{in1} \cdot E_1$$

$$C_{convolution2} = C_{in2} \cdot E_1 + C_{in1} \cdot E_2$$

$$C_{convolution3} = C_{in3} \cdot E_1 + C_{in2} \cdot E_2 + C_{in1} \cdot E_3$$

$$C_{convolution4} = C_{in4} \cdot E_1 + C_{in3} \cdot E_2 + C_{in2} \cdot E_3 + C_{in1} \cdot E_4$$

$$C_{convolution5} = C_{in5} \cdot E_1 + C_{in4} \cdot E_2 + C_{in3} \cdot E_3 + C_{in2} \cdot E_4 + C_{in1} \cdot E_5$$

$$C_{convolution6} = \qquad C_{in5} \cdot E_2 + C_{in4} \cdot E_3 + C_{in3} \cdot E_4 + C_{in2} \cdot E_5$$

$$C_{convolution7} = \qquad C_{in5} \cdot E_3 + C_{in4} \cdot E_4 + C_{in3} \cdot E_5$$

$$C_{convolution8} = \qquad C_{in5} \cdot E_4 + C_{in5} \cdot E_5$$

$$C_{convolution9} = \qquad C_{in5} \cdot E_5$$

Tanking the values of $C_{in}$ = tracer input and $E$ = RTD of the reactor, we can easily calculate:

$$C_{convolution1} = 0 \cdot 0$$

$$C_{convolution2} = 8 \cdot 0 + 0 \cdot 0.05$$

$$C_{convolution3} = 4 \cdot 0 + 8 \cdot 0.05 + 0 \cdot 0.5$$

$$C_{convolution4} = 6 \cdot 0 + 4 \cdot 0.05 + 8 \cdot 0.5 + 0 \cdot 0.35$$

$$C_{convolution5} = 0 \cdot 0 + 6 \cdot 0.05 + 4 \cdot 0.5 + 8 \cdot 0.35 + 0 \cdot 0$$

$$C_{convolution6} = \quad 0 \cdot 0.05 + 6 \cdot 0.5 + 4 \cdot 0.35 + 8 \cdot 0$$

$$C_{convolution7} = \quad\quad 0 \cdot 0.5 + 6 \cdot 0.35 + 4 \cdot 0$$

$$C_{convolution8} = \quad\quad\quad 0 \cdot 0.35 + 6 \cdot 0$$

$$C_{convolution9} = \quad\quad\quad\quad 0 \cdot 0$$

And finally:

$$C_{convolution} \ (\text{mol/l}) = \quad [0 \quad 0 \quad 0.4 \quad 4.2 \quad 5.2 \quad 4.4 \quad 2.1 \quad 0 \quad 0]$$

**Problem 4.2**  In an experiment to determine the $E(t)$ of a certain reactor, the tracer is introduced into the reactor following the equation $C_{in} = t \cdot \exp(-t/2)$. The response of the non-ideal tank reactor is given in the following table:

| $t_{out}$ (min) | 4 | 5 | 6 | 7 | 8 | 9 | 10 | 11 | 12 | 13 |
|---|---|---|---|---|---|---|---|---|---|---|
| $C_{out}$ (mg/l) | 0.9 | 1.8 | 2.1 | 5.2 | 3.6 | 4.5 | 1.7 | 0.8 | 0.7 | 0.5 |

Calculate the $E(t)$ of the reactor, together with the time vector associated with the $E(t)$.

**Solution to Problem 4.2**

For doing this deconvolution, the easiest is to calculate discrete values of $C_{in}$ and then deconvolute. We can choose the number of points we use for the calculation (with limits), and also the time values, but the time increment should be the same as the $C_{out}$ signal (i.e. one minute), and the $C_{in}$ cannot be zero at all points. Another restriction is that we will need enough data for $C_{in}$. We have $v = 10$ points on the convoluted curve ($C_{out}$). For example, if we use $m = 10$ points to calculate $C_{in}$, as we have $v = 10$ values of $C_{out}$, we can only calculate $n = v - m + 1 = 1$ point of the RTD. If we use $m = 9$ points in $C_{in}$, we will be able to calculate two values of $E(t)$. Let us choose to use $m = 5$ points in $C_{in}$, so we will calculate six points in $E(t)$. Choosing $t_{0,in} = 0$, then:

| $t_{in}$ (min) | $T$ | 0 | 1 | 2 | 3 | 4 |
|---|---|---|---|---|---|---|
| $C_{in}$ (mg/l) | $t \cdot \exp(-t/2)$ | 0 | 0.61 | 0.74 | 0.67 | 0.54 |

Following the reasoning presented before, the deconvoluted signal will have $(10 - 5 + 1) = 6$ points. The time vector of these points is easy to calculate, as $t_{E,0} = t_{out,0} - t_{in,0}$ and the time increment is one minute, so:

| $t_E$ (min) | 4 | 5 | 6 | 7 | 8 | 9 |
|---|---|---|---|---|---|---|

For calculating the $E(t)$, we should solve the following equations system:

$$
\begin{pmatrix} 0.9 \\ 1.8 \\ 2.1 \\ 5.2 \\ 3.6 \\ 4.5 \\ 1.7 \\ 0.8 \\ 0.7 \\ 0.5 \end{pmatrix}
=
\begin{bmatrix}
E_1 & 0 & 0 & 0 & 0 \\
E_2 & E_1 & 0 & 0 & 0 \\
E_3 & E_2 & E_1 & 0 & 0 \\
E_4 & E_3 & E_2 & E_1 & 0 \\
E_5 & E_4 & E_3 & E_2 & E_1 \\
E_6 & E_5 & E_4 & E_3 & E_2 \\
0 & E_6 & E_5 & E_4 & E_3 \\
0 & 0 & E_6 & E_5 & E_4 \\
0 & 0 & 0 & E_6 & E_5 \\
0 & 0 & 0 & 0 & E_6
\end{bmatrix}
\cdot
\begin{pmatrix} 0 \\ 0.61 \\ 0.74 \\ 0.67 \\ 0.54 \end{pmatrix}
$$

As we can see, we have 6 unknowns and 10 equations, so the system is overdetermined. An optimization is needed. This can be done by optimizing the objective function:

$$\text{O.F.} = [C_{out(v \times 1)} - \mathbf{A}_{(v \times m)} \cdot C_{in(m \times 1)}]^2$$

that accounts for the differences between the known $C_{out}$ and that predicted by the convolution.

The optimized values of $E(t)$ are:

| $t_E$ (min) | 4 | 5 | 6 | 7 | 8 | 9 |
|---|---|---|---|---|---|---|
| $E(t)$ | 1.79 | 2.33 | 2.58 | 0.00 | 1.24 | 0.00 |

The next figure graphically presents the results.

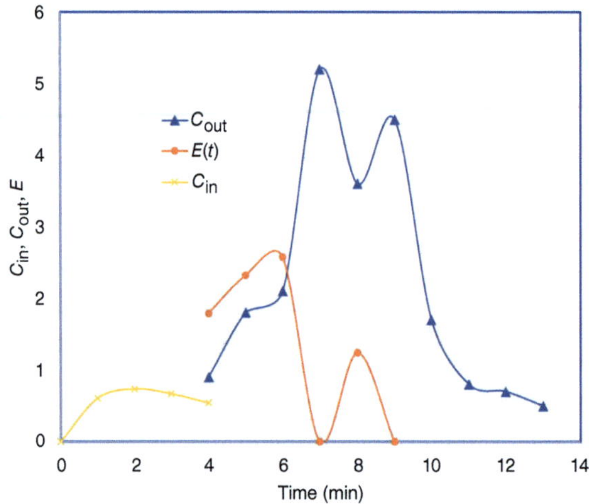

**Problem 4.3**   Given the curves for the tracer input signal and residence time distribution of the recipient, we want to calculate the signal at the exit of the system.

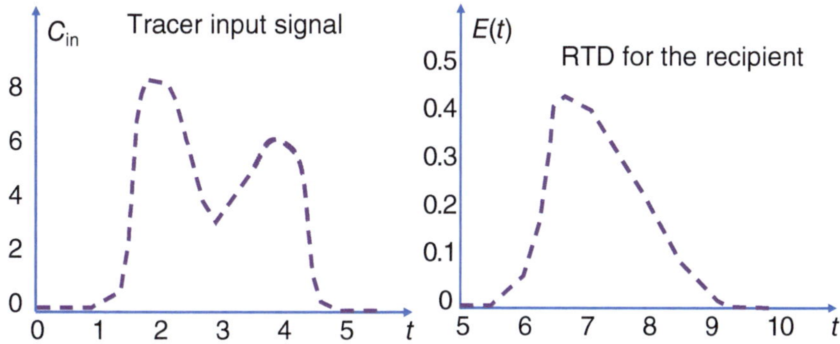



**Solution to Problem 4.3**

We must calculate both the concentration and the time vectors. First, we must choose a number of points in the plots that would make a good curve. For example, we can choose $m = 5$ (from the $C_{in}$ curve) and $n = 4$ (from the $E$ curve), so we will have $v = m + n - 1 = 8$ points in the calculated $C_{out}$. Hence:

| $t_{in}$ | 1 | 2 | 3 | 4 | 5 |
|---|---|---|---|---|---|
| $C_{in}$ | 0 | 8 | 4 | 6 | 0 |

| $t_E$ | 6 | 7 | 8 | 9 |
|---|---|---|---|---|
| $E$ | 0.05 | 0.5 | 0.3 | 0.1 |

Note that we should choose points with the same time interval in both curves. The time vector is easy to calculate. The first value of time is $(1 + 6) = 7$, and the last one is $(5 + 9) = 14$. The corresponding $\Delta t_{convolution}$ is 1, so:

| $t_{out}$ | 7 | 8 | 9 | 10 | 11 | 12 | 13 | 14 |
|---|---|---|---|---|---|---|---|---|

The calculation of $C_{out}$ is done by:

$$
C_{out} =
\begin{bmatrix}
0.05 & 0 & 0 & 0 & 0 \\
0.5 & 0.05 & 0 & 0 & 0 \\
0.3 & 0.5 & 0.05 & 0 & 0 \\
0.1 & 0.3 & 0.5 & 0.05 & 0 \\
0 & 0.1 & 0.3 & 0.5 & 0.05 \\
0 & 0 & 0.1 & 0.3 & 0.5 \\
0 & 0 & 0 & 0.1 & 0.3 \\
0 & 0 & 0 & 0 & 0.1
\end{bmatrix}_{8\times 5}
\cdot
\begin{bmatrix}
0 \\ 8 \\ 4 \\ 6 \\ 0
\end{bmatrix}_{5\times 1}
=
\begin{bmatrix}
0 \\ 0.4 \\ 4.2 \\ 4.7 \\ 5 \\ 2.2 \\ 0.6 \\ 0
\end{bmatrix}
$$

Graphically:

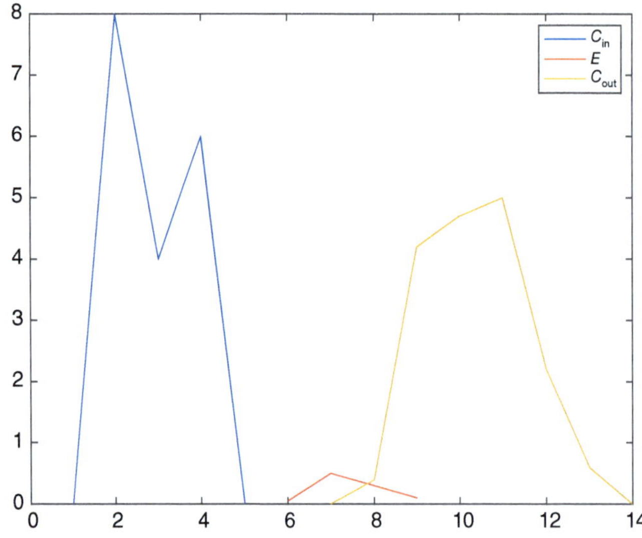


In Matlab®:

```
C=[0 8 4 6 0]
E=[0.05 0.5 0.3 0.1]
tC=1:5
tE=6:9
to=7:14
Co=conv(C,E)
plot(tC,C,tE,E,to,Co)
legend('C_i_n','E','C_o_u_t')
```

**Problem 4.4**   Let us have the following data for the input signal to a continuous stirred tank reactor (CSTR):

| $t$ (s) | 0 | 1 | 2 | 3 | 4 | 5 | 6 | 7 | 8 | 9 | 10 | 11 |
|---|---|---|---|---|---|---|---|---|---|---|---|---|
| $C_{in}$ | 0 | 0 | 8 | 6 | 4 | 5 | 6 | 3 | 1 | 0 | 0 | 0 |

In the CSTR, the residence time is known to be 5 seconds. What is the signal at the exit of the reactor?

**Solution to Problem 4.4**
In the CSTR the $E(t)$ is:

$$E(t) = \frac{1}{\bar{t}} \exp\left(-\frac{t}{\bar{t}}\right) = 0.2 \exp(-0.2t)$$

For doing the convolution of a signal of $m = 12$ values, we can choose $n = 12$ values of $E(t)$ at times separated by one second (as is in the $C_{in}$ signal).

In this way:

| $t_E$ (s) | 0 | 1 | 2 | 3 | 4 | 5 | 6 | 7 | 8 | 9 | 10 | 11 |
|---|---|---|---|---|---|---|---|---|---|---|---|---|
| $E$ (s$^{-1}$) | 0.200 | 0.164 | 0.134 | 0.110 | 0.090 | 0.074 | 0.060 | 0.049 | 0.040 | 0.033 | 0.027 | 0.022 |

We will obtain $v = 12 + 12 - 1 = 23$ points in the convoluted signal. And we have that:

$$
\begin{pmatrix} C_{out,1} \\ C_{out,2} \\ C_{out,3} \\ C_{out,4} \\ \cdots \\ \cdots \\ \cdots \\ C_{out,23} \end{pmatrix}
=
\begin{bmatrix}
E_1 & 0 & 0 & \cdots & 0 \\
E_2 & E_1 & 0 & \cdots & 0 \\
E_3 & E_2 & E_1 & \cdots & 0 \\
\cdots & \cdots & \cdots & \cdots & \cdots \\
E_n & E_{n-1} & \cdots & \cdots & \cdots \\
0 & E_n & E_{n-1} & \cdots & \cdots \\
& & E_n & & \\
& & \cdots & & \\
& & \cdots & & \\
0 & 0 & 0 & 0 & E_n
\end{bmatrix}_{23\times12}
\cdot
\begin{pmatrix} C_{in,1} \\ C_{in,2} \\ \cdots \\ C_{in,12} \end{pmatrix}_{12\times1}
$$

$$
\begin{pmatrix} C_{out,1} \\ C_{out,2} \\ C_{out,3} \\ C_{out,4} \\ \cdots \\ \cdots \\ \cdots \\ C_{out,23} \end{pmatrix}
=
\begin{bmatrix}
0.200 & 0 & 0 \\
0.164 & 0.200 & 0 \\
0.134 & 0.164 & 0.200 \\
0.110 & 0.134 & 0.164 \\
0.090 & 0.110 & 0.134 \\
0.074 & 0.090 & 0.110 \\
0.060 & 0.074 & 0.090 \\
0.049 & 0.060 & 0.074 \\
0.040 & 0.049 & 0.060 \\
0.033 & 0.040 & 0.049 \\
0.027 & 0.033 & 0.040 \\
0.022 & 0.027 & 0.033 \\
0 & 0.022 & 0.027 \\
0 & 0 & 0.022 \\
0 & 0 & 0 \\
0 & 0 & 0 \\
\cdots & \cdots & \cdots
\end{bmatrix}
\cdots \quad \cdots
\cdot
\begin{pmatrix} 0 \\ 0 \\ 8 \\ 6 \\ 4 \\ 5 \\ 6 \\ 3 \\ 1 \\ 0 \\ 0 \\ 0 \end{pmatrix}
$$

Finally, we can obtain:

$$
C_{\text{out}} = \begin{vmatrix}
0 \\
0 \\
1.6 \\
2.512 \\
2.856 \\
3.340 \\
3.936 \\
3.826 \\
3.330 \\
2.724 \\
2.228 \\
1.824 \\
1.491 \\
1.218 \\
0.852 \\
0.590 \\
0.411 \\
0.246 \\
0.093 \\
0.022 \\
0 \\
0 \\
0
\end{vmatrix}
$$

And the time vector is $t_{\text{out}} = (0, 1, 2, 3, \dots 22)$.

**Problem 4.5**   The following signal is introduced to a system of three reactors such as that shown in figure ($C_{\text{in}}$ in mmol/l).

$$C_{\text{in}} = 0.02 * t^2 \quad \text{for } 0 < t < 12\,\text{s}$$

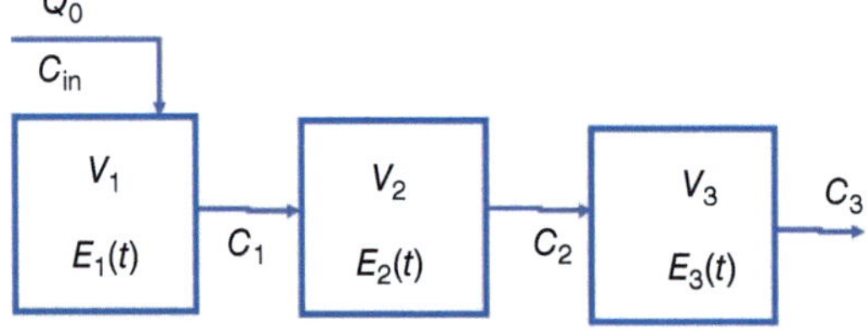


At the exit of the first reactor, the signal is measured at times 4–25 seconds, obtaining a single peak of height 7 mmol/l at $t = 13$ s.

The second reactor has a CSTR behavior with $t_{m2} = 10$ s, and the third reactor has $E(t)$ given by the following data:

| $t_{r3}$ (s) | 0 | 1 | 2 | 3 | 4 | 5 | 6 | 7 | 8 | 9 |
|---|---|---|---|---|---|---|---|---|---|---|
| $E_3(t)$ | 0.5319 | 0.2660 | 0.1064 | 0.0532 | 0.0266 | 0.0106 | 0.0053 | 0 | 0 | 0 |

Calculate the corresponding $E(t)$ of all three reactors, and the signal evolution at the exit of all three reactors. Plot them correspondingly.

**Solution to Problem 4.5**

For details refer the Wiley website at http://www.wiley-vch.de/ISBN9783527354115

This problem is better solved by using a program such as Matlab® or Octave. We can do the mentioned tasks by using the following:

```
clear all
close all

global Cin C1

tin=0:12;
Cin=0.02*tin.^2;

m=length(Cin);

%%% At the exit of the first reactor:

t1=4:25;
C1=[0 0 0 0 0 0 0 0 0 7 0 0 0 0 0 0 0 0 0 0 0 0];

v=length(C1);
n=v+1-m;

% RTD of the first reactor:
Cout=C1;
Esup=ones(n,1)/2;
E1=abs(fminsearch(@minimal,Esup)); % Minimization of...
the O.F.
trl=(t1(1)-tin(1)):(t1(end)-tin(end));

%%%% RTD of the second reactor:
tr2=0:50;
tau2=10;
E2=1/tau2*exp(-tr2/tau2);

%%% At the exit of the second reactor:
C2=conv(C1,E2);
t2=(tr2(1)+t1(1)):(tr2(end)+t1(end));

%%%% RTD of the third reactor:
tr3=0:9;
E3=[0.5319    0.2660    0.1064    0.0532    0.0266...
0.0106    0.0053    0    0    0];

%%% At the exit of the third reactor:
C3=conv(C2,E3);
t3=(tr3(1)+t2(1)):(tr3(end)+t2(end));
```

```
%% Figures:

figure(1)
plot(tin,Cin,t1,C1,t2,C2,t3,C3)
legend('Cin','C1','C2','C3')
title('Concentrations at the input of the system and at...
the exit of each reactor')
xlabel('Time (s)')
ylabel('Conc.')

figure(2)
plot(tr1,E1,tr2,E2,tr3,E3)
legend('E1','E2','E3')
title('RTD of the three reactors')
xlabel('Time (s)')
ylabel('E(t)')
```

We also need the function "minimal":

```
function OF=minimal(f0)  % f0 is the E curve being...
optimised
global Cin C1
Cout=C1;

n=length(f0); % Dimmension of the resulting E vector
v=length(Cout); % Dimmension of vector Cout
m=length(Cin); % Dimmension of vector Cin
for i=1:m
    A(i:i+n-1,i)=abs(f0);% Constructing the convolution...
matrix
end
OF=sum((Cin'-A'*Cout').^2);
```

The following is obtained:

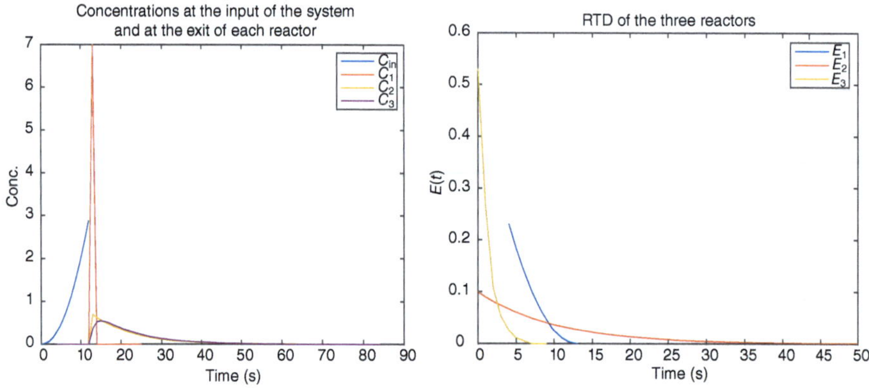

**Problem 4.6**  Two pulse inputs are fed to a reactor of unknown RTD, with a difference of 10 seconds. The response of the reactor to such feed is given in the following graph. Determine the $E(t)$ of the reactor.

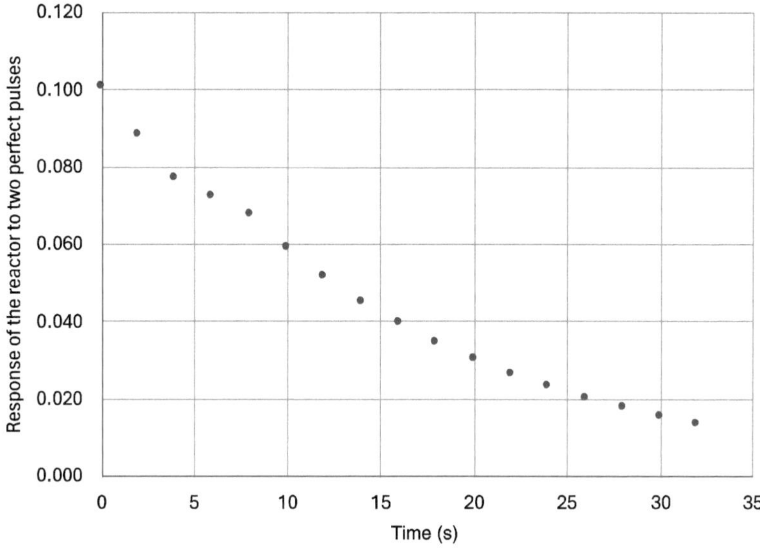

## Solution to Problem 4.6

From the graph, we can read the following data:

| $t$ (s)  | 0     | 2     | 4     | 6     | 8     | 10    | 12    | 14    | 16    |
|----------|-------|-------|-------|-------|-------|-------|-------|-------|-------|
| Response | 0.101 | 0.088 | 0.077 | 0.072 | 0.068 | 0.059 | 0.052 | 0.045 | 0.040 |

| $t$ (s)  | 18    | 20    | 22    | 24    | 26    | 28    | 30    | 32    |
|----------|-------|-------|-------|-------|-------|-------|-------|-------|
| Response | 0.035 | 0.030 | 0.027 | 0.023 | 0.020 | 0.018 | 0.016 | 0.014 |

As the response is due to two different pulses (at 0 and 10 seconds), the input can be represented by:

| $t$ (s)         | 0 | 2 | 4 | 5 | 6 | 8 | 10 | 12 |
|-----------------|---|---|---|---|---|---|----|----|
| $C_{\text{input}}$ | 1 | 0 | 0 | 0 | 0 | 0 | 1  | 0  |

Following the deconvolution operation:

$$
\begin{pmatrix} C_{\text{out},1} \\ C_{\text{out},2} \\ C_{\text{out},3} \\ C_{\text{out},4} \\ \cdots \\ \cdots \\ C_{\text{out},v} \end{pmatrix}
=
\begin{pmatrix}
E_1 & 0 & 0 & \cdots & 0 \\
E_2 & E_1 & 0 & \cdots & 0 \\
E_3 & E_2 & E_1 & \cdots & 0 \\
\cdots & \cdots & \cdots & \cdots & \cdots \\
E_n & E_{n-1} & \cdots & \cdots & \cdots \\
0 & E_n & E_{n-1} & \cdots & \cdots \\
 & & E_n & & \\
 & & \cdots & & \\
 & & \cdots & & \\
0 & 0 & 0\ 0 & 0 & E_n
\end{pmatrix}
\cdot
\begin{pmatrix} C_{\text{in},1} \\ C_{\text{in},2} \\ \cdots \\ C_{\text{in},m} \end{pmatrix}
$$

We have that $v = 17$ (number of points in $C_{out}$), $m = 7$ (number of points in $C_{in}$), and so $n = v - m + 1 = 11$ points can be determined in the $E(t)$:

$$
\begin{pmatrix} 0.101 \\ 0.088 \\ 0.077 \\ 0.072 \\ 0.068 \\ 0.059 \\ 0.052 \\ 0.045 \\ 0.040 \\ 0.035 \\ 0.030 \\ 0.027 \\ 0.023 \\ 0.020 \\ 0.018 \\ 0.016 \\ 0.014 \end{pmatrix}_{17\times1}
=
\begin{bmatrix}
E_1 & 0 & 0 & 0 & \cdots \\
E_2 & E_1 & 0 & 0 & \cdots \\
E_3 & E_2 & E_1 & 0 & \cdots \\
E_4 & E_3 & E_2 & E_1 & \cdots \\
E_5 & E_4 & E_3 & E_2 & \cdots \\
E_6 & E_5 & E_4 & E_3 & \cdots \\
E_7 & E_6 & E_5 & E_4 & \cdots \\
E_8 & E_7 & E_6 & E_5 & \cdots \\
E_9 & E_8 & E_7 & E_6 & \cdots \\
E_{10} & E_9 & E_8 & E_7 & \cdots \\
E_{11} & E_{10} & E_9 & E_8 & \cdots \\
0 & E_{11} & E_{10} & E_9 & \cdots \\
 & & \cdots & & \\
0 & 0 & 0 & 0 & \cdots
\end{bmatrix}_{17\times7}
\begin{pmatrix} 1 \\ 0 \\ 0 \\ 0 \\ 0 \\ 1 \\ 0 \end{pmatrix}_{7\times1}
$$

Let us use Matlab® to solve the problem, just like in Problem 4.5. The problem is then to deconvolute the output signal, bearing in mind that the function $E(t)$ in the reactor is superposed at different times.

```
clear all
close all

global Cin C1

tin=0:2:12;
Cin=[1 0 0 0 0 1 0];

m=length(Cin);

%%% At the exit of the reactor:

t1=0:2:32;
C1=[0.101      0.088    0.077    0.072    0.068    0.059...
0.052    0.045    0.040 0.035 0.030    0.027    0.023    0.020...
0.018    0.016    0.01];

v=length(C1);
n=v+1-m;

% RTD of the reactor:
Cout=C1;
Esup=ones(n,1)/2;
E1=abs(fminsearch(@minimal,Esup)); % Minimization of...
the O.F.
tr1=(t1(1)-tin(1)):2:(t1(end)-tin(end));
```

```
%% Figures:

figure(1)
plot(tin,Cin,t1,C1)
legend('Cin','C1')
title('Concentrations at the input of the system and at...
the exit of the reactor')
xlabel('Time (s)')
ylabel('Conc.')

figure(2)
plot(tr1,E1)
legend('E1')
title('RTD')
xlabel('Time (s)')
ylabel('E(t)')
```

We obtain:

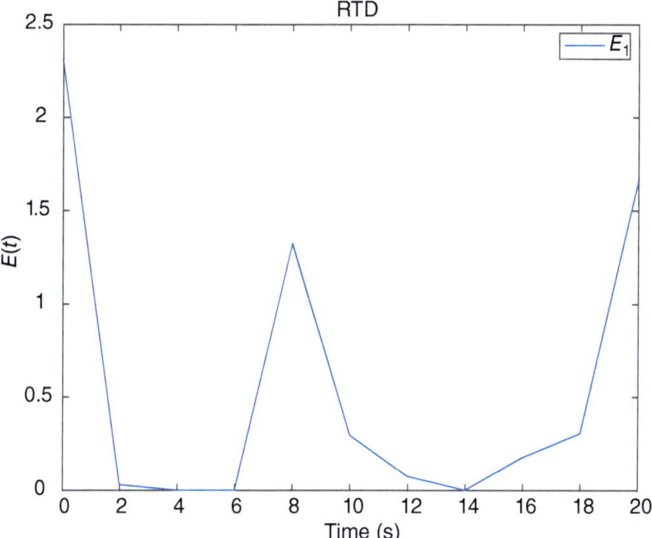

# 5

# Partial Differential Equations in Chemical Reactor Engineering

## Summary

### Finite Differences Method (FDM)

#### First Derivative

$$\text{Front differences: } f'(x_i) = \frac{f(x_{i+1}) - f(x_i)}{\Delta x}$$

$$\text{Rear differences: } f'(x_i) = \frac{f(x_i) - f(x_{i-1})}{\Delta x}$$

#### Second Derivative

$$\text{Front differences: } f''(x_i) = \frac{f(x_i) - 2f(x_{i+1}) + f(x_{i+2})}{\Delta x^2}$$

$$\text{Central differences: } f''(x_i) = \frac{f(x_{i-1}) - 2f(x_i) + f(x_{i+1})}{\Delta x^2}$$

$$\text{Rear differences: } f''(x_i) = \frac{f(x_{i-2}) - 2f(x_{i-1}) + f(x_i)}{\Delta x^2}$$

### Stability of the FDM

One way to ensure stability is through the following theorem. Given a scheme like the following:

$$f(x)_i^{t+1} = A \cdot f(x)_{i+1}^t + B \cdot f(x)_i^t + C \cdot f(x)_{i-1}^t$$

If $A$, $B$, and $C$ are all positive ($A \geq 0$, $B \geq 0$, and $C \geq 0$) and also $(A + B + C) \leq 1$, the scheme is stable, i.e., the errors in the solution will not increase.

### Ideal Reactors Working in Unsteady State

#### CSTR Working in Unsteady State

During an unsteady state, we will have:

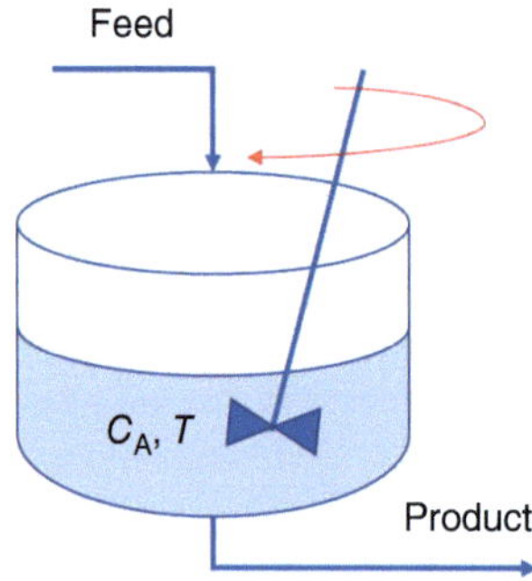


Input $= n_{A0}$ (moles/time)

Output $= n_A$ (moles/time)

Accumulation $= V \cdot dC_A/dt$ (volume $\cdot$ (moles/volume)/time)

Generation $= r_A \cdot V$ ((moles/volume $\cdot$ time) $\cdot$ volume)

Mole balance:

$$n_{A0} + r_A V = n_A + V\frac{dC_A}{dt}$$

#### PFR Working in Unsteady State (No Dispersion)



Mole balance:

$$\frac{\partial C_A}{\partial t} = r_A - Q\frac{\partial C_A}{\partial V}$$

## PFR Working in Dynamic Regime (With Dispersion)

Mole balance:

$$D_e \cdot \frac{\partial^2 C_A}{\partial z^2} - u \cdot \frac{\partial C_A}{\partial z} + r_A = \frac{\partial C_A}{\partial t}$$

For more details, please consult Conesa (2019).

**Problem 5.1**   Acetic anhydride undergoes hydrolysis in a continuous stirred tank reactor (CSTR) at a temperature of 40 °C. Initially, the reactor contains $0.57\,\mathrm{m}^3$ of an aqueous solution with a concentration of $0.487\,\mathrm{kmol/m}^3$ of anhydride. The reactor is rapidly heated to 350 K, and at that point, a feed solution with a concentration of $0.985\,\mathrm{kmol/m}^3$ of anhydride is introduced into the reactor at a rate of 9.55 l/s. Simultaneously, the product pump is activated, and the product is withdrawn from the reactor at the same rate of 9.55 l/s. The reaction follows first-order kinetics with a rate constant of $6.35 \cdot 10^{-3}\,\mathrm{s}^{-1}$.

(a) Perform an unsteady state balance and solve to obtain the transient behavior during startup in terms of the concentrations of anhydride and acid in the product stream.
(b) Determine the concentrations of anhydride and acid in the product stream after 120 seconds.

**Solution to Problem 5.1**

For details refer the Wiley website at http://www.wiley-vch.de/ISBN9783527354115

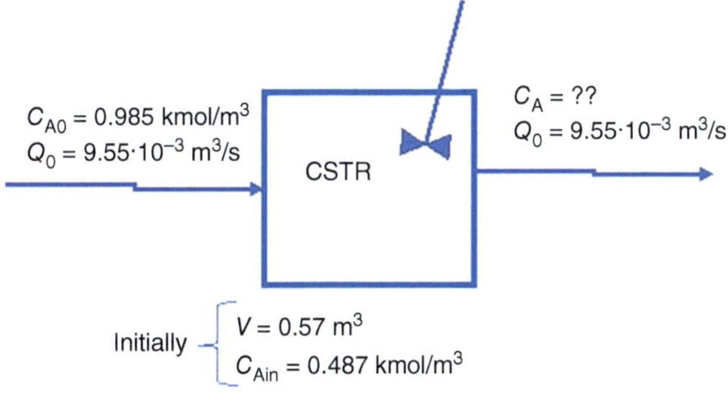


The reaction is: $A + H_2O \rightarrow B$

In this problem, we have a CSTR working in a non-steady state. A mass balance in the reactor would be:

$$n_{A0} - n_A + (-r_A) \cdot V = \frac{dn_A}{dt}$$

With $r_A = k \cdot C_A$

Obtaining:

$$\frac{1}{t} \cdot C_{A0} - \frac{1}{t} \cdot C_A - k \cdot C_A = \frac{dC_A}{dt}$$

Furthermore, the initial value of the concentration of reactant in the vessel is not zero, as in the reactor there is an amount of anhydride. Using the finite differences method (FDM), we have:

$$\frac{C_A^{t+1} - C_A^t}{\Delta t} = \frac{1}{t} \cdot \left(C_{A0} - C_A^t\right) - k \cdot C_A^t$$

For "B":

$$n_{B0} - n_B + (-r_B)V = \frac{dn_B}{dt}$$

$$-Q_0 C_B + k C_A V = Q_0 \frac{dC_B}{dt}$$

$$\frac{-Q_0 C_B^t \Delta t + k C_A^t V \Delta t}{Q_0} = C_B^{t+1} - C_B^t$$

With initial values of $C_A = 0.487\,\text{kmol/m}^3$ and $C_B = 0$. Note the differences between $C_{A0}$ and the initial value of $C_A$ in the reactor. These two values can be equal or, in other cases, different, as in the present one. We can now simulate the behavior of the tank until a steady state is observed.

| *t* (s) | $\Delta t$ (s) | $C_A$ (kmol/m³) | $C_B$ (kmol/m³) |
| --- | --- | --- | --- |
| 0 | — | 0.487 | 0 |
| 2 | 2 | 0.482 | 0.006 |
| 5 | 3 | 0.476 | 0.015 |
| 10 | 5 | 0.465 | 0.030 |
| 20 | 10 | 0.444 | 0.059 |
| 40 | 20 | 0.406 | 0.114 |
| 80 | 40 | 0.342 | 0.209 |
| 120 | 40 | 0.298 | 0.282 |
| 160 | 40 | 0.268 | 0.339 |
| 200 | 40 | 0.248 | 0.384 |
| 300 | 100 | 0.214 | 0.477 |
| 400 | 100 | 0.207 | 0.533 |
| 500 | 100 | 0.206 | 0.576 |
| 700 | 200 | 0.205 | 0.644 |
| 900 | 200 | 0.206 | 0.689 |
| 1100 | 200 | 0.206 | 0.720 |
| 1500 | 400 | 0.206 | 0.759 |
| 2000 | 500 | 0.205 | 0.777 |
| 2500 | 500 | 0.207 | 0.777 |

Plotting $C_A$ and $C_B$ versus time:

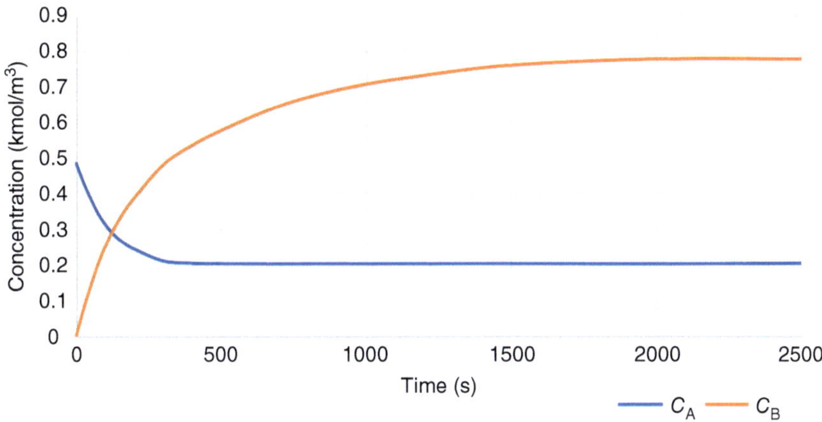

In the table, we see that the concentration of anhydride at 120 seconds is $0.298 \, \text{kmol/m}^3$.

**Problem 5.2** The mass transfer of substance A between solvent I and solvent II, which will be considered totally immiscible, is being studied. It is further assumed that the concentration of A is small enough that Fick's second law can be used to describe the diffusion in both regions. It is desired to solve the diffusion equations:

$$\frac{\partial C_{\mathrm{I}}}{\partial t} = D_{\mathrm{I}} \frac{\partial^2 C_{\mathrm{I}}}{\partial z^2} \text{ from } z = -\infty \text{ to } z = 0$$

$$\frac{\partial C_{\mathrm{II}}}{\partial t} = D_{\mathrm{II}} \frac{\partial^2 C_{\mathrm{II}}}{\partial z^2} \text{ from } z = 0 \text{ to } z = +\infty$$

Being:

$C_{\mathrm{I}}$    concentration of A in phase I
$C_{\mathrm{II}}$    concentration of A in phase II
$D_{\mathrm{I}}$    diffusion coefficient of A in phase I
$D_{\mathrm{II}}$    diffusion coefficient of A in phase II

The initial and boundary conditions are:

$$
\begin{aligned}
&\text{for } t = 0 &&\rightarrow C_{\mathrm{I}} = C_{\mathrm{I}}^0 &&\text{for } -\infty < z < 0 \\
&\text{for } t = 0 &&\rightarrow C_{\mathrm{II}} = C_{\mathrm{II}}^0 &&\text{for } 0 < z < +\infty \\
&\text{for } z = 0 &&\rightarrow C_{\mathrm{II}} = m C_{\mathrm{I}} &&\text{for } t < 0 \\
&\text{for } z = 0 &&\rightarrow -D_{\mathrm{I}} \frac{\partial C_{\mathrm{I}}}{\partial z} = -D_{\mathrm{II}} \frac{\partial C_{\mathrm{II}}}{\partial z} \\
&\text{for } z = -\infty &&\rightarrow C_{\mathrm{I}} = C_{\mathrm{I}}^0 \\
&\text{for } z = +\infty &&\rightarrow C_{\mathrm{II}} = C_{\mathrm{II}}^0
\end{aligned}
$$

The first limit condition for $z = 0$ reveals the existence of equilibrium at the interface, where $m$ is a distribution coefficient. The second states that the molar flux density calculated for $z = 0$ is the same on both sides of the fluid.

(a) Plot diagrams of the system, indicating the conditions mentioned at $t = 0$ and $t > 0$. Include the initially known concentration values and the boundary conditions of the problem.

(b) Apply the FDM to this system, studying the stability of the solution and clearly indicating how to take into account the limit and initial conditions.

(c) Explain how to carry out the calculation of the concentration profile that is established in the areas adjacent to the interface at different times.

**Solution to Problem 5.2**

(a)

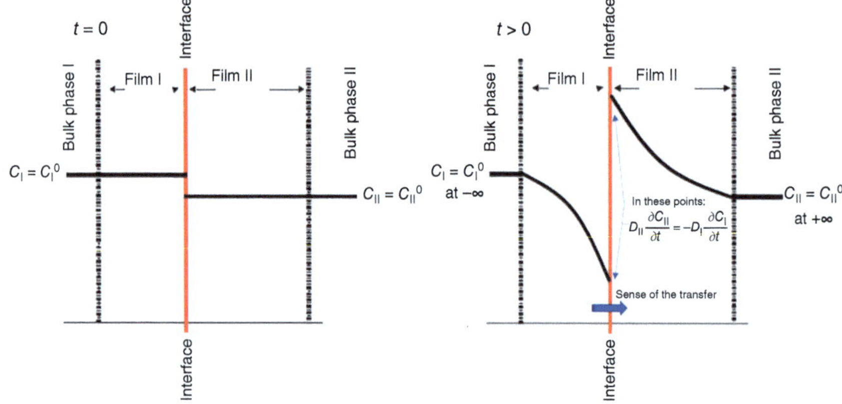


(b) From:

$$\frac{\partial C_{\mathrm{I}}}{\partial t} = D_{\mathrm{I}} \frac{\partial^2 C_{\mathrm{I}}}{\partial z^2}$$

We can do:

$$\frac{C_{\mathrm{I}_i}^{t+1} - C_{\mathrm{I}_i}^{t}}{\Delta t} = D_{\mathrm{I}} \frac{C_{\mathrm{I}_{i+1}}^{t} - 2C_{\mathrm{I}_i}^{t} + C_{\mathrm{I}_{i-1}}^{t}}{\Delta z^2}$$

$$C_{\mathrm{I}_i}^{t+1} = \left( D_{\mathrm{I}} \frac{C_{\mathrm{I}_{i+1}}^{t} - 2C_{\mathrm{I}_i}^{t} + C_{\mathrm{I}_{i-1}}^{t}}{\Delta z^2} \right) \Delta t + C_{\mathrm{I}_i}^{t}$$

$$= \left( \frac{D_{\mathrm{I}} \Delta t}{\Delta z^2} \right) C_{\mathrm{I}_{i+1}}^{t} + \left( 1 - 2\frac{D_{\mathrm{I}} \Delta t}{\Delta z^2} \right) C_{\mathrm{I}_i}^{t} + \left( \frac{D_{\mathrm{I}} \Delta t}{\Delta z^2} \right) C_{\mathrm{I}_{i-1}}^{t}$$

And equally for phase II:

$$C_{\mathrm{II}_i}^{t+1} = \left( \frac{D_{\mathrm{II}} \Delta t}{\Delta z^2} \right) C_{\mathrm{II}_{i+1}}^{t} + \left( 1 - 2\frac{D_{\mathrm{II}} \Delta t}{\Delta z^2} \right) C_{\mathrm{II}_i}^{t} + \left( \frac{D_{\mathrm{I}} \Delta t}{\Delta z^2} \right) C_{\mathrm{II}_{i-1}}^{t}$$

For the equation being held in phase I, the stability conditions are:

$$\left( \frac{D_{\mathrm{I}} \Delta t}{\Delta z^2} \right) > 0$$

$$\left( 1 - 2\frac{D_{\mathrm{I}} \Delta t}{\Delta z^2} \right) > 0$$

$$\left[\left(1 - 2\frac{D_{\mathrm{I}}\Delta t}{\Delta z^2}\right) + \left(\frac{D_{\mathrm{I}}\Delta t}{\Delta z^2}\right) + \left(\frac{D_{\mathrm{I}}\Delta t}{\Delta z^2}\right)\right] \leq 1$$

The first and third conditions are always true, and from the second one:

$$\Delta t < \frac{\Delta z^2}{2D_{\mathrm{I}}}$$

And equivalent reasoning for phase II:

$$\Delta t < \frac{\Delta z^2}{2D_{\mathrm{II}}}$$

So the condition is:

$$\Delta t = \min\left(\frac{\Delta z^2}{2D_{\mathrm{I}}}, \frac{\Delta z^2}{2D_{\mathrm{II}}}\right)$$

In the interface:

$$-D_{\mathrm{I}}\frac{\partial C_{\mathrm{I}}}{\partial z} = -D_{\mathrm{II}}\frac{\partial C_{\mathrm{II}}}{\partial z}$$

$$-D_{\mathrm{I}}\left[\frac{C_{\mathrm{I}_{i+1}} - C_{\mathrm{I}_i}}{\Delta z}\right]_{z=0} = -D_{\mathrm{II}}\left[\frac{C_{\mathrm{II}_{i+1}} - C_{\mathrm{II}_i}}{\Delta z}\right]_{z=0}$$

$$\frac{D_{\mathrm{I}}}{D_{\mathrm{II}}} = \left[\frac{C_{\mathrm{II}_{i+1}} - C_{\mathrm{II}_i}}{C_{\mathrm{I}_{i+1}} - C_{\mathrm{I}_i}}\right]_{z=0}$$

(c) We begin with the initial condition, which is:

$$C_{\mathrm{I}} = C_{\mathrm{I}}^0 \quad \text{for all positions "$i$" in phase I } (-\infty < z < 0)$$

$$C_{\mathrm{II}} = C_{\mathrm{II}}^0 \quad \text{for all positions "$i$" in phase II } (0 < z < +\infty)$$

Choosing a value of $\Delta z$ (or a number of increments $N$ in the length variable), we should choose a $\Delta t$ according to the stability condition. In the intervals 2 to $(N-1)$, i.e., all increments except the extremes, we apply the equation corresponding to $C_{\mathrm{I}_i}^{t+1}$ or $C_{\mathrm{II}_i}^{t+1}$.

In $z = +\infty$: $C_{\mathrm{II}} = C_{\mathrm{II}}^0$
In $z = -\infty$: $C_{\mathrm{I}} = C_{\mathrm{I}}^0$
In $z = 0$: $\dfrac{D_{\mathrm{I}}}{D_{\mathrm{II}}} = \left[\dfrac{C_{\mathrm{II}_{i+1}} - C_{\mathrm{II}_i}}{C_{\mathrm{I}_{i+1}} - C_{\mathrm{I}_i}}\right]_{z=0}$

**Problem 5.3** Perform a mass balance in a dispersed plug flow reactor (PFR) in which a first-order reaction occurs. We will assume open and closed boundary conditions at both ends and steady state regime. Apply the FDM and indicate the equations that result in the calculation of the reactant concentration. Indicate the stability conditions and how to consider the boundary conditions of this problem.

**Solution to Problem 5.3**

As pointed out previously, the mass balance in the PFR system can be expressed as:

$$n_{\mathrm{A}} + r_{\mathrm{A}} \cdot dV = (n_{\mathrm{A}} + dn_{\mathrm{A}})$$

The accumulation is zero because the problem works in a steady state. But in this case:

$$Q \cdot dC_A \neq dn_A$$

As:

$$n_A = Q \cdot C_A - D_e \cdot S \cdot \frac{dC_A}{dz}$$

In the previous equation, two terms are present, the first one corresponding to the convection (general movement of the flow) and the second one corresponding to the diffusion by Fick's law. Deriving and doing $Q = (u \cdot S)$, then:

$$\frac{dn_A}{dz} = u \cdot S \cdot \frac{dC_A}{dz} - D_e \cdot S \cdot \frac{d^2C_A}{dz^2}$$

And introducing the mole balance:

$$D_e \cdot \frac{d^2C_A}{dz^2} - u \cdot \frac{dC_A}{dz} + r_A = 0$$

If we assume first-order kinetics: $r_A = -k \cdot C_A$
And then:

$$D_e \cdot \frac{d^2C_A}{dz^2} - u \cdot \frac{dC_A}{dz} - k \cdot C_A = 0$$

For a closed–closed system, boundary conditions are given by the Danckwerts boundary conditions, so:

$$\text{At } z = 0 \rightarrow u \cdot C_A|_{z=0} - D_e \cdot \left.\frac{dC_A}{dz}\right|_{z=0} = u \cdot C_{A0}$$

$$\text{At } z = L \rightarrow \left.\frac{dC_A}{dz}\right|_{z=L} = 0$$

On the other hand, if the system is open:

$$\text{At } z = 0 \rightarrow C_A|_{z=0} = C_{A0}$$

$$\text{At } z = L \rightarrow \left.\frac{dC_A}{dz}\right|_{z=L} = 0$$

As we can check, the balance equation is not dependent on time (steady state), only on distance, but the derivative is second-order. In this situation, a change in variables is needed. Let us name $C_A'$ as a new variable:

$$C_A' = \frac{dC_A}{dz}$$

$$\frac{d^2C_A}{dz^2} = \frac{dC_A'}{dz}$$

The mass balance can be written as follows:

$$D_e \cdot \frac{dC_A'}{dz} - u \cdot C_A' - k \cdot C_A = 0$$

$$C_A' = \frac{dC_A}{dz}$$

And the boundary conditions are, for a closed system:

$$\text{At } z = 0 \rightarrow u \cdot C_A\big|_{z=0} - D_e \cdot C'_A\big|_{z=0} = u \cdot C_{A0}$$

$$\text{At } z = L \rightarrow C'_A\big|_{z=L} = 0$$

If the system were open:

$$\text{At } z = 0 \rightarrow C_A\big|_{z=0} = C_{A0}$$

$$\text{At } z = L \rightarrow C'_A\big|_{z=L} = 0$$

Using FDM: Let us apply FDM to the mass balance equations:

$$D_e \cdot \frac{C'_{A_{i+1}} - C'_{A_i}}{\Delta z} - u \cdot C'_{A_i} - k \cdot C_{A_i} = 0$$

$$C'_{A_i} = \frac{C_{A_{i+1}} - C_{A_i}}{\Delta z}$$

In this way:

$$C'_{A_{i+1}} = \frac{\frac{k}{u} C_{A_i} + C'_{A_i}\left(1 + \frac{D_e}{u \cdot \Delta z}\right)}{\frac{D_e}{u \cdot \Delta z}}$$

$$C_{A_{i+1}} = C'_{A_i} \cdot \Delta z + C_{A_i}$$

Values of $C_A$ and $C'_A$ at $z = 0$: For the integration of the former equations, we need a starting point.

*For an open–open recipient*, the value of $C_A\big|_{z=0}$ is obviously given by $C_{A0}$. Contrarily, the value of the derivative $C'_A\big|_{z=0}$ is not known, as we only know that the value at $z = L$ is zero.

In order to solve this problem, it is needed an estimation of the value of $C'_A\big|_{z=0}$ that, when integrating, meets in the most accurate way the boundary condition $C'_A\big|_{z=L} = 0$. The system requires, in this way, an optimization of the value of the derivative at the input, minimizing the value of $C'_A\big|_{z=L}$:

$$\text{Objective Function} = \text{O.F.} = \min\left(\text{abs}(C'_A\big|_{z=L})\right)$$

where the absolute value (abs) is introduced to avoid negative values.

*For a closed–closed recipient*, the value of $C_A\big|_{z=0}$ is not given by $C_{A0}$. The condition we know is:

$$u \cdot C_A\big|_{z=0} - D_e \cdot C'_A\big|_{z=0} = u \cdot C_{A0}$$

i.e.,:

$$u \cdot C_A\big|_{z=0} - D_e \cdot C'_A\big|_{z=0} - u \cdot C_{A0} = 0$$

Also, as in the open case, the value of the derivative $C'_A\big|_{z=0}$ is not known, as we only know that the value at $z = L$ is zero.

For solving the system, it is needed an estimation of the value of $C_A\big|_{z=0}$ and $C'_A\big|_{z=0}$ that, when integrating, meets in the most accurate way the boundary conditions

$C'_A\big|_{z=L} = 0$ and previous equation. The system requires, in this way, an optimization of:

Objective Function = O.F.

$$= \min \left( \text{abs} \left( C'_A\big|_{z=L} \right) + \text{abs} \left[ u \cdot C_A\big|_{z=0} - D_e \cdot C'_A\big|_{z=0} - u \cdot C_{A0} \right] \right)$$

**Problem 5.4**  A reactor for cracking ethylbenzene (gas) behaves like a CSTR. It is submerged in a bath of molten KCl salts in order to keep the external temperature constant and equal to $T_{ext} = 770\,°C$. The evolution with time of the temperature in the reactor wall can be calculated by solving the heat balance in a non-steady state, which leads to the following expression:

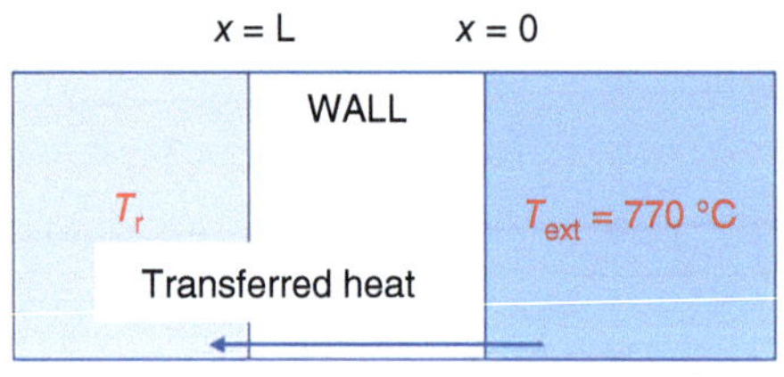


$$\rho \cdot C_p \cdot \frac{\delta T}{\delta t} = k \cdot \frac{\delta^2 T}{\delta x^2}$$

where $\rho\, C_p$ is the product of the density of the wall and its heat capacity, $k$ is its thermal conductivity, $T$ is the temperature of the wall, "$t$" the time, and "$x$" the distance measured on the reactor wall of total width "$L$."

It is intended to simulate the heating of the wall of this reactor, which starts from a temperature $T_0$ at $t = 0$. The boundary conditions of the problem indicate that outside ($x = 0$), the temperature of the wall is the same as that of the molten salt, while inside there is heat transfer, such that:

$$k \cdot \frac{\delta T}{\delta x}\bigg|_{x=L} = U \cdot A \cdot (T - T_r)$$

Being "$U$" the global heat transfer coefficient, "$A$" the area, and "$T_r$" the reaction temperature. Applying the finite difference method, indicate the equations that result in the calculation of the temperature along the wall. Indicate the stability conditions and how to consider the boundary conditions of this problem.

**Solution to Problem 5.4**
The equation:

$$\rho \cdot C_p \cdot \frac{\delta T}{\delta t} = k \cdot \frac{\delta^2 T}{\delta x^2}$$

Using the FDM can be written as:

$$\rho \cdot C_p \cdot \frac{T_i^{t+1} - T_i^t}{\Delta t} = k \cdot \frac{T_{i+1}^t - 2T_i^t + T_{i-1}^t}{\Delta x^2}$$

that, rearranging:

$$T_i^{t+1} = \frac{k\Delta t}{\rho C_p \Delta x^2} T_i^t + \left(1 - 2\frac{k\Delta t}{\rho C_p \Delta x^2}\right) T_i^t + \frac{k\Delta t}{\rho C_p \Delta x^2} T_i^t$$

With $A = C = \frac{k\Delta t}{\rho C_p \Delta x^2}$ and $B = 1 - 2\frac{k\Delta t}{\rho C_p \Delta x^2}$ the stability conditions are $A > 0$, $B > 0$, and $(A + B + C) \leq 1$. The first and last conditions are always fulfilled. From $B > 0$, we obtain:

$$\left(1 - 2\frac{k\Delta t}{\rho C_p \Delta x^2}\right) > 0$$

$$\Delta t < \frac{\rho C_p \Delta x^2}{2k}$$

The boundary conditions are:

$$\text{in } x = 0 \quad T = \text{Text}$$

$$\text{in } x = L: \quad k \cdot \frac{T_N - T_{N-1}}{\Delta x} = U \cdot A \cdot (T_{N-1} - T_r) \text{ and so:}$$

$$T_N = \frac{\Delta x}{k} \cdot U \cdot A \cdot (T_{N-1} - T_r) + T_{N-1}$$

With $N$ being the last point of the mesh.

**Problem 5.5** A reactive gas ("A") is diffusing into flat catalyst particles, forming a concentration gradient as shown in the figure. The diffusion in the solid layer can be modeled with Fick's law, resulting in:

$$\frac{\delta C_A}{\delta t} = D_{AB} \frac{\delta^2 C_A}{\delta x^2} - kC_A$$

With the conditions shown in the figure, indicate how the concentration profiles can be determined using the finite difference method. Indicates the boundary, initial, and stability conditions of the solution.

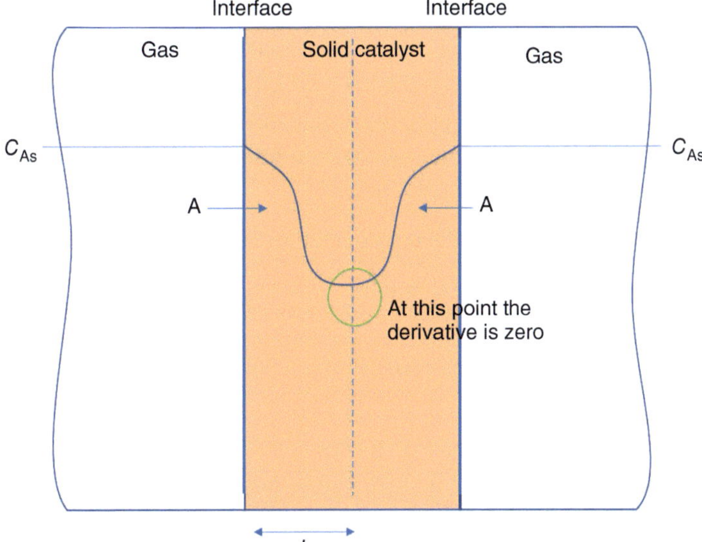

**Solution to Problem 5.5**

Let's consider the interior point of the catalyst. Since all the values at time "$t$" are assumed to be known, the spatial derivatives can be approximated using central differences:

$$\frac{\partial C_A}{\partial x} = \frac{C^t_{Ai+1} - C^t_{Ai-1}}{2 \cdot \Delta x}$$

$$\frac{\partial C^2_A}{\partial x^2} = \frac{C^t_{Ai+1} - 2 \cdot C^t_{Ai} + C^t_{Ai-1}}{\Delta x^2}$$

However, the first-order time derivative can be approximated by the forward difference:

$$\frac{\partial C_A}{\partial t} = \frac{C^{t+1}_{Ai} - C^t_{Ai}}{\Delta t}$$

The resolution procedure consists of substituting these approximations in the differential equation that governs the problem and solving explicitly for $C^{t+1}_{Ai}$. This equation will contain, on the right side, only terms of $C^t_A$. The boundary conditions will be used to calculate the limit points, and the problem can be solved forward in time. As a consideration, we must point out that the error in the resolution of this problem is proportional to $(\Delta t + \Delta x^2)$.

Applying the procedure to the previous example and using the indices "$i$" for the position and "$t$" for the time:

$$\frac{C^{t+1}_{Ai} - C^t_{Ai}}{\Delta t} = D_{AB} \frac{C^t_{Ai+1} - 2 \cdot C^t_{Ai} + C^t_{Ai-1}}{\Delta x^2} - k \cdot C^t_{Ai}$$

Rearranging:

$$C^{t+1}_{Ai} = \frac{\Delta t \cdot D_{AB}}{\Delta x^2} C^t_{Ai+1} - 2 \cdot \frac{\Delta t \cdot D_{AB}}{\Delta x^2} C^t_{Ai} - \frac{\Delta t \cdot D_{AB}}{\Delta x^2} C^t_{Ai-1} + k \cdot \Delta t \cdot C^t_{Ai} + C^t_{Ai}$$

We finally get that:

$$C^{t+1}_{Ai} = \frac{\Delta t \cdot D_{AB}}{\Delta x^2} C^t_{Ai+1} + \left(1 - k \cdot \Delta t - 2 \cdot \frac{\Delta t \cdot D_{AB}}{\Delta x^2}\right) C^t_{Ai} + \frac{\Delta t \cdot D_{AB}}{\Delta x^2} C^t_{Ai-1}$$

With the above equation, we can obtain the values of the concentration in the nodes in time $(t + \Delta t)$ depending on the value at time "$t$", which, together with the initial and boundary conditions, allows us to solve the problem.

For the problem, a valid initial condition is:

$t = 0 \rightarrow C_A = 0$, for all points but the one contacting the other phase, where $C_A = C_{As}$

These initial conditions can be written as:

$$t = 0; \quad C_A = 0; \quad \text{for } \forall x \neq 0$$
$$t = 0; \quad C_A = C_{AS}; \quad \text{for } \forall x = 0$$

The boundary conditions are:

At the interface ($x = 0$): $C_A = C_{As}$ (known and constant value), $\forall t$

At the center of the particle ($x = L$), by symmetry: $\frac{\partial C_A}{\partial x} = 0$

The boundary condition at the center of the catalyst, $x = L$, is similar to other conditions used in chemical engineering and reflects the fact that, in the present example, the derivative in the center is zero because it represents a minimum in the concentration of the species reacting.

$$C_A = C_{As}; \text{ for } x = 0$$

$$\left[ \frac{C^t_{Ai} - C^t_{Ai-1}}{\Delta x} \right]_{x=L} = 0 \quad \text{i.e., } C^t_{AN} = C^t_{AN-1}$$

Note that the derivative in the last point must be done in the form of a rear derivative, i.e., using the points "$i-1$" and "$i$". If central or front differences are tried to be used, the information in the point "$N+1$" is needed, and the system is not solvable.

We can also use a second-order approximation of the first derivative, and in this case, for the last node (rear differences), we will have:

$$\left[ \frac{3C^{t+1}_{Ai} - 4C^t_{Ai-1} + C^t_{Ai-2}}{2 \cdot \Delta x} \right]_{i=N} = 0 \quad \text{i.e., } C^{t+1}_{AN} = \frac{1}{3}\left(4C^t_{AN-1} + C^t_{AN-2}\right)$$

The equation represents the relationship between the concentrations at the final node and the two nodal points closest to it.

**Problem 5.6**  Perform a mass balance in a reactor:

(a) CSTR, to which a tracer is injected in pulse.
(b) CSTR, to which a reagent that decomposes with first-order kinetics (non-steady regime) begins to be fed.
(c) PFR without dispersion, to which tracer is injected in pulse.
(d) PFR with dispersion to which tracer is injected in pulse, assuming closed–closed vessel.
(e) CSTR–PFR combination working in steady state, first-order kinetics.

**Solution to Problem 5.6**

(a) CSTR, to which a tracer is injected in pulse.
    During the unsteady state, we will have:

Input $= 0$ (no tracer input during mass balance, as it is a pulse input)

Output $= n_A$ (moles/time)

Accumulation $= V \cdot dC_A/dt$ (volume $\cdot$ (moles/volume)/time)

Generation $= 0$ (tracer does not decompose nor form)

So that, in the mass balance:

Input $+$ Generation $=$ Output $+$ Accumulation

$$0 + 0 = n_A + \frac{dC_A}{dt} \cdot V$$

$$Q_0 C_A + \frac{dC_A}{dt} \cdot V = 0$$

$$\frac{dC_A}{C_A} = \frac{-Q_0}{V} dt = -\frac{dt}{\bar{t}}$$

$$\int_{C_{A0}}^{C_A} \frac{dC_A}{C_A} = -\frac{t}{\bar{t}}$$

$$\ln \frac{C_A}{C_{A0}} = -\frac{t}{\bar{t}}$$

$$C_A = C_{A0} \exp\left(-\frac{t}{\bar{t}}\right)$$

(b) CSTR, to which a reagent that decomposes with first-order kinetics (non-steady regime) begins to be fed.

During the unsteady state, we will have:

$$\text{Input} = n_{A0} \ (\text{moles/time})$$

$$\text{Output} = n_A \ (\text{moles/time})$$

$$\text{Accumulation} = V \cdot dC_A/dt \ (\text{volume} \cdot (\text{moles/volume})/\text{time})$$

$$\text{Generation} = r_A \cdot V \ ((\text{moles/volume} \cdot \text{time}) \cdot \text{volume})$$

So that, in the mass balance:

$$\text{Input} + \text{Generation} = \text{Output} + \text{Accumulation}$$

$$n_{A0} + r_A \cdot V = n_A + \frac{dC_A}{dt} \cdot V$$

If we assume a first-order reaction:

$$Q \cdot C_{A0} - k \cdot C_A \cdot V = Q \cdot C_A + \frac{dC_A}{dt} \cdot V$$

$$\frac{dC_A}{dt} = \frac{C_{A0} - C_A}{V/Q} - k \cdot C_A = \frac{C_{A0} - C_A}{\bar{t}} - k \cdot C_A$$

This equation is an ordinary differential equation (ODE), not in partial derivatives, and can be solved numerically with no problem using the FDM:

$$\frac{C_A^{t+1} - C_A^t}{\Delta t} = \frac{C_{A0} - C_A^t}{\bar{t}} - k \cdot C_A^t$$

And solving for the concentration:

$$C_A^{t+1} = \Delta t \left[ \frac{C_{A0} - C_A^t}{\bar{t}} - k \cdot C_A^t \right] + C_A^t$$

For simulating this behavior, a simple Matlab® program can be used. For example, in the case we want to simulate a CSTR reactor with $\bar{t} = 10\,\text{s}$ and $C_{A0} = 1\,\text{mol/m}^3$, where $k = 1\,\text{s}^{-1}$, the corresponding Matlab program can be:

```
clear all
tresid=10; % sec, average residence time
k=1; % sec^-1, kinetic first-order constant
CA0=1; % Input concentration, mol/m^3
t=0;
i=1;
CA(1)=CA0; % Boundary condition at the input
tfinal=8; % sec, final time
Nt=100; % Number of intervals of time
inct=tfinal/Nt; % Sec, increment of time
tv=0:inct:tfinal; % Values of time to be used in the calculation
%CSTR concentrations are the same in all points of the reactor
while t<tfinal
   i=i+1
   t=t+inct;
   CAn(i)=inc?((CA0-CA(i-1))/tresid-k*CA(i-1))+CA(i-1);
End
CA=CAn;
plot(tv,CA(1:Nt+1))
xlabel('Time (sec)')
ylabel('C_A (mol/m^3)')
```

(c) PFR without dispersion, to which tracer is injected in pulse.

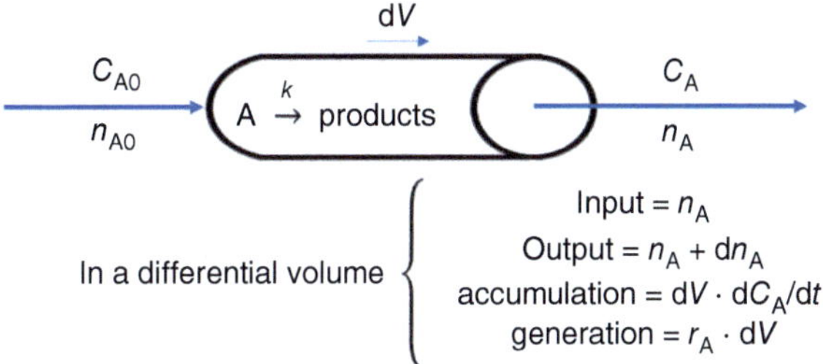


The pulse of tracer is injected at the reactor input, and it moves along the reactor as time passes. We can do a mass balance in a volume differential:

$$\text{Input} + \text{Generation} = \text{Output} + \text{Accumulation}$$

$$n_A + r_A \cdot \partial V = (n_A + \partial n_A) + \frac{\partial C_A}{\partial t} \cdot \partial V$$

Bearing in mind, for a constant-density system:

$$Q \cdot \partial C_A = \partial n_A$$

What means that:

$$\frac{\partial C_A}{\partial t} = r_A - Q\frac{\partial C_A}{\partial V}$$

If we assume first-order kinetics:

$$\frac{\partial C_A}{\partial t} = -kC_A - Q\frac{\partial C_A}{\partial V}$$

Using the finite differences approximation, first-order:

$$\frac{C_{Ai}^{t+1} - C_{Ai}^{t}}{\Delta t} = -kC_{Ai}^{t} - Q\frac{C_{Ai+1}^{t} - C_{Ai}^{t}}{\Delta V}$$

Rearranging:

$$C_{Ai}^{t+1} = \left(1 - k \cdot \Delta t + Q\frac{\Delta t}{\Delta V}\right) C_{Ai}^{t} - Q\frac{\Delta t}{\Delta V} C_{Ai+1}^{t}$$

Applying the stability criteria, the system is stable (errors do not increase), if:

$$Q\frac{\Delta t}{\Delta V} > 0$$

$$\left(1 - k \cdot \Delta t - Q\frac{\Delta t}{\Delta V}\right) > 0$$

$$\left(1 - k \cdot \Delta t - Q\frac{\Delta t}{\Delta V}\right) + Q\frac{\Delta t}{\Delta V} \leq 1$$

The first and third conditions are always fulfilled, and from the second one:

$$\Delta t < \frac{1}{\left(k + \frac{Q}{\Delta V}\right)}$$

In the initial state:

$$\forall V \text{ if } t = 0 \rightarrow C_A = 0$$

The boundary conditions of this system are:

- At the input ($V = 0$): $C_A = C_{A0}$ (known concentration)
- At the exit ($V = V_{\text{total}}$) $\rightarrow C_A(V^-) = C_A(V^+)$ (closed vessel), so that at this section:

$$\left[\frac{\partial C}{\partial V}\right]_{V_{\text{total}}} = 0$$

(d) PFR with dispersion to which tracer is injected in impulse, assuming closed–closed vessel.

The mass balance in the PFR system in the dynamic regime can be expressed as:

$$n_A + r_A \cdot \partial V = (n_A + \partial n_A) + \frac{\partial C_A}{\partial t} \cdot \partial V$$

But in this case:

$$Q \cdot \partial C_A \neq \partial n_A$$

As:

$$n_A = Q \cdot C_A - D_e \cdot S \cdot \frac{\partial C_A}{\partial z}$$

Deriving and doing $Q = (u \cdot S)$, then:

$$\frac{\partial n_A}{\partial z} = u \cdot S \cdot \frac{\partial C_A}{\partial z} - D_e \cdot S \cdot \frac{\partial^2 C_A}{\partial z^2}$$

And introducing the mole balance:

$$D_e \cdot \frac{\partial^2 C_A}{\partial z^2} - u \cdot \frac{\partial C_A}{\partial z} + r_A = \frac{\partial C_A}{\partial t}$$

Using the finite differences approximation, first-order, and assuming a first-order reaction:

$$D_e \cdot \frac{C_{Ai+1}^t - 2C_{Ai}^t + C_{Ai-1}^t}{\Delta z^2} - u \cdot \frac{C_{Ai+1}^t - C_{Ai}^t}{\Delta z} - k \cdot C_{Ai}^t = \frac{C_{Ai}^{t+1} - C_{Ai}^t}{\Delta t}$$

And rearranging:

$$C_{Ai}^{t+1} = D_e \frac{\Delta t}{\Delta z^2} \cdot C_{Ai-1}^t + \left(1 - 2\frac{\Delta t}{\Delta z^2}D_e + u\frac{\Delta t}{\Delta z} - k\Delta t\right)C_{Ai}^t$$
$$+ \left(\frac{\Delta t}{\Delta z^2}D_e - u\frac{\Delta t}{\Delta z}\right)C_{Ai+1}^t$$

The system is stable (errors do not increase) if:

$$D_e\frac{\Delta t}{\Delta z^2} > 0$$

$$\left(1 - 2\frac{\Delta t}{\Delta z^2}D_e + u\frac{\Delta t}{\Delta z} - k\Delta t\right) > 0$$

$$\left(\frac{\Delta t}{\Delta z^2}D_e - u\frac{\Delta t}{\Delta z}\right) > 0$$

$$\left[D_e\frac{\Delta t}{\Delta z^2} + \left(1 - 2\frac{\Delta t}{\Delta z^2}D_e + u\frac{\Delta t}{\Delta z} - k\Delta t\right) + \left(\frac{\Delta t}{\Delta z^2}D_e - u\frac{\Delta t}{\Delta z}\right)\right] \leq 1$$

As in other equations previously discussed, the first and last conditions are always fulfilled, but for the third one:

$$\Delta z < \frac{D_e}{u}$$

And from the second one:

$$\Delta t < \frac{1}{\left(\frac{2}{\Delta z^2}D_e - \frac{u}{\Delta z} + k\right)}$$

Obviously, all these two conditions must be fulfilled.

(e) CSTR–PFR combination working in steady state, first-order kinetics.

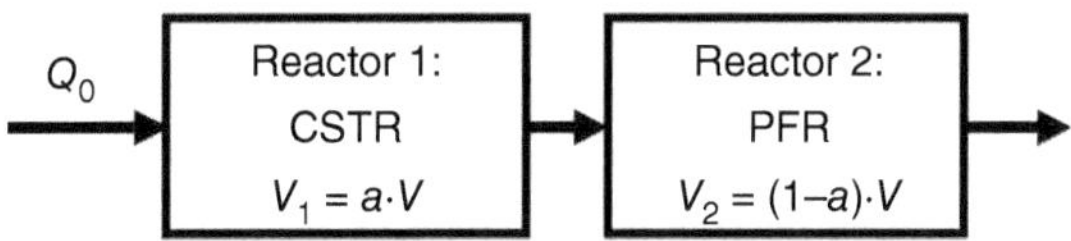


In the Reactor 1, a mass balance is:

$$Q_0 C_{A0} + r_A \cdot V_1 = Q_0 C_{A1}$$

$$Q_0 C_{A0} - k \cdot C_{A1} \cdot V_1 = Q_0 C_{A1}$$

$$C_{A1} = \frac{Q_0 C_{A0}}{Q_0 + kV_1} = \frac{1}{1 + k\frac{V_1}{Q_0}}C_{A0}$$

In Reactor 2:

$$C_{A2} = C_{A1} \cdot \exp\left(-\frac{kV_2}{Q_0}\right)$$

And then:

$$C_{A2} = \frac{C_{A0}}{1 + k\frac{V_1}{Q_0}} \cdot \exp\left(-\frac{kV_2}{Q_0}\right)$$

**Problem 5.7**  A tubular reactor is provided with rectangular metal fins in order to facilitate the removal of the heat produced by the reaction (see figure).

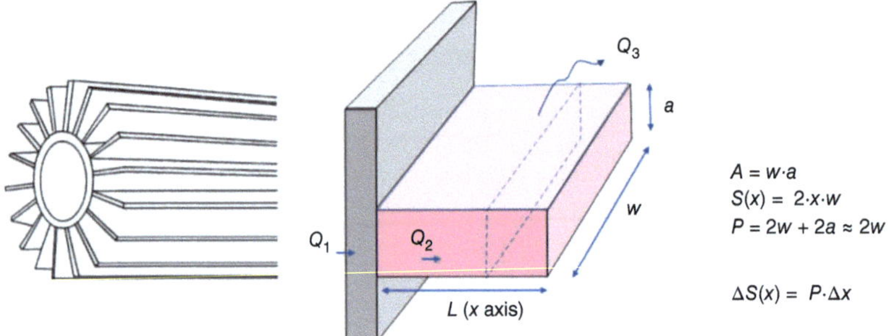


S = Surface area of the fin (heat loss by convection)
A = cross-sectional area of the fin (heat transmision by conduction)

In the steady state, the heat produced in a certain section ($Q_1$) will be given by the sum of the heat that is transmitted by conduction ($Q_2$) and convection to the medium ($Q_3$). This balance is given by:

$$Q_2 - Q_1 + Q_3 = 0$$

$$-k \cdot A \cdot \frac{dT}{dx}\bigg]_{x+\Delta x} - k \cdot A \cdot \frac{dT}{dx}\bigg]_{x} + h \cdot S \cdot (T - T_{amb}) = 0$$

where $T_{amb}$ is the temperature of the fluid surrounding the reactor, "$A$" is the cross section, and "$S$" is the surface area of the paddle. If we express the area $S(x)$ as a function of the width $x$ multiplied by the perimeter $P(x)$, we will have, taking a limit when $\Delta x$ tends to zero:

$$\frac{d}{dx}\left(k \cdot A \cdot \frac{dT}{dx}\right) - h \cdot P \cdot (T - T_{amb}) = 0$$

For a paddle of uniform cross section, it follows that $A(x) = A$ is constant and $P(x) = P$ is also constant. The boundary conditions for this problem generally assume that the heat reaching the tip of the paddle (of length "$L$") is lost by convection, so:

$$\begin{cases} T = T_0 & \text{for } x = 0 \\ -k \cdot A \cdot \dfrac{dT}{dx} = h \cdot P \cdot (T - T_{amb}) & \text{for } x = L \end{cases}$$

Applying the finite difference method, indicate the equations that result in the calculation of the temperature along the fin. Indicate the stability conditions and how to consider the boundary conditions of this problem.

**Solution to Problem 5.7**

For details refer the Wiley website at http://www.wiley-vch.de/ISBN9783527354115

The equation governing the heat transfer:

$$\frac{d}{dx}\left(kA\frac{dT}{dx}\right) - hP(T - T_{amb}) = 0$$

is an ODE, but it is of second-order. We can solve the system by changing a variable:

$$T' = \frac{dT}{dx}$$

And in this way, the second-order equation is transformed into a system of two first-order equations:

$$\begin{cases} T' = \dfrac{dT}{dx} \\ kA\dfrac{dT'}{dx} = hP(T - T_{amb}) \end{cases}$$

The boundary conditions are:

$$\begin{cases} \text{in } x = 0 \rightarrow T_i = T_0 \\ \text{in } x = L \rightarrow T'_N = \dfrac{hP}{kA}(T_{amb} - T_N) \end{cases}$$

with $N$ being the last point of the mesh. There are no stability conditions.

The issue at hand stems from having known conditions for the equations at different points (0 and $L$), which makes numerical integration unfeasible without additional steps. To resolve this, an optimization process is required to determine the value of the derivative $T'$ at $x = 0$ in a way that ensures the derivative at $x = L$ matches the desired value.

We can also try to apply the FDM as usual, beginning with the equation governing the heat transfer:

$$\frac{d}{dx}\left(kA\frac{dT}{dx}\right) - hP(T - T_{amb}) = 0$$

$$kA\left(\frac{d^2T}{dx^2}\right) - hP(T - T_{amb}) = 0$$

is an ODE, but it is of second-order. We can try to solve the problem by applying FDM. Let us use rear differences in this case:

$$kA\left(\frac{T_i - 2T_{i-1} + T_{i-2}}{\Delta x^2}\right) - hP(T_i - T_{amb}) = 0$$

Solving for $T_i$:

$$T_i = \frac{\Delta x^2 hP(T_i - T_{amb})}{kA} + 2T_{i-1} - T_{i-2} = -\frac{\Delta x^2 hP}{kA}T_{amb} - \frac{\Delta x^2 hP}{kA}T_i + 2T_{i-1} - T_{i-2}$$

$$T_i = \frac{-\frac{\Delta x^2 hP}{kA}T_{amb} + 2T_{i-1} - T_{i-2}}{1 + \frac{\Delta x^2 hP}{kA}}$$

This equation relates the temperature at one point to the temperatures at the two points preceding it. We must consider that the initial condition only gives us a temperature (at $x = 0$), so the problem cannot be solved without assuming a derivative value in the first increment.

In the spreadsheet corresponding to this problem, it has been applied to a particular case, where we assume the following values:

| | |
|---|---|
| $T_{amb}$ (K) | 298 |
| $T_0$ (K) | 373 |
| $k$ (W/m·K) | 100 |
| $h$ (W/m²·K) | 10 |
| $A$ (m²) | 0.0001 |
| $P$ (m) | 0.01 |
| $L$ (m) | 0.6 |

First, temperature in $x = 0$ is fixed at $T = T_0$. Then, a value of $T'$ at the first interval is assumed, calculating the corresponding temperature in the second interval:

$$T_{i=2} = T_{i=1} + T'_{i=1} \cdot \Delta x$$

The remaining values of $T$ are calculated using the equation relating $T_i$ to $T_{i-1}$ and $T_{i-2}$. The derivative at the point $x = L$ ($i = N$) is calculated and compared to the expected value. An optimization of the initial value of the derivative is needed. In this sense, a value of the derivative at $x = 0$ is chosen, and the corresponding value at $x = L$ is calculated ($T'_{x=L}$). An objective function is defined based on the expected value of derivative at this point:

$$\text{O.F.} = \left( T'_{x=L} - \frac{hP}{kA}(T_{amb} - T_N) \right)^2$$

The value of the $T'_{x=0}$ chosen should minimize the O.F. mentioned. In another way:

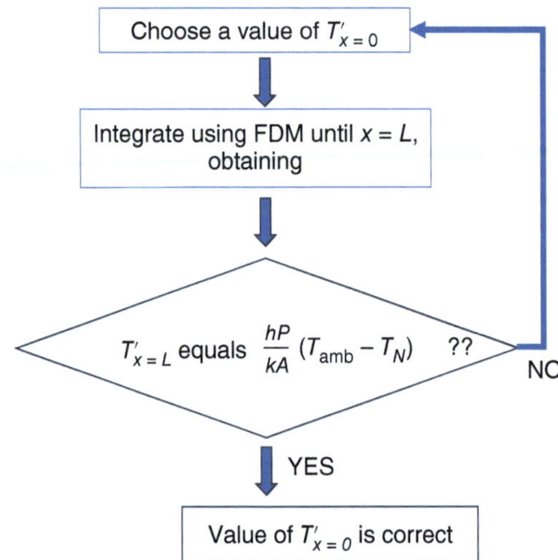

Finally, the behavior of this system is given by the graph:

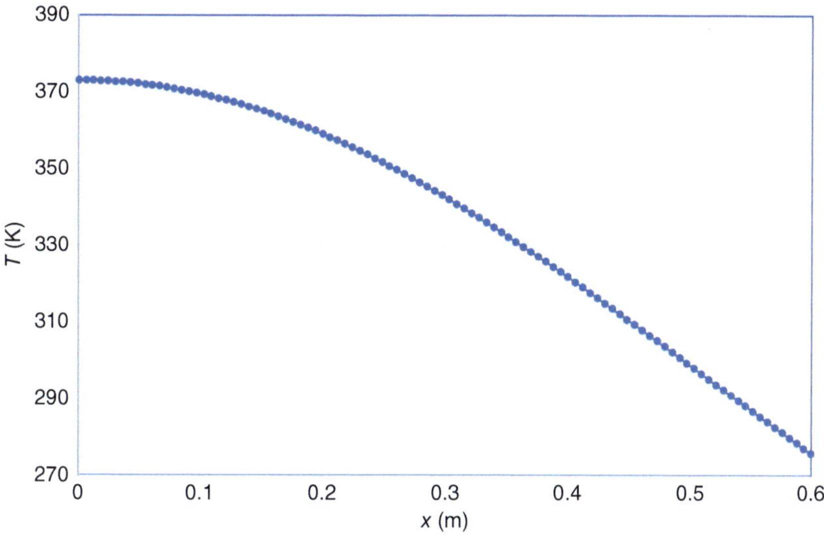


**Problem 5.8**   In a one-meter-long reactor, a mixture of two substances, A and B, is introduced through one of its ends. The concentration at the inlet end for each of the substances is given by the following expression:

$$(C_i) = \frac{1}{\bar{t}_i} \cdot \exp\left(-\frac{(t-1.5)^2}{0.2}\right)$$

with $\bar{t}_A = 0.2$ and with $\bar{t}_B = 0.4$ and where $t =$ time (min); $C_i =$ concentration (mol/m³) of species A or B.

In the reactor, dispersion of each of the two substances occurs, as well as a reaction of order 1 with respect to each of the reactants (global order 2).

Determine the differential equations that describe the concentration of each species with distance if the reactor operates in a non-steady state with the feed. Specify the boundary conditions. Apply the method of finite differences to solve the system and study its stability.

**Solution to Problem 5.8**

The mass balance in a differential volume is:

$$n_A + r_A \cdot \partial V = (n_A + \partial n_A) + \frac{\partial C_A}{\partial t} \cdot \partial V$$

$$S \cdot r_A \cdot \partial z = (\partial n_A) + \frac{\partial C_A}{\partial t} \cdot S \partial z$$

$$\left( \frac{\partial n_A}{\partial z} \right) = S \cdot r_A - S \frac{\partial C_A}{\partial t}$$

In the dispersion model:

$$n_A = Q \cdot C_A - D_e \cdot S \cdot \frac{\partial C_A}{\partial z}$$

So:

$$\frac{\partial n_A}{\partial z} = u \cdot S \cdot \frac{\partial C_A}{\partial z} - D_e \cdot S \cdot \frac{\partial^2 C_A}{\partial z^2}$$

And, equating:

$$u \frac{\partial C_A}{\partial z} - D_e \frac{\partial^2 C_A}{\partial z^2} = r_A - \frac{\partial C_A}{\partial t}$$

In the same way:

$$u \cdot \frac{\partial C_B}{\partial z} - D_e \frac{\partial^2 C_B}{\partial z^2} = r_B - \frac{\partial C_B}{\partial t}$$

Applying FDM:

$$u \cdot \frac{C_{Ai+1}^t - C_{Ai}^t}{\Delta z} - D_e \frac{C_{Ai+1}^t - 2C_{Ai}^t + C_{Ai-1}^t}{\Delta z^2} = k C_{Ai}^t C_{Bi}^t - \frac{C_{Ai}^{t+1} - C_{Ai}^t}{\Delta t}$$

$$C_{Ai}^{t+1} = \frac{D_e \Delta t}{\Delta z^2} C_{Ai-1}^t + \left( 1 - 2\frac{D_e \Delta t}{\Delta z^2} + u\frac{\Delta t}{\Delta z} - k C_{Bi}^t \Delta t \right) C_{Ai}^t + \left( \frac{D_e \Delta t}{\Delta z^2} - u\frac{\Delta t}{\Delta z} \right) C_{Ai-1}^t$$

And, in the same way for $C_B$, the boundary conditions are:

$$\begin{cases} \text{in } z = 0 - - \rightarrow \ C_A]_{z=0} = \frac{1}{0.2} \exp\left( -\frac{(t-1.5)^2}{0.2} \right) \\[2ex] \text{in } z = L - - \rightarrow \ .C_A(L^+) = C_A(L^-) - \rightarrow \ \left. \frac{dC_A}{dz} \right]_{z=L} = 0 \end{cases}$$

Stability:

$$\frac{D_e \Delta t}{\Delta z^2} > 0 \text{ always}$$

$$\left( \frac{D_e \Delta t}{\Delta z^2} - u\frac{\Delta t}{\Delta z} \right) > 0 \rightarrow \Delta z < \frac{D_e}{u}$$

$$\left( 1 - 2\frac{D_e \Delta t}{\Delta z^2} + u\frac{\Delta t}{\Delta z} - k C_{Bi}^t \Delta t \right) > 0 \rightarrow \Delta t < \left( \frac{1}{\frac{2D_e}{\Delta z^2} - \frac{u}{\Delta z} + k C_{Bi}^t} \right)$$

And the reasoning is completely analogous for species "B".

**Problem 5.9** Explain the difference between closed–closed and open–open boundary conditions, indicating the mathematical treatment of such situations.

**Solution to Problem 5.9**

The molar flow of a component in a differential reactor is:

$$n_j = u \cdot S \cdot C_j - D_e \cdot S \cdot \frac{\partial C_j}{\partial z}$$

where the first term is due to the convection, and the second to the dispersion. If the vessel is closed–closed, there is neither dispersion nor radial variation in the concentration both upstream and downstream of the reaction zone. In open–open boundary conditions, this is not true.

For closed–closed system:

$$n_j(0^-, t) = n_j(0^+, t)$$

$$uSC_j(0^-, t) = -SD_e \left. \frac{\partial C_j}{\partial z} \right|_{z=0^+} + uSC_j(0^+, t)$$

If a known concentration of $C_{j0}$ is fed:

$$C_{j0} = -\frac{D_e}{u} \left. \frac{\partial C_j}{\partial z} \right|_{z=0^+} + C_j(0^+, t)$$

And, at the exit:

$$C_j(0^-, t) = C_j(0^+, t)$$

So:

$$\left. \frac{\partial C_j}{\partial z} \right|_{z=L} = \frac{C_j(0^+, t) - C_j(0^-, t)}{\Delta z} = 0$$

If the vessel were open–open:

$$n_j(0^-, t) = n_j(0^+, t)$$

$$uSC_j(0^-, t) - SD_e \left. \frac{\partial C_j}{\partial z} \right|_{z=0^-} = -SD_e \left. \frac{\partial C_j}{\partial z} \right|_{z=0^+} + uSC_j(0^+, t)$$

that, if $D_e$ does not vary, gives:

$$C_j(0^-, t) = C_j(0^+, t)$$

At the exit, with the same assumption:

$$C_j(L^-, t) = C_j(L^+, t)$$

**Problem 5.10** On a liquid film, a gas (A) is absorbed that reacts to give a compound B. This compound, in turn, reacts to give another C:

$$A + L \rightarrow B \qquad (-r_A) = k_1 C_A$$
$$B \rightarrow C \qquad (-r_B) = -k_1 C_A + k_2 C_B$$

An unsteady state of mass balance is given by:

$$\frac{\partial C_A}{\partial t} = D_A \frac{\partial^2 C_A}{\partial x^2} + r_A$$

$$\frac{\partial C_B}{\partial t} = D_B \frac{\partial^2 C_B}{\partial x^2} + r_B$$

With initial and boundary conditions:

$$t = 0 \rightarrow (C_A = 0 \quad \forall x \neq 0; \quad C_B = 0)$$

$$x = 0 \rightarrow (C_A = C_{A0} \quad C_B = C_{B0} \quad \forall t)$$

$$x = L \rightarrow \left( \frac{dC_A}{dx} = 0; \frac{dC_B}{dx} = 0 \right)$$

Using the finite difference method, indicate the resulting equations and how you would proceed to calculate the concentration profiles at different times. What happens to the points located at the extremes?

**Solution to Problem 5.10**

From the balances and the rate laws:

$$\frac{\partial C_A}{\partial t} = D_A \frac{\partial^2 C_A}{\partial x^2} - k_1 C_A$$

$$\frac{\partial C_B}{\partial t} = D_B \frac{\partial^2 C_B}{\partial x^2} + k_1 C_A - k_2 C_B$$

Applying FDM:

$$\frac{C_{A_i}^{t+1} - C_{A_i}^{t}}{\Delta t} = D_A \frac{C_{A_i+1}^{t} - 2C_{A_i}^{t} + C_{A_{i-1}}^{t}}{\Delta x^2} - k_1 C_{A_i}^{t}$$

$$\frac{C_{B_i}^{t+1} - C_{B_i}^{t}}{\Delta t} = D_B \frac{C_{B_i+1}^{t} - 2C_{B_i}^{t} + C_{B_{i-1}}^{t}}{\Delta x^2} - k_1 C_{A_i}^{t} - k_2 C_{B_i}^{t}$$

Rearranging:

$$C_{A_i}^{t+1} = D_A \Delta t \frac{C_{A_i+1}^{t} - 2C_{A_i}^{t} + C_{A_{i-1}}^{t}}{\Delta x^2} - k_1 \Delta t C_{A_i}^{t} + C_{A_i}^{t} =$$

$$= \frac{D_A \Delta t}{\Delta x^2} C_{A_{i-1}}^{t} + \left( 1 - 2\frac{D_A \Delta t}{\Delta x^2} - k_1 \Delta t \right) C_{A_i}^{t} + \frac{D_A \Delta t}{\Delta x^2} C_{A_{i+1}}^{t}$$

$$C_{B_i}^{t+1} = \frac{D_B \Delta t}{\Delta x^2} C_{B_{i-1}}^{t} + \left( 1 - 2\frac{D_B \Delta t}{\Delta x^2} - k_2 \Delta t \right) C_{B_i}^{t} + \frac{D_B \Delta t}{\Delta x^2} C_{B_{i+1}}^{t} - k_1 \Delta t C_{A_i}^{t}$$

The stability conditions are:

$$\frac{D_A \Delta t}{\Delta x^2} > 0 \text{ (always)}$$

$$\left( 1 - 2\frac{D_A \Delta t}{\Delta x^2} - k_1 \Delta t \right) > 0 \rightarrow \Delta t < \frac{1}{k + \frac{2D_A}{\Delta x^2}}$$

$$\left[ \frac{D_A \Delta t}{\Delta x^2} + \left( 1 - 2\frac{D_A \Delta t}{\Delta x^2} - k_1 \Delta t \right) + \frac{D_A \Delta t}{\Delta x^2} \right] \leq 1 \text{ (always)}$$

And for "B":

$$\left(1 - 2\frac{D_{\mathrm{B}}\Delta t}{\Delta x^2} - k_2\Delta t\right) > 0 \rightarrow \Delta t < \frac{1}{k + \frac{2D_{\mathrm{B}}}{\Delta x^2}}$$

Respect to the boundary conditions:

- In $x = 0$:

$$C_{\mathrm{A}} = C_{\mathrm{A0}}, C_{\mathrm{B}} = C_{\mathrm{B0}}$$

- In $x = L$:

$$\left.\frac{C^t_{\mathrm{A}i} - C^t_{\mathrm{A}(i-1)}}{\Delta x}\right]_{x=L} = 0$$

$$\left.\frac{C^t_{\mathrm{B}i} - C^t_{\mathrm{B}(i-1)}}{\Delta x}\right]_{x=L} = 0$$

this is equivalent to:

$$C^t_{\mathrm{A}_N} = C^t_{\mathrm{A}_{N-1}}$$

$$C^t_{\mathrm{B}_N} = C^t_{\mathrm{B}_{N-1}}$$

being $N$, the last increment of length (corresponding to $x = L$).

**Problem 5.11**   Consider a first-order reaction occurring in a CSTR where the inlet concentration of reactant has been held constant at $C_{\mathrm{A0}}$ for $t < 0$. At time $t = 0$, the inlet concentration is changed to $C_{\mathrm{A1}}$. Find the outlet response for $t > 0$ assuming isothermal, constant-volume, and constant-density operations.

**Solution to Problem 5.11**

For details refer the Wiley website at http://www.wiley-vch.de/ISBN9783527354115

In the CSTR, a mass balance in a non-steady state is:

$$n_{\mathrm{Ain}} + r_{\mathrm{A}}V = n_{\mathrm{Aout}} + V\frac{dC_{\mathrm{Aout}}}{dt}$$

$$C_{\mathrm{Ain}} + (-kC_{\mathrm{Aout}})\bar{t} = C_{\mathrm{Aout}} + \bar{t}\frac{dC_{\mathrm{Aout}}}{dt}$$

$$C_{\mathrm{A1}} + (-kC_{\mathrm{Aout}})\bar{t} = C_{\mathrm{Aout}} + \bar{t}\frac{dC_{\mathrm{Aout}}}{dt}$$

$$C_{\mathrm{A1}} = C_{\mathrm{Aout}}(1 + k\bar{t}) + \bar{t}\frac{dC_{\mathrm{Aout}}}{dt}$$

A differential equation of the form:

$$y'(t) = \frac{a - y(t)(1 + k)}{b}$$

has the solution:

$$y(t) = \frac{a}{k + 1} + \chi\exp\left(-\frac{kt}{b} - \frac{t}{b}\right)$$

Solving for the concentration at the output:

$$C_{\text{Aout}} = \frac{C_{A1}}{1 + k\bar{t}} + \chi \exp\left(-\frac{(1 + k\bar{t})t}{\bar{t}}\right) \quad t \geq 0$$

where $\chi$ is an integration constant. At $t < 0$, a steady-state balance gives:

$$C_{\text{Ain}} + r_A\bar{t} = C_{\text{Aout}}$$
$$C_{A0} + r_A\bar{t} = C_{\text{Aout}}$$

$$C_{\text{Aout}} = \frac{C_{A0}}{1 + k \cdot \bar{t}} \quad t < 0$$

This will be the initial condition for the unsteady balance. Applying this initial condition, we obtain:

$$C_{\text{Aout}} = \frac{C_{A1} + (C_{A0} - C_{A1})\exp\left(-\frac{(1+k\bar{t})t}{\bar{t}}\right)}{1 + k\bar{t}}$$

We can simulate a curve just to see the result. For example, with the following parameters, we get the following plot:

| | |
|---|---|
| $t_m$ (min) | 1 |
| $C_{A0}$ (mol/l) | 1 |
| $C_{A1}$ (mol/l) | 2 |
| $k$ (min$^{-1}$) | 1 |

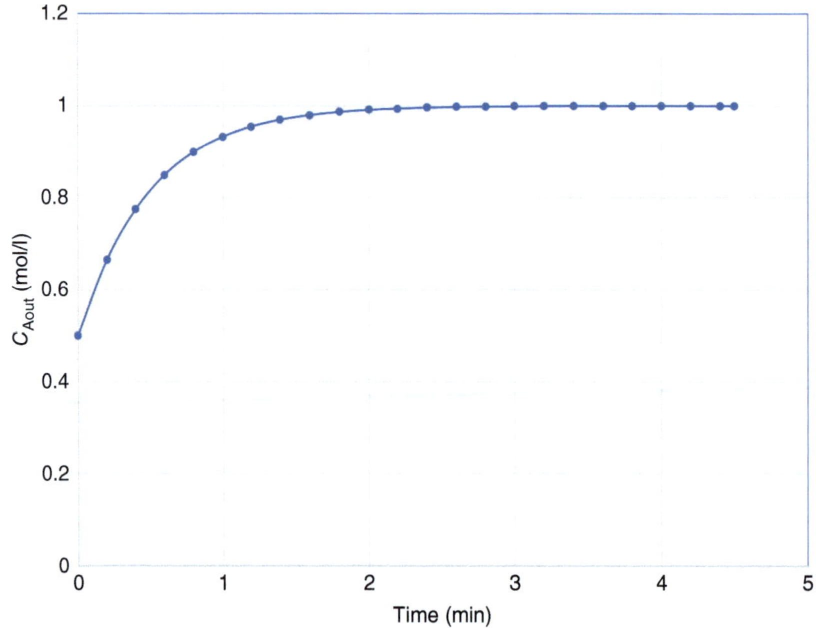


If $C_{A1}$ is smaller than $C_{A0}$:

| | |
|---|---|
| $t_m$ (min) | 1 |
| $C_{A0}$ (mol/l) | 1 |
| $C_{A1}$ (mol/l) | 0.1 |
| $k$ (min$^{-1}$) | 1 |

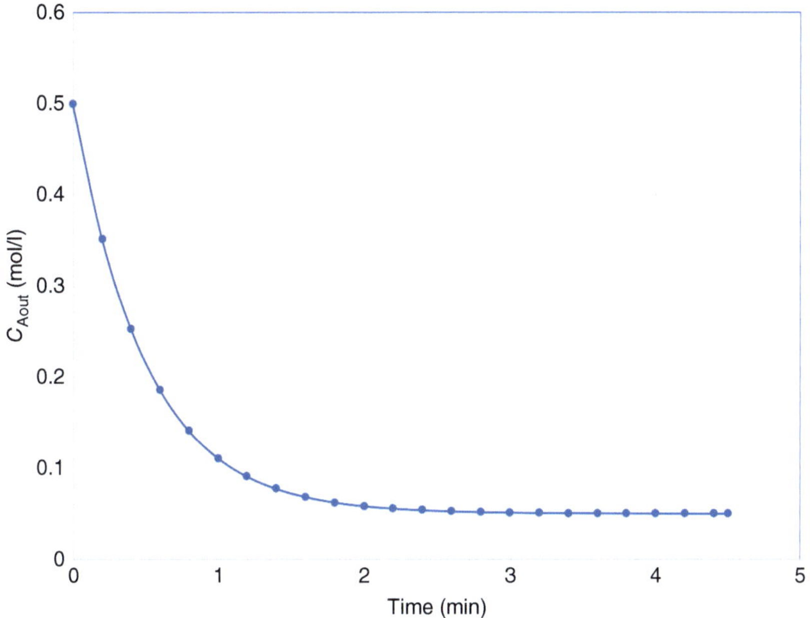



**Problem 5.12** The initial portion of a reactor startup is usually fed-batch. Determine the fed-batch startup transient for an isothermal, constant-density stirred tank reactor. Suppose the tank is initially empty and is filled at a constant rate $Q_0$ with fluid having concentration $C_{A0}$. A first-order reaction begins immediately. Find the concentration within the tank, $C_A$, as a function of time $t$.

**Solution to Problem 5.12**

For details refer the Wiley website at http://www.wiley-vch.de/ISBN9783527354115

A balance in the reactor gives:

$$\text{Accumulation} = \text{Input} + \text{Generation}$$

$$\frac{d(C_A V)}{dt} = C_A \frac{dV}{dt} + V \frac{dC_A}{dt} = Q_0 C_{A0} + V \cdot r_A$$

$$C_A Q_0 + V \frac{dC_A}{dt} = Q_0 C_{A0} + V \cdot (-kC_A)$$

In a fed-batch, during the filling period: $V = Q_0 t \rightarrow t = V/Q_0$

$$C_A + t \frac{dC_A}{dt} = C_{A0} + (-ktC_A)$$

$$\frac{dC_A}{dt} = \frac{C_{A0} - C_A(1 + kt)}{t}$$

A differential equation of the form:

$$y'(t) = \frac{a - y(t)(1 + kt)}{t}$$

has the solution:

$$y(t) = \frac{a}{kt} + \frac{\chi \exp(-kt)}{t}$$

Solving for the concentration at the reactor:

$$C_A = \frac{C_{A0}}{kt} + \frac{\chi \exp(-kt)}{t} \qquad t \geq 0$$

that can also be written as:

$$tC_A = \frac{C_{A0}}{k} + \chi \exp(-kt)$$

The initial condition is $C_A = C_{A0}$ at $t = 0$, so:

$$0C_{A0} = \frac{C_{A0}}{k} + \chi \exp(-k0)$$

Hence:

$$\chi = -C_{A0}$$

and:

$$C_A = \frac{C_{A0}}{kt} + \frac{-C_{A0}\exp(-kt)}{kt} = \frac{C_{A0}(1 - \exp(-kt))}{kt}$$

We can simulate a curve just to see the result. For example, with the following parameters, we get the following plot:

| | |
|---|---|
| $Q_0$ (l/min) | 1 |
| $C_{A0}$ (mol/l) | 1 |
| $V_0$ (l) | 0 |
| $k$ (min$^{-1}$) | 1 |

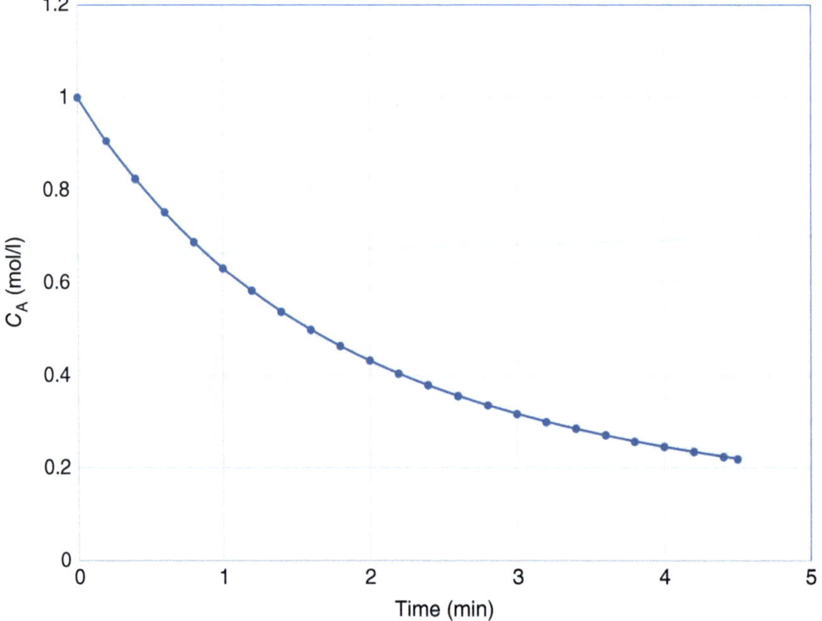

**Problem 5.13**  A plastics company has two products in a CSTR. Product P, a homopolymer, is made by the reaction:

$$A \rightarrow P \quad r = k_1 C_A$$

Product Q is a modified version of the homopolymer made by a second reaction:

$$P + D \rightarrow Q \quad r = k_2 C_P C_D$$

The reactor operates at constant volume, constant density, constant flow rate, and isothermally. The only difference between the two products is the addition of component D to the feed when product Q is made.

Explore methods for making a running transition from product P to product Q. There is no P or Q in the reactor feed.

**Solution to Problem 5.13**

For details refer the Wiley website at http://www.wiley-vch.de/ISBN9783527354115

We can do a mass balance for each reactant:

$$\bar{t}\frac{dC_{Aout}}{dt} = C_{Ain} - C_{Aout} - \bar{t}k_1 C_{Aout}$$

$$\bar{t}\frac{dC_{Pout}}{dt} = -C_{Pout} - \bar{t}(k_2 C_{Pout} C_{Dout} - k_1 C_{Aout})$$

$$\bar{t}\frac{dC_{Qout}}{dt} = -C_{Qout} + \bar{t}(k_2 C_{Pout} C_{Dout})$$

$$\bar{t}\frac{dC_{Dout}}{dt} = C_{Din} - C_{Dout} - \bar{t}(k_2 C_{Pout} C_{Dout})$$

The analysis, once we have the differential equations, could be numerical. Suppose the following values:

$$C_{Ain} = 20 \, \text{mol/m}^3 \qquad\qquad C_{Din} = 10 \, \text{mol/m}^3$$
$$\bar{t} = 1 \, \text{h} \qquad\qquad k_1 = 4 \, \text{h}^{-1} \qquad k_2 = 1 \, \text{m}^3/\text{mol·h}$$

This kinetic system allows only one steady state. It is stable and can be found by solving the ODEs starting from any initial condition. For example, starting from: $C_{Aout} = 1 \, \text{mol/m}^3$, $C_{Pout} = 1 \, \text{mol/m}^3$, $C_{Dout} = C_{Qout} = 0 \, \text{mol/m}^3$, we find:

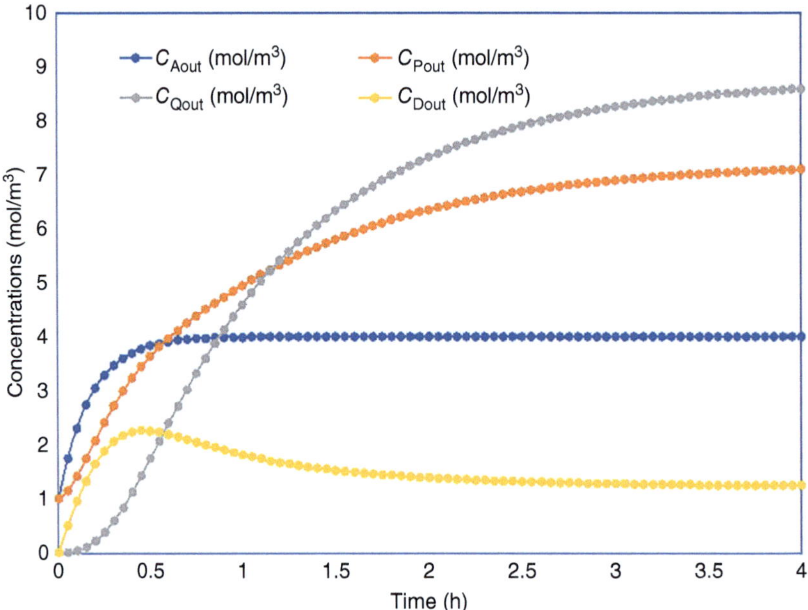

If we start from: $C_{\text{Aout}} = 10 \, \text{mol/m}^3$, $C_{\text{Pout}} = 10 \, \text{mol/m}^3$, $C_{\text{Dout}} = C_{\text{Qout}} = 1 \, \text{mol/m}^3$, we find:

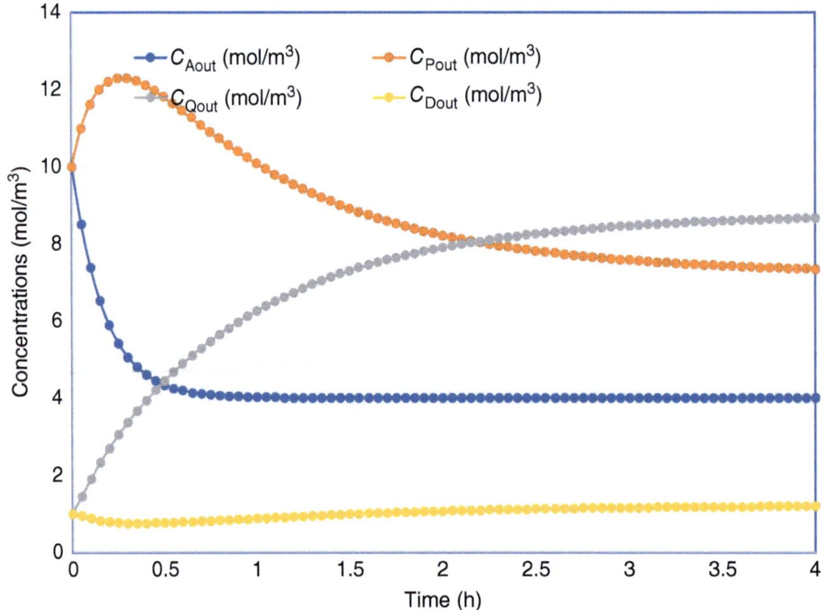

in such a way that the steady state response when making product P is always $C_{\text{Aout}} = 4 \, \text{mol/m}^3$ and $C_{\text{Pout}} = 7.23 \, \text{mol/m}^3$ if reactant D is present.

For considering the transition from product P to Q, the simplest way is to add component D to the feed at the required steady concentration. The governing equations are solved subject to the initial condition that the reactor initially contains the steady-state composition corresponding to product P. With $C_{\text{Din}} = 0$, we find that:

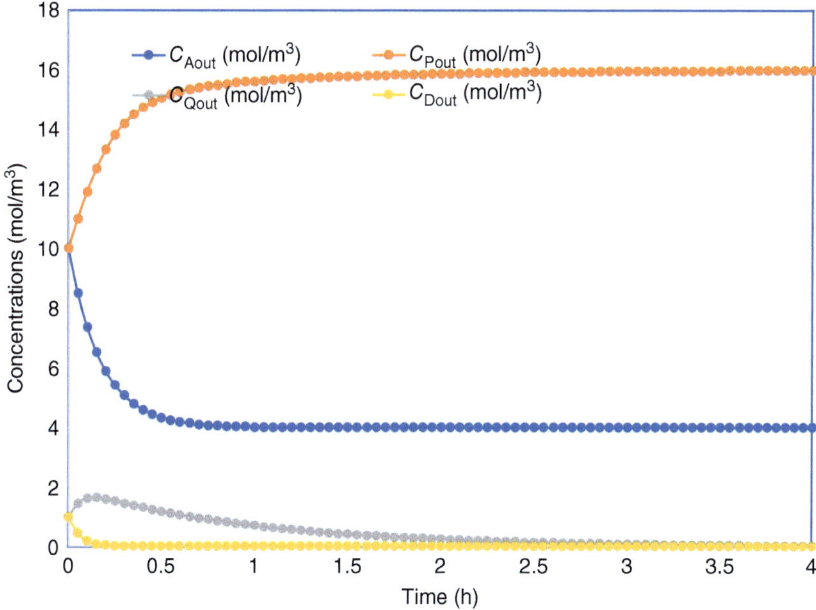


this means that the steady-state outflow will contain 4 mol/m$^3$ of A and 16 mol/m$^3$ of P if D is not present. Running from this condition and introducing $C_{\text{Din}} = 10 \, \text{mol/m}^3$ at $t = 1$ h, we finally find:

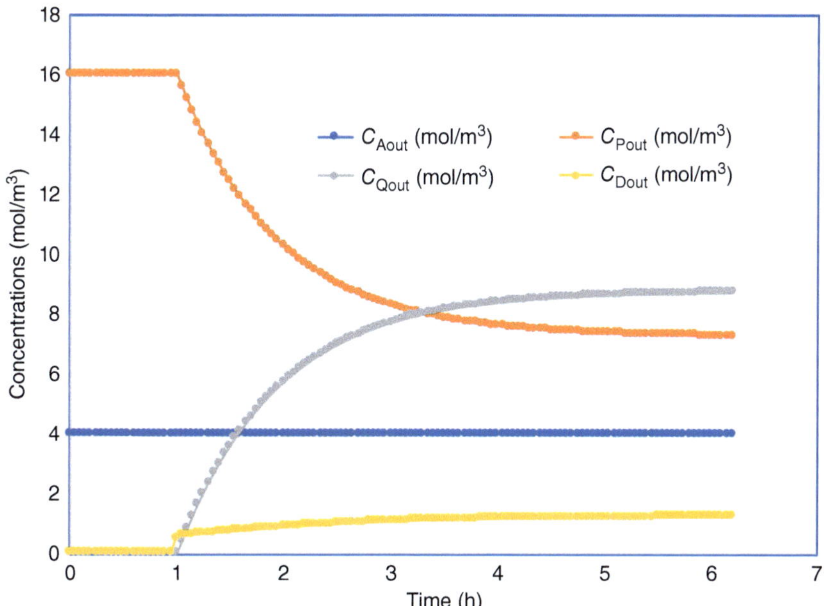

**Problem 5.14**   The vessel of a tubular reactor is receiving a constant flow of $100\,l/min$ of benzene with concentration $C_{A1} = 10^{-1}$ mol/l. At a certain moment $(t = 0)$, this current is replaced by another, in which $C_{A2} = 10^{-2}$ mol/l. Determine how the outlet concentration of benzene and chlorobenzene changes, given that $Cl_2$ is injected into the reactor and the reaction takes place:

$$\text{Benzene} + Cl_2(g) \rightarrow \text{Chlorobenzene} + HCl\,(g)$$

The volume of the reactor is $100\,l$, and the rate of the reaction is first-order (chlorine is injected in excess) and given by:

$$r_A = -kC_A = -0.035\,s^{-1}\,C_A$$

being "A" the benzene reacting.

**Solution to Problem 5.14**

For details refer the Wiley website at http://www.wiley-vch.de/ISBN9783527354115

In the initial state, the reactor is performing a first-order reaction in a stationary system, so at the exit of the reactor:

$$X_{Aout} = 1 - \exp(-k\bar{t}) = 1 - \exp(-0.035 \cdot 60) = 0.8775$$

$$C_{Aout} = C_{A0}(1 - X_{Aout}) = 0.012\,24\,\frac{mol}{l}$$

So the system can be simulated, bearing in mind that the conversion at each point can be calculated by:

$$X_A = 1 - \exp\left(-\frac{kV}{Q_0}\right)$$
$$C_A = C_{A0}(1 - X_A) = C_{A0}\exp\left(-\frac{kV}{Q_0}\right)$$

where $V$ is the volume of the reactor until it reaches the desired position. Using the data given, we find the following plot as an initial condition:

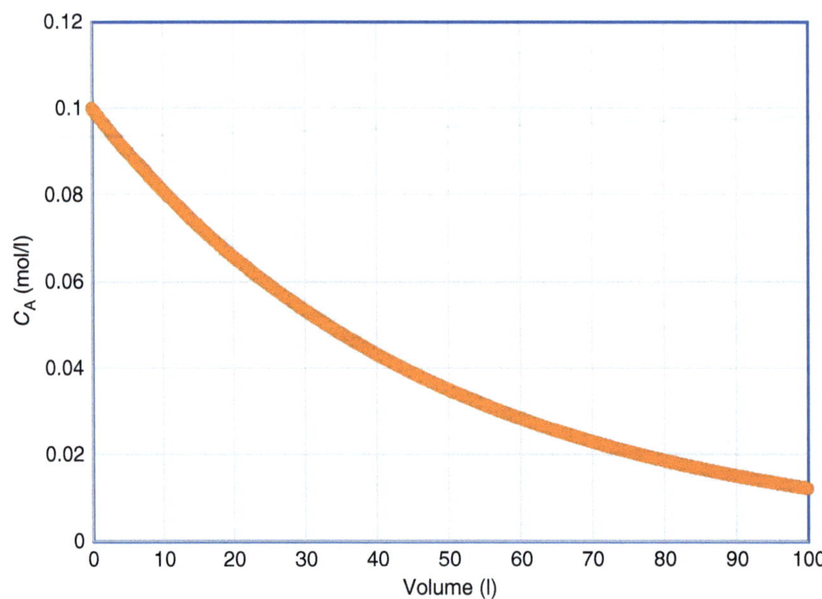

During the transient state, the mass balance in a PFR done over a differential volume, $dV$, is as shown in figure:



As different variables are used, differential forms are partial differential, and introducing into the balance:

$$\text{Input} + \text{Generation} = \text{Output} + \text{Accumulation}$$

$$n_A + r_A \cdot \partial V = (n_A + \partial n_A) + \frac{\partial C_A}{\partial t} \cdot \partial V$$

Bearing in mind, for a constant-density system:

$$Q \cdot \partial C_A = \partial n_A$$

What means that:

$$\frac{\partial C_A}{\partial t} = r_A - Q\frac{\partial C_A}{\partial V}$$

With first-order kinetics:

$$\frac{\partial C_A}{\partial t} = -kC_A - Q\frac{\partial C_A}{\partial V}$$

Using the finite differences approximation, first-order:

$$\frac{C_{Ai}^{t+1} - C_{Ai}^{t}}{\Delta t} = -kC_{Ai}^{t} - Q\frac{C_{Ai+1}^{t} - C_{Ai}^{t}}{\Delta V}$$

Rearranging:

$$C_{Ai}^{t+1} = \left(1 - k \cdot \Delta t + Q\frac{\Delta t}{\Delta V}\right) C_{Ai}^{t} - Q\frac{\Delta t}{\Delta V} C_{Ai+1}^{t}$$

Applying the stability criteria, the system is stable (errors do not increase) if:

$$Q\frac{\Delta t}{\Delta V} > 0$$

$$\left(1 - k \cdot \Delta t - Q\frac{\Delta t}{\Delta V}\right) > 0$$

$$\left(1 - k \cdot \Delta t - Q\frac{\Delta t}{\Delta V}\right) + Q\frac{\Delta t}{\Delta V} \leq 1$$

The first and third conditions are always fulfilled, and from the second one:

$$\Delta t < \frac{1}{\left(k + \frac{Q}{\Delta V}\right)}$$

The boundary conditions of this system are:

- At the input ($V = 0$): $C_A = C_{A2}$ (change in the concentration)
- At the exit ($V = V_{total}$) $\rightarrow C_A(V^-) = C_A(V^+)$ (closed vessel), so that at this section:

$$\left[\frac{\partial C}{\partial V}\right]_{V_{total}} = 0$$

The procedure can be applied using a Matlab program. For example:

```
clear all
close all
CA1=0.1; % Input concentration mol/L (t<0)
CA2=0.01; % Input concentration mol/L (t>=0)
Qv0=100; % Volumetric flow rate L/min
Vtotal=100; % L Total volume
k=0.035; % Reaction Kinetic constant, s-1
N=50; % Number of increment in x (position)

incV=Vtotal/(N-1); % Increment V
inct=0.95*incV/(Qv0+k*incV); % Must be lesser than defined
V=0:incV:Vtotal; % Values of V that Will be used in the...
calculation

% Initial values:
tend=[];
CAend=[];
t=0;
CA=CA1*exp(-k*V/Qv0*60);
CA(1)=CA2;
tfinal=2; % min, Final time
A=1-k*inct-Qv0*inct/incV;

while t<tfinal
    t=t+inct
    CAn(1)=CA2; % Boundary condition for the first increment
    CAn(N)=CA(N-1); % Boundary condition for the last increment
    for i=2:N-1
        CAn(i)=A*CA(i)+(Qv0*inct/incV)*CA(i-1); % PDE mass balance
    end
    CA=CAn;
    CAend=[CAend CA(end)]; % Saving the concentration at the exit
    tend=[tend t]; % Saving the values of time used
    plot(V,CA)
    xlabel('Volume (L)')
    ylabel('C_A (mol/L)')
    drawnow
end

figure
plot(tend,CAend)
xlabel('Time (s)')
ylabel('C_A at the exit (mol/L)')
```

Using the former program, we can check for the concentration profile at different times and also plot the exit concentration versus time:

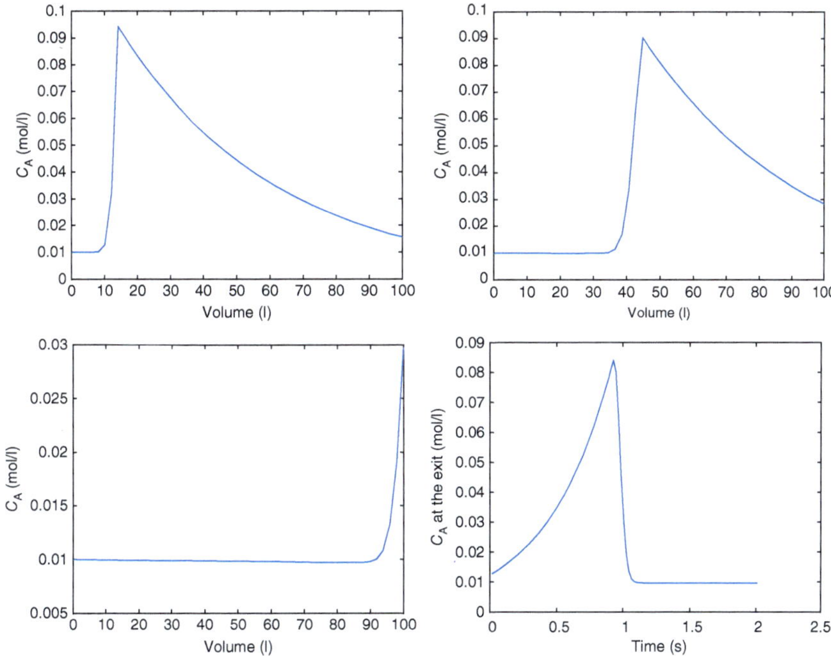

**Problem 5.15**   Heat conduction with a heat source of viscous origin: Simulate the temperature behavior of an incompressible Newtonian fluid through the space between two coaxial cylinders, as the outside temperature $(T_b)$ changes from 300 to 310 K and the outer surface of the cylinder moves at a rate of $2\pi$ rad/s. Determine the initial situation as the approximation to a steady state with $T_b = 300$ K, and then do the necessary calculations to introduce a change in $T_b$ to 310 K.

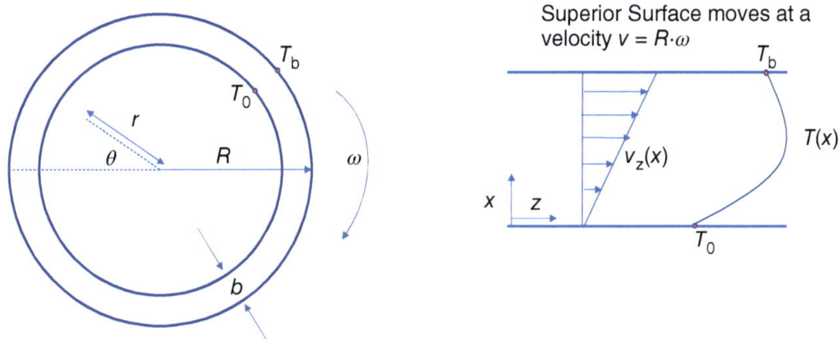

Use the following data:

| | |
|---|---|
| $T_0$ (K) | 273 |
| $\mu$ (Pa·s) | $10^{-3}$ |
| $k$ (W/m·K) | $10^{-3}$ |
| $b$ (m) | $10^{-3}$ |
| $\omega$ (rad/s) | $2\pi$ |
| $R$ (m) | 1 |
| $\rho$ (kg/m$^3$) | 1000 |
| $C_p$ (J/kg·K) | $10^{-3}$ |

**Solution to Problem 5.15**

For details refer the Wiley website at http://www.wiley-vch.de/ISBN9783527354115

Let us consider the flow of an incompressible Newtonian fluid through the space between two coaxial cylinders, as indicated in the figure. As the outer cylinder rotates, the cylindrical fluid layers rub against the adjacent fluid layers, resulting in heat production. In other words, mechanical energy is invariably degraded into heat energy. The heat source per unit volume resulting from this "viscous dissipation" will be designated by $S_v$. This magnitude depends on the local velocity gradient: the faster a fluid layer moves relative to an adjacent one, the greater the heating produced by viscous dissipation. The internal and external surfaces of the fluid are maintained, respectively, at temperatures $T = T_0$ and $T = T_b$. Obviously, $T$ will be an exclusive function of "$r$".

If the thickness $b$ of the slit is small compared to the radius $R$ of the outer cylinder, the problem can be approximately solved using the simplified system indicated in the second part of the figure.

A heat energy balance applied to an envelope of thickness $\Delta x$, width $W$, and length $L$ in a steady state leads to:

$$WL\,q_x\big]_x - WL\,q_x\big]_{x+\Delta x} + WL\Delta x S_v = 0$$

Using the simplified system, the value of $S_v$ can be expressed as:

$$S_v = -\tau_{xz}\left(\frac{dv_z}{dz}\right) = \mu\left(\frac{dv_z}{dz}\right)^2$$

Since $v_z = x \cdot v/b$, we will have to:

$$S_v = \mu\left(\frac{v}{b}\right)^2$$

And therefore, using the heat balance:

$$\frac{dq_x}{dx} = \mu\left(\frac{v}{b}\right)^2$$

Also, using Fourier's law for heat transfer, we have:

$$q_x = -k\frac{dT}{dx}$$

which is introduced in the heat balance:

$$-k\frac{d^2T}{dx^2} = \mu\left(\frac{v}{b}\right)^2$$

This balance will be subject to the following limit conditions:

$T = T_0 \text{ para } x = 0$

$T = T_b \text{ para } x = L$

Integrating once:

$$-k\frac{dT}{dx} = \mu\left(\frac{v}{b}\right)^2 x + C_1$$

And again:

$$-k \cdot T = \mu\left(\frac{v}{b}\right)^2 \frac{x^2}{2} + C_1 x + C_2$$

Using the boundary conditions to determine $C_1$ and $C_2$, we finally arrive at:

$$\frac{T - T_0}{T_b - T_0} = \left(\frac{x}{b}\right) + \frac{1}{2}\text{Br}\left(\frac{x}{b}\right)\left(1 - \frac{x}{b}\right)$$

where $\text{Br} = \mu\frac{v^2}{k(T_b-T_0)}$ is the Brinkman number.

Let us take the values given in the statement (with $T_b = 300\,\text{K}$), and we will simulate the stationary profile:

$$\text{Br} = \mu\frac{v^2}{k(T_b - T_0)} = 1.46$$

Given the $v = R \cdot w = 6.28\,\text{m/s}$
We obtain:

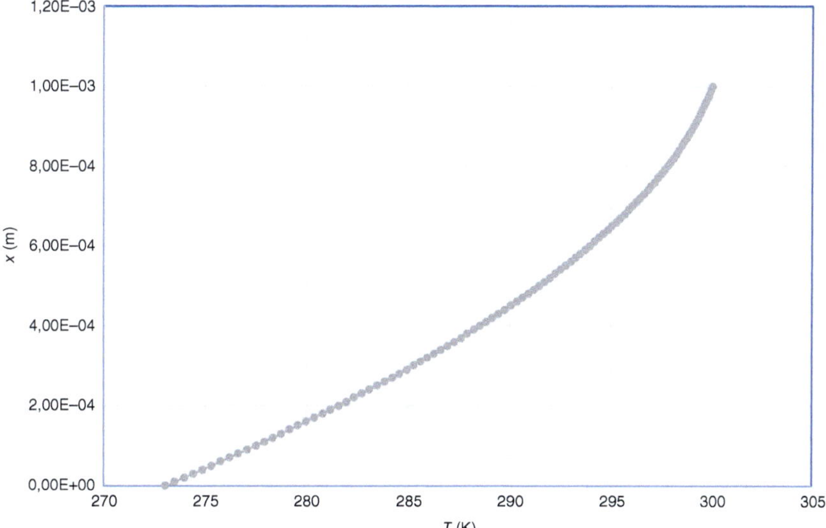

The lower the conductivity ($k$), the greater the thickness ($b$), the greater the viscosity ($\mu$), the greater the speed ($w$), or the greater the radius ($R$), the more curved the profile will be.

To simulate the system in a non-steady state, we start with the expression of the heat balance:

$$WL\,q_x]_x - WL\,q_x]_{x+\Delta x} + WL\Delta x S_v = WL\Delta x \rho C_p \frac{\mathrm{d}T}{\mathrm{d}t}$$

That means dividing by $WL\Delta x$ and taking differential elements:

$$\frac{q_x]_x - q_x]_{x+\Delta x}}{\Delta x} + S_v = \rho C_p \frac{\mathrm{d}T}{\mathrm{d}t}$$

$$-\frac{\mathrm{d}q_x}{\mathrm{d}x} + S_v = \rho C_p \frac{\mathrm{d}T}{\mathrm{d}t}$$

That is to say:

$$-\frac{\mathrm{d}q_x}{\mathrm{d}x} + \mu\left(\frac{v}{b}\right)^2 = \rho C_p \frac{\mathrm{d}T}{\mathrm{d}t}$$

Using Fourier's law:

$$-\frac{\mathrm{d}}{\mathrm{d}x}\left(-k\frac{\mathrm{d}T}{\mathrm{d}x}\right) + \mu\left(\frac{v}{b}\right)^2 = \rho C_p \frac{\mathrm{d}T}{\mathrm{d}t}$$

$$k\frac{\mathrm{d}^2 T}{\mathrm{d}x^2} + \mu\left(\frac{v}{b}\right)^2 = \rho C_p \frac{\mathrm{d}T}{\mathrm{d}t}$$

Applying the FDM:

$$k\frac{\left(T_{i+1}^t - 2T_i^t + T_{i-1}^t\right)}{\Delta x^2} + \mu\left(\frac{v}{b}\right)^2 = \frac{\rho C_p \left(T_i^{t+1} - T_i^t\right)}{\Delta t}$$

Therefore:

$$\frac{k}{\rho C_p}\frac{\Delta t}{\Delta x^2}\left(T_{i+1}^t - 2T_i^t + T_{i-1}^t\right) + \frac{\mu}{\rho C_p}\Delta t\left(\frac{v}{b}\right)^2 = \left(T_i^{t+1} - T_i^t\right)$$

And clearing:

$$T_i^{t+1} = \frac{k}{\rho C_p}\frac{\Delta t}{\Delta x^2}\left(T_{i+1}^t - 2T_i^t + T_{i-1}^t\right) + \frac{\mu}{\rho C_p}\Delta t\left(\frac{v}{b}\right)^2 + T_i^t$$

Taking:

$$\alpha = \frac{k}{\rho C_p}\frac{\Delta t}{\Delta x^2}$$

$$T_i^{t+1} = \alpha\left(T_{i+1}^t + T_{i-1}^t\right) + (1 - 2\alpha)T_i^t + \frac{\mu}{\rho C_p}\Delta t\left(\frac{v}{b}\right)^2$$

Starting from the profile obtained in steady state ($t = 0$), we will obtain the variation of the profile with time when the outside temperature ($T_b$) changes to $310\,\mathrm{K}$, that is, $T_b = 310\,\mathrm{K}$ for $t > 0$.

The stability conditions are:

$$\frac{k}{\rho C_{\mathrm{p}}} \frac{\Delta t}{\Delta x^2} > 0$$

$$1 - 2\frac{k}{\rho C_{\mathrm{p}}} \frac{\Delta t}{\Delta x^2} > 0$$

Therefore, the latter indicates that:

$$1 > 2\frac{k}{\rho C_{\mathrm{p}}} \frac{\Delta t}{\Delta x^2}$$

$$\frac{1}{2} \frac{\rho C_{\mathrm{p}}}{k} > \frac{\Delta t}{\Delta x^2}$$

That is to say:

$$\Delta t < \frac{1}{2} \frac{\rho C_{\mathrm{p}}}{k} \Delta x^2$$

taking $\Delta x = 10^{-5}$ m we have to

$$\Delta t < 5 \cdot 10^{-8} \text{ s}$$

At the extremes, we will assume that the transmission is zero; therefore, at $x = 0$ and $x = b$:

$$\frac{\mathrm{d}T}{\mathrm{d}x} = 0$$

And we will have to:

$$T_{N-1}^{t+1} = T_N^t = T_b$$
$$T_1^{t+1} = T_0^t$$

Taking $\Delta t = 4.9 \cdot 10^{-8}$ s we get:

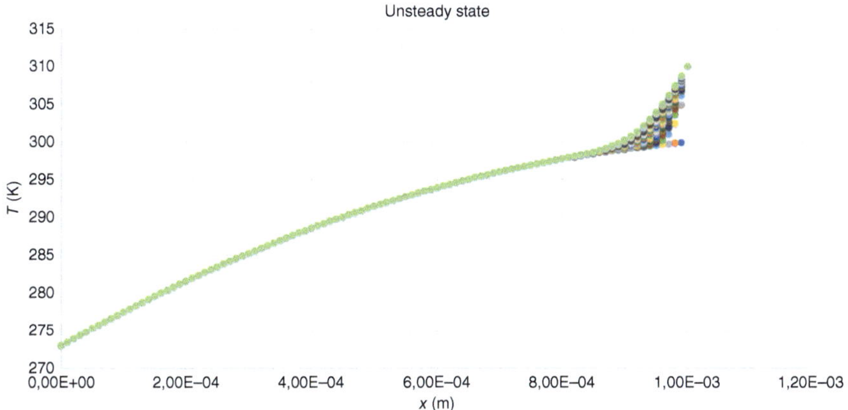


Here you can see how the temperature changes along the axis, with time, up to a total time of $2 \cdot 10^{-6}$ s.

**Problem 5.16**

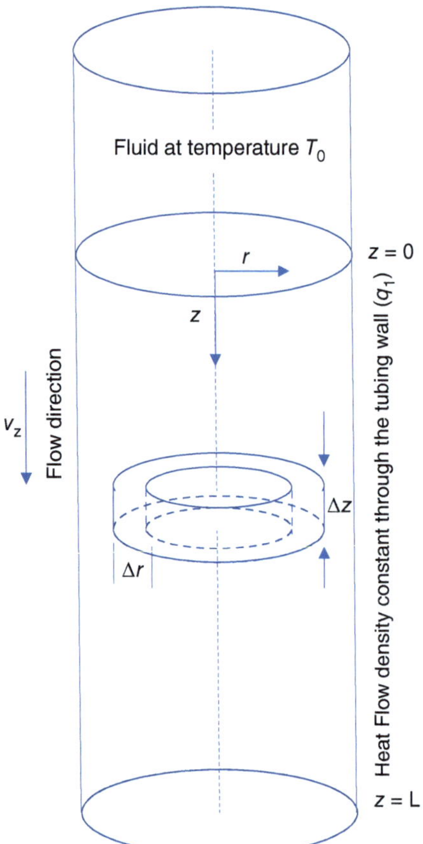


A fluid flows through a circular tube, like the one shown in the figure, at a temperature $T_0$. Through the walls of the tube, there is a flow of heat whose density can be considered constant. Determine the expression for the heat balance that allows the calculation of the temperature profiles that will occur in this system in both the axial ($z$) and radial ($r$) coordinates. We will assume that the following data are valid to carry out the simulation:

| | |
|---|---|
| $T_0$ (K) | 400 |
| $q_1$ (W/s) | 1 |
| $k$ (W/m·K) | 1 |
| $v_z$ (m/s) | 0.1 |
| $R$ (m) | 1 |
| $\rho$ (kg/m³) | 1000 |
| $C_p$ (J/kg·K) | 1 |
| $L$ (m) | 1 |

**Solution to Problem 5.16**

For details refer the Wiley website at http://www.wiley-vch.de/ISBN9783527354115

The heat balance can be done in the circular ring shown in the figure, in such a way that:

| | |
|---|---|
| Input of energy through "$r$" by conduction | $q_r]_r \cdot 2\pi r \Delta z$ |
| Output of energy through "$r + \Delta r$" by conduction | $q_r]_{r+\Delta r} \cdot 2\pi (r + \Delta r)\Delta z$ |
| Input of energy through "$z$" by conduction | $q_z]_z \cdot 2\pi r \Delta z$ |
| Output of energy through "$z + \Delta z$" by conduction | $q_z]_{z+\Delta z} \cdot 2\pi r \Delta z$ |
| Input of energy through "$z$" due to the flow of fluid | $\rho C_p v_z (T - T_0)]_z \cdot 2\pi r \Delta z$ |
| Output of energy through "$z + \Delta z$" due to the flow of fluid | $\rho C_p v_z (T - T_0)]_{z+\Delta z} \cdot 2\pi r \Delta z$ |

By equating the energy inputs and outputs, the energy balance applied to the circular ring is established, which will be:

$$\frac{(rq_r)]_{r+\Delta r} - (rq_r)]_r}{\Delta r} + r\frac{(q_z)]_{z+\Delta z} - (q_z)]_z}{\Delta z} + r\rho C_p v_z \frac{T]_{z+\Delta z} - T]_z}{\Delta z} = 0$$

If $\Delta z$ and $\Delta r$ tend to zero, we obtain:

$$\rho C_p v_z \frac{\partial T}{\partial z} = -\frac{1}{r}\frac{\partial}{\partial z}(rq_r) - \frac{\partial q_z}{\partial z}$$

Assuming a constant velocity:

$$v_z = v_{z\,max}$$

And the Fourier law for both directions $z$ and $r$:

$$q_z = -k\frac{\partial T}{\partial z}$$

$$q_r = -k\frac{\partial T}{\partial r}$$

We finally obtain:

$$\rho C_p v_{z\,max} \frac{\partial T}{\partial z} = k\left[\frac{1}{r}\frac{\partial}{\partial r}\left(r\frac{\partial T}{\partial r}\right) + \frac{\partial^2 T}{\partial z^2}\right]$$

Generally, the conduction of heat in the $z$-direction (the term containing the second derivative) is small compared to the convective transmission (the term containing $\frac{\partial T}{\partial z}$), so second derivative can be omitted from the equation. One case where this cannot be done is the slow flow of substances that have high heat conductivity, such as liquid metals. When the conductive term in the $z$-direction can be removed, it finally remains:

$$\rho C_p v_{z\,max} \frac{\partial T}{\partial z} = k\frac{1}{r}\frac{\partial}{\partial r}\left(r\frac{\partial T}{\partial r}\right)$$

The second term can be expressed (by using the chain rule) as follows:

$$\frac{1}{r}\frac{\partial}{\partial r}\left(r\frac{\partial T}{\partial r}\right) = \frac{1}{r}\left(r\frac{\partial^2 T}{\partial r^2}\right) + \frac{1}{r}\left(\frac{\partial T}{\partial r}\right) = \frac{\partial^2 T}{\partial r^2} + \frac{1}{r}\left(\frac{\partial T}{\partial r}\right)$$

So:

$$v_{z\,max}\frac{\partial T}{\partial z} = k\left[\frac{\partial^2 T}{\partial r^2} + \frac{1}{r}\left(\frac{\partial T}{\partial r}\right)\right]$$

This equation will give the temperature as a function of both dimensions $z$ and $r$ in the tube. The boundary conditions will be:

for $r = 0$        $\frac{\partial T}{\partial r} = 0$ (by symmetry $T$ should present a maximum)

for $r = R$        $-k\frac{\partial T}{\partial r} = q_1$ (a constant)

for $z = 0$        $T = T_0$ (for any value of "$r$")

For applying the FDM, we will use superscripts for positions in $z$ ($T^j$) and subscripts for positions in $r$ ($T_i$). From the heat balance, we can write:

$$\rho C_p v_{z\,max} \cdot \frac{T_i^{j+1} - T_i^j}{\Delta z} = k\left[\frac{T_{i+1}^j - 2T_i^j + T_{i-1}^j}{\Delta r^2} + \frac{1}{r_i}\left(\frac{T_{i+1}^j - T_i^j}{\Delta r}\right)\right]$$

$$T_i^{j+1} = \frac{k\Delta z\left[\frac{T_{i+1}^j - 2T_i^j + T_{i-1}^j}{\Delta r^2} + \frac{1}{r_i}\left(\frac{T_{i+1}^j - T_i^j}{\Delta r}\right)\right]}{\rho C_p v_{z\,max}} + T_i^j$$

Rearranging:

$$T_i^{j+1} = \frac{k\Delta z}{\Delta r^2 \rho C_p v_{z\,max}}\left(T_{i+1}^j - 2T_i^j + T_{i-1}^j\right) + \frac{k\Delta z}{r_i \rho \Delta r C_p v_{z\,max}}\left(T_{i+1}^j - T_i^j\right) + T_i^j$$

Using:

$$\alpha = \frac{k\Delta z}{\Delta r^2 \rho C_p v_{z\,max}}$$

$$\beta = \frac{k\Delta z}{r_i \rho \Delta r C_p v_{z\,max}}$$

We have that:

$$T_i^{j+1} = \alpha\left(T_{i+1}^j - 2T_i^j + T_{i-1}^j\right) + \beta\left(T_{i+1}^j - T_i^j\right) + T_i^j$$

$$T_i^{j+1} = (\alpha + \beta)T_{i+1}^j + (1 - 2\alpha - \beta)T_i^j + \alpha T_{i-1}^j$$

For the stability of the FDM solution, the only condition giving information is:

$$(1 - 2\alpha - \beta) > 0$$

From this, we can check that:

$$\Delta z < \frac{\Delta r \rho C_p v_{z\,max}}{k\left(\frac{2}{\Delta r} + \frac{1}{r_i}\right)}$$

That will be used to calculate the stable condition of the solution. Bearing in mind that $r_i$ varies between 0 and $R$, we have:

$$\Delta z < \frac{\Delta r \rho C_p v_{z\,\mathrm{max}}}{k\left(\frac{2}{\Delta r} + \frac{1}{R}\right)}$$

The initial condition of the system will be that temperature equals $T_0$ at any point of the fluid at $z = 0$. Boundary conditions can be written using the FDM:

* for $r = 0$    $\frac{T^{j+1}_{i=1} - T^{j}_{i=0}}{\Delta r} = 0$        So: $T^{j+1}_{i=1} = T^{j}_{i=0}$

* for $r = R$    $-k\frac{T^{j}_{i} - T^{j}_{i-1}}{\Delta r} = q_1$        So: $T^{j}_{i} = T^{j}_{i-1} - \frac{\Delta r}{k}q_1$ in the last interval for $r$-integration

* for $z = 0$    $T = T_0$ (for any value of "$r$")

Using the previous equations, we finally get:

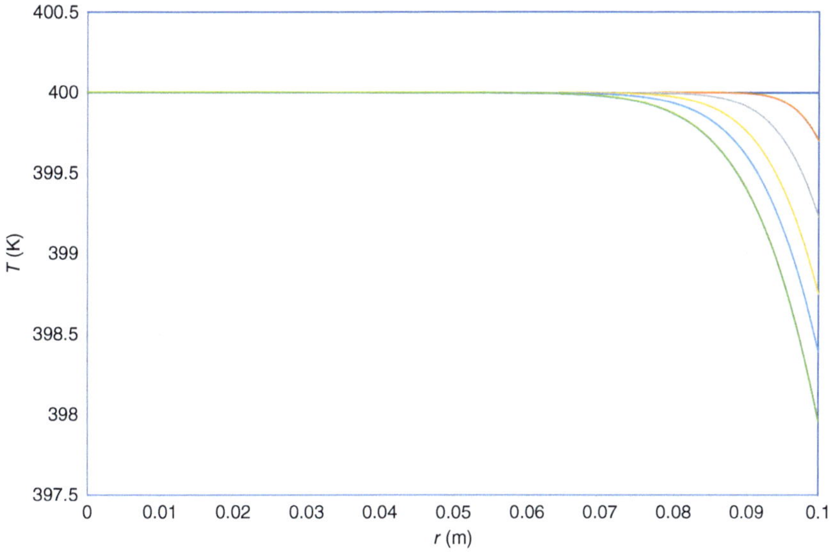


with the different lines representing the different temperature profiles at the various values of $z$.

**Problem 5.17** A cylindrical tank capable of holding $28\,\mathrm{m}^3$ of liquid is provided with a sufficiently powerful agitator to maintain the liquid at a uniform temperature (see figure). Heat is transmitted to the liquid through a coil arranged in such a way that the area available for heat transfer is proportional to the amount of liquid in the tank. The heating coil consists of 10 turns of 125 cm in diameter, built with a tube of 2.5 cm in external diameter. The tank is fed continuously with 10 kg/min of water at

200 °C, starting with the tank empty at time $t = 0$. Water vapor at 105 °C is introduced inside the tube, and the global transmission coefficient of heat is 500 kcal/h·m²·K. How does the temperature of the water change as the tank fills?

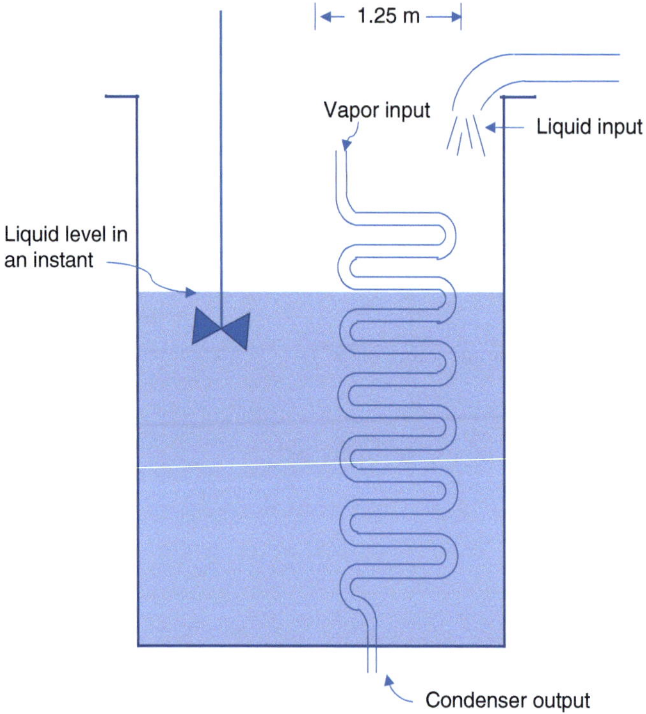


### Solution to Problem 5.17

For details refer the Wiley website at http://www.wiley-vch.de/ISBN9783527354115

These types of problems are best solved by using symbols and entering the numerical values at the end. For this, the following notation is used:

$A_0$ = total area available for heat transfer
$A(t)$ = heat transfer area at an instant
$V_0$ = volume of liquid in the full tank
$V(t)$ = volume of liquid in the tank at an instant
$T$ = temperature of the liquid at an instant
$T_1$ = temperature of the liquid at the inlet
$T_s$ = temperature of the water vapor
$U_0$ = overall heat transfer coefficient
$t$ = time elapsed from the moment when water is turned on
$Q$ = flow velocity of the water entering the tank
$\rho$ = density of the liquid

For the problem statement, the following assumptions are made:

* The temperature of the steam in the coil is constant.
* The density and specific heat of water do not vary much with temperature.
* Since the fluid is approximately incompressible, $C_p = C_v$.
* The stirrer keeps the temperature even throughout the mass.
* The heat transmission coefficient is independent of position and time.
* The walls of the tank are adiabatic.

The fluid contained in the tank is taken as the system. For the system we are considering:

$$\rho C_p \frac{d}{dt} V(T - T_1) = U_0 A (T_s - T)$$

The area and volume in an instant are:

$$V(t) = Qt/\rho$$

$$A(t) = QtA_0/(\rho V_0)$$

Introducing these expressions in the balance:

$$\rho C_p \left( \frac{dV}{dt}(T - T_1) + \frac{V d(T - T_1)}{dt} \right) = QC_p(T - T_1) + QC_p t \left( \frac{dT}{dt} \right)$$

$$= U_0 Qt \frac{A_0}{\rho V_0}(T_s - T)$$

The initial value of this equation is $T = T_1$ for $t = 0$.

This is an ODE simple to solve by numerical methods. Applying FDM to the equation:

$$QC_p(T^t - T_1) + QC_p t \left( \frac{(T^{t+1} - T^t)}{\Delta t} \right) = U_0 Qt \frac{A_0}{\rho V_0}(T_s - T^t)$$

Rearranging:

$$T^{t+1} = \frac{U_0 Qt \frac{A_0}{\rho V_0}(T_s - T^t) - QC_p(T^t - T_1)}{QC_p t} \Delta t + T^t$$

$$T^{t+1} = \left( \frac{U_0 A_0}{\rho V_0 C_p}(T_s - T^t) - \frac{(T^t - T_1)}{t} \right) \Delta t + T^t$$

The total area for heat transmission can be calculated as follows:

$$A_0 = 10 * 1.25 * (2\pi 0.025) = 1.96 \text{ m}^2$$

Using a spreadsheet, it is easy to do the calculation, obtaining:

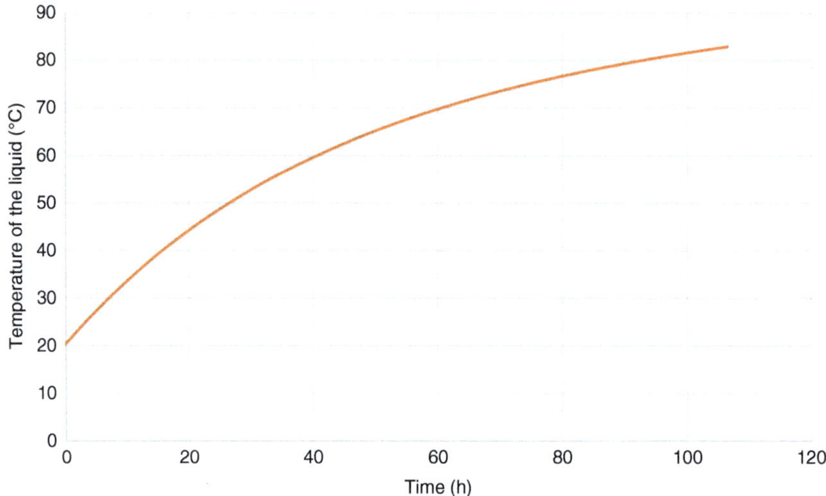


**Problem 5.18**  It is desired to derive an equation for the speed at which a liquid A evaporates within a vapor B in a tube of infinite length (see figure). The liquid level is maintained at all times at position $z = 0$. The entire system is at constant temperature and pressure, and vapors A and B are assumed to form an ideal gas mixture; therefore, the molar density c is constant throughout the gas phase. Finally, it is assumed that B is insoluble in A.

In the system, $D_{AB} = 10^{-4}$ m$^2$/s, and the equilibrium molar fraction of A is $X_{A0} = 0.1$. Let us simulate the system for $z_{total} = 0.3$ m.

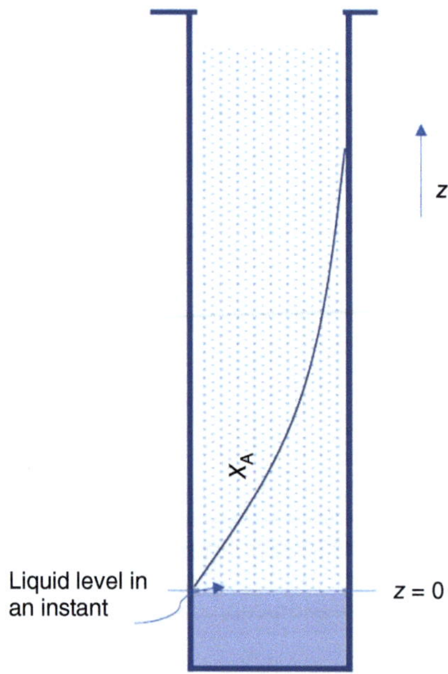

**Solution to Problem 5.18**

For details refer the Wiley website at http://www.wiley-vch.de/ISBN9783527354115

The continuity equations for components A and B are:

$$\frac{\partial C_A}{\partial t} = -\frac{\partial N_{Az}}{\partial z}$$

$$\frac{\partial C_B}{\partial t} = -\frac{\partial N_{Bz}}{\partial z}$$

Adding these two equations, we get:

$$\frac{\partial c}{\partial t} = -\frac{\partial}{\partial z}(N_{Az} + N_{Bz})$$

Since $c$ is constant, it follows that $\frac{\partial c}{\partial t} = 0$, and $N_{Az} + N_{Bz}$ is an exclusive function of time. But since it is known that for $z = 0$, there is no movement of B, it turns out that $N_{B0} = 0$. For the same position, the flux density of A is obtained:

$$N_{A0} = -c\frac{D_{AB}}{1 - X_{A0}} \frac{\partial X_A}{\partial z}\bigg]_{z=0}$$

in which the subscript 0 indicates that the quantities are evaluated for the $z = 0$ plane. Therefore, for any value of $z$:

$$N_{Az} + N_{Bz} = N_{A0} + N_{B0} = -c\frac{D_{AB}}{1 - X_{A0}} \frac{\partial X_A}{\partial z}\bigg]_{z=0}$$

and the flux density of A for any value of $z$ is given by:

$$N_{Az} = -cD_{AB}\frac{\partial X_A}{\partial z} - X_A\left(c\frac{D_{AB}}{1 - X_{A0}}\right)\frac{\partial X_A}{\partial z}\bigg]_{z=0}$$

Substituting, being $D_{AB}$ constant, we obtain:

$$\frac{\partial X_A}{\partial t} = D_{AB}\frac{\partial^2 X_A}{\partial z^2} + \frac{D_{AB}}{1 - X_{A0}}\frac{\partial X_A}{\partial z}\bigg]_{z=0}\frac{\partial X_A}{\partial z}$$

This equation is solved with the following initial and limit conditions:

| | | |
|---|---|---|
| Cond. initial: | for $t = 0$ | $X_A = 0$ |
| Cond. limit 1: | for $z = 0$ | $X_A = X_{A0}$ |
| Cond. limit 2: | for $z = \infty$ | $X_A = 0$ |

being $X_{A0}$ the equilibrium concentration in the gas phase, that for an ideal gas mixture corresponds exactly to the ratio between the vapor pressure of pure A and the total pressure.

Let us apply the FDM:

$$\frac{X_{Ai}^{t+1} - X_{Ai}^{t}}{\Delta t} = D_{AB}\frac{X_{Ai+1}^{t} - 2X_{Ai}^{t} + X_{Ai-1}^{t}}{\Delta z^2} + \frac{D_{AB}}{1 - X_{A0}}X_A'\bigg]_{z=0}\frac{X_{Ai+1}^{t} - X_{Ai}^{t}}{\Delta z}$$

$$X_{Ai}^{t+1} = \left( D_{AB} \frac{X_{Ai+1}^t - 2X_{Ai}^t + X_{Ai-1}^t}{\Delta z^2} + \frac{D_{AB}}{1 - X_{A0}} X_A' \Big]_{z=0} \frac{X_{Ai+1}^t - X_{Ai}^t}{\Delta z} \right) \Delta t + X_{Ai}^t$$

$$X_{Ai}^{t+1} = \left( \frac{D_{AB}}{\Delta z^2} \Delta t + \frac{D_{AB}\Delta t}{1 - X_{A0}} X_A' \Big]_{z=0} \right) X_{Ai+1}^t$$

$$+ \left( 1 - 2\frac{D_{AB}}{\Delta z^2} \Delta t - \frac{D_{AB}\Delta t}{1 - X_{A0}} X_A' \Big]_{z=0} \right) X_{Ai}^t + \frac{D_{AB}}{\Delta z^2} X_{Ai-1}^t$$

The only stability condition giving information is:

$$\left( 1 - 2\frac{D_{AB}\Delta t}{\Delta z^2} - \frac{D_{AB}\Delta t}{1 - X_{A0}} X_A' \Big]_{z=0} \right) > 0$$

Giving:

$$1 > \left( 2\frac{D_{AB}}{\Delta z^2} + \frac{D_{AB}}{1 - X_{A0}} X_A' \Big]_{z=0} \right) \Delta t$$

$$\Delta t < \frac{1}{\left( 2\frac{D_{AB}}{\Delta z^2} + \frac{D_{AB}}{1-X_{A0}} X_A' \Big]_{z=0} \right)}$$

For its part, the derivative at the first increment is:

$$X_A' \Big]_{z=0}^{t+1} = \frac{X_{A0}^t - X_{A1}^t}{\Delta z}$$

The required calculation is quite hard to do in a spreadsheet (as the time increment should be very low), and a Matlab program is better to use. In this case, the following is valid:

```
clear all
close all

DAB=      1.00E-04;%        m2/s
XA0=      0.1;
Ztotal=            0.2;%    m
N=100; % number of points
incz=   Ztotal/(N-1);
derivatz0=-2.47;

alpha=DAB/incz^2;
beta=DAB*derivatz0/(1-XA0);

maxinct= (1/(2*alpha+beta));
inct=0.9*maxinct;
alpha=alpha*inct;

Z=0:incz:Ztotal; % Values of z that Will be used in the...
calculation

% Initial values:

t=0;
XA=zeros(1,N);
```

```
XA(1)=XA0;
tfinal=0.2; % sec, Final time

while t<tfinal
    t=t+inct
    XAn(1)=XA0; % Boundary condition for the first increment
    XAn(N)=0; % Boundary condition for the last increment
    derivatz0=(XA(2)-XA(1))/inct
    beta= DAB*derivatz0/(1-XA0);
    for i=2:N-1
        XAn(i)=(1-2*alpha-beta)*XA(i)+(alpha+beta)*XA(i+1)+...
(alpha)*XA(i-1); % PDE mass balance
    end
    XA=XAn;
    plot(Z,XA)
    xlabel('Z (m)')
    ylabel('X_A (-)')
    drawnow
end
```

We can stop the calculation at any time and see the corresponding profile. For example:

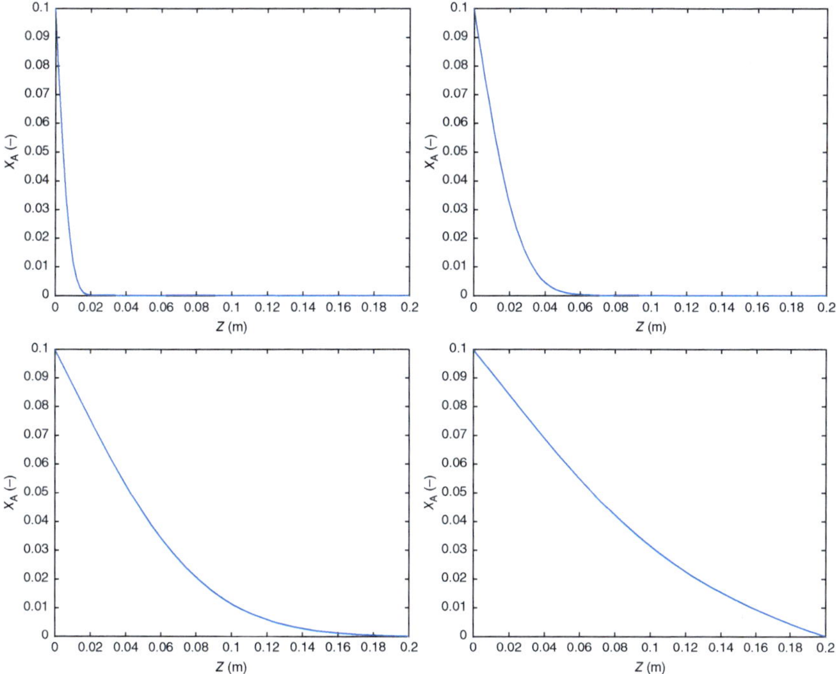

**Problem 5.19** Repeat Problem 5.7, but consider a non-steady state approximation. In the non-steady regime, the fin is started at room temperature ($T_{amb}$), and there is a temperature increase inside the reactor up to $T_0$. The energy balance in the system

takes into account the heat transmitted by conduction and lost by convection to the medium. This balance is given by:

$$\text{Heat input} = (q_x \cdot A)_x$$

$$\text{Heat exit} = (q_x \cdot A)_{x+dx} + q_{conv} \cdot \Delta S$$

$$\text{Heat accumulation} = A \cdot \rho C_p \cdot \frac{dT}{dt}$$

With:

$q_x = -k \cdot \frac{dT}{dx}$   (conduction of heat within the fin)

$q_{conv} = h(T - T_\infty)$   (heat loss by convection on the surface of the fin).

This gives:

$$k \cdot A \cdot \frac{d^2 T}{dx^2} - h \cdot P \cdot (T - T_{amb}) = \rho C_p \cdot A \cdot \frac{dT}{dt}$$

where $T_{amb}$ is the temperature of the fluid surrounding the reactor, "$A$" is the cross section, "$S$" is the surface area of the fin, and "$P$" is its perimeter.

The boundary conditions of this problem generally consider that at one end the temperature is constant, and the heat loss that reaches the end of the fin (of length "$L$") is zero, so:

$$\begin{cases} T = T_0 & \text{in } x = 0 \\ -k \cdot A \cdot \dfrac{dT}{dx} = 0 & \text{in } x = L \end{cases}$$

Applying the finite difference method, indicate the equations that result in the calculation of the temperature along the fin. Indicates the stability conditions and how to take into account the boundary conditions of this problem.

**Solution to Problem 5.19**

For details refer the Wiley website at http://www.wiley-vch.de/ISBN9783527354115

$$\frac{d^2 T}{dx^2} - \frac{hP}{kA} \cdot (T - T_\infty) = \frac{\rho C_p}{k} \cdot \frac{dT}{dt}$$

Applying FDM:

$$\frac{T_{i+1}^t + T_{i-1}^t - 2T_i^t}{\Delta x^2} - \frac{hP}{kA} \cdot \left(T_i^t - T_\infty\right) = \frac{\rho C_p}{k} \cdot \frac{T_i^{t+1} - T_i^t}{\Delta t}$$

Solving for $T_i^{t+1}$:

$$T_i^{t+1} = \frac{k \cdot \Delta t}{\rho C_p \Delta x^2} \cdot \left(T_{i+1}^t + T_{i-1}^t - 2T_i^t\right) - \frac{hP}{kA} \cdot \frac{k \cdot \Delta t}{\rho C_p} \cdot \left(T_i^t - T_\infty\right) + T_i^t$$

$$T_i^{t+1} = \frac{k \cdot \Delta t}{\rho C_p \Delta x^2} \cdot T_{i+1}^t + \frac{k \cdot \Delta t}{\rho C_p \Delta x^2} \cdot T_{i-1}^t + \left(1 - 2\frac{k \cdot \Delta t}{\rho C_p \Delta x^2} - \frac{hP}{A} \cdot \frac{\Delta t}{\rho C_p}\right)$$
$$\cdot T_i^t + \frac{hP}{A} \cdot \frac{\Delta t}{\rho C_p} \cdot T_\infty$$

Stability conditions are:

$$\frac{k \cdot \Delta t}{\rho C_\mathrm{p} \Delta x^2} > 0 \ldots \text{always is fulfilled}$$

$$\left(1 - 2\frac{k \cdot \Delta t}{\rho C_\mathrm{p} \Delta x^2} + \frac{hP}{A} \cdot \frac{\Delta t}{\rho C_\mathrm{p}}\right) > 0 \ldots \rightarrow \Delta t < \left(\frac{2k}{\rho C_\mathrm{p} \Delta x^2} - \frac{hP}{\rho A C_\mathrm{p}}\right)^{-1}$$

Also:

$$\frac{k \cdot \Delta t}{\rho C_\mathrm{p} \Delta x^2} + \frac{k \cdot \Delta t}{\rho C_\mathrm{p} \Delta x^2} + \left(1 - 2\frac{k \cdot \Delta t}{\rho C_\mathrm{p} \Delta x^2} - \frac{hP}{A} \cdot \frac{\Delta t}{\rho C_\mathrm{p}}\right) \leq 1$$

And we see that:

$$\left(1 - \frac{hP}{A} \cdot \frac{\Delta t}{\rho C_\mathrm{p}}\right) \leq 1$$

So:

$$\frac{hP}{A} \cdot \frac{\Delta t}{\rho C_\mathrm{p}} > 0$$

that is always fulfilled.

Boundary conditions: heat loss from the tip of the fin is negligible

$$T = T_0 \text{ in } x = 0$$

$$\frac{\mathrm{d}T}{\mathrm{d}x} = 0 \text{ in } x = L$$

So:

$$\frac{T_N - T_{N-1}}{\Delta x} = 0$$

And: $T_N = T_{N-1}$

Let us calculate the evolution of the temperature in a particular case. For example, with a fin made of brass:

$$A = 2 \cdot 10^{-4}\,\mathrm{m^2} \qquad S = 0.03\,\mathrm{m^2} \qquad P = 0.05\,\mathrm{m} \qquad L = 0.6\,\mathrm{m}$$
$$k = 100\,\mathrm{W/m \cdot K} \qquad \rho = 873\,\mathrm{kg/m^3} \qquad C_\mathrm{p} = 38\,\mathrm{J/K \cdot kg}$$
$$T_0 = 100\,^{\circ}\mathrm{C} \qquad T_\mathrm{amb} = 25\,^{\circ}\mathrm{C} \qquad h = 10\,\mathrm{W/m^2 \cdot K}$$

In Matlab, following the scheme of calculation of previous problems, we can do:

```
clc
clear all
close all


N = 50;
A = 2e-4; %m2
S = 0.03; % m2
P =0.05; %m
k = 100; %W/mK
ro = 873; % kg/m3
Cp = 38; % [ J / K · kg ]
L = 0.6;    % [ m ]
```

```
TO = 100;  % [ °C ]
Tair = 25;   % [ °C ]
h = 10;    % [ W / m2 · K ]


Dx = L / (N-1);    % [ cm ]

max_Dt =   (2*k*ro/(Dx^2)-(h*P/A/ro/Cp))^-1 ;
Dt = max_Dt*0.9;


t=0;
T(1) = TO;
T(2:N) = Tair;

Tout = Tair;


%————————————————————————————————————
x = 0: Dx : L ;

Alfa =   k*Dt/ro/Cp/Dx^2;
B = 1 - 2 * Alfa+h*P*Dt/k/ro/Cp;

while t<50
     t = t + Dt
    Taux(1) = TO;
    Taux(N) = T(N-1);
    for i = 2:N-1
         Taux(i) = Alfa*T(i+1) + B * T(i) + Alfa*T(i-1)-h*P*Dt*...
Tair/A/ro/Cp;
     end
     T = Taux;
      plot (x,T)
      drawnow
   end
```

The program can be used to follow the evolution of the temperature profile with
time along the fin.


**Problem 5.20**  A reactive spherical particle of diameter $D = 0.1$ m with a temperature $(T_0)$ of 90 °C, initially equal in all its points, is placed in a reactor maintained at
20 °C $(T_r)$.

(a) Performing a global analysis, that is, assuming that all points on the sphere are
always at the same temperature $(T)$, if an energy balance is proposed, we obtain
that:

$$\text{Input} + \text{Generation} = \text{Output} + \text{Accumulation}$$

$$0 + 0 = h(4\pi R^2)(T - T_r)\Delta t + \left(\frac{4\pi^3}{3}\right)\rho_S C_p(T^{t+1} - T^t)$$

Dividing by $\Delta t$ and taking limits, we get:

$$\frac{dT}{dt} = -\frac{3h(T - T_\mathrm{r})}{R\rho_\mathrm{S}C_\mathrm{p}}$$

with initial value $T = T_0$.

(a.1) Calculate how the temperature of the reactive varies during the first 100 seconds and plot $T$ versus $t$.

(a.2) Calculate what the temperature of the sphere would be after 40 seconds.

Admit in all cases that the temperature of the reactor does not vary.

(b) Performing a local analysis, that is, if it is not accepted that the temperature of all points on the sphere of reactive varies at the same rate but rather that the temperature depends on the radius ($r$), following a procedure analogous to the previous one, we arrive at the following expression:

$$\frac{\delta T}{\delta t} = \frac{1}{r^2 \rho_\mathrm{S} C_\mathrm{p}} \frac{\delta}{\delta r}\left(k_\mathrm{term} r^2 \frac{\delta T}{\delta r}\right)$$

With boundary conditions:

$$-k_\mathrm{term} \frac{\delta T}{\delta r}\bigg]_{r=R} = h(T - T_\mathrm{r})$$

$$-k_\mathrm{term} \frac{\delta T}{\delta r}\bigg]_{r=0} = 0$$

(b.1) Calculate, using an explicit finite difference method, the temperature profile at 40 seconds. (Divide the radius into 30 points), and deduce the stability conditions.

(b.2) Represent the profile obtained in the previous section together with the profile at 40 seconds from section (a).

Data:

$k_\mathrm{term} = 10\,\mathrm{W/m\cdot^\circ C}$, thermal conductivity
$C_\mathrm{p} = 410\,\mathrm{J/kg\cdot K}$, heat capacity
$h = 220\,\mathrm{W/m^2\cdot K}$, heat transfer coefficient with the exterior
$\rho_\mathrm{s} = 2000\,\mathrm{kg/m^3}$, density of the sphere
$T_\mathrm{r} = 20\,^\circ C$, reactor temperature
$T_0 = 90\,^\circ C$, initial temperature of the sphere

**Solution to Problem 5.20**

For details refer the Wiley website at http://www.wiley-vch.de/ISBN9783527354115

(a) In the first part, an ODE should be solved. In the system:

$$\frac{T^{t+1} - T^t}{\Delta t} = -\frac{3h(T^t - T_\mathrm{r})}{R\rho_\mathrm{S}C_\mathrm{p}}$$

$$T^{t+1} = T^t - \frac{3h(T^t - T_\mathrm{r})}{R\rho_\mathrm{S}C_\mathrm{p}}\Delta t$$

Using a time increment of 20 seconds, we can do:

```
clear all
close all

R=0.1% m
kterm = 10 % W/(m °C), thermal conductivity
Cp = 410 %J/(kg K), heat capacity
h = 220 %W/(m2 K), heat transfer coefficient with the exterior
ds = 2000 %kg/m3, density of the sphere
Tr  = 20 %°C, reactor temperature
T0 = 90 %&°C, initial temperature of the sphere

deltat=20; % seconds

T(1)=T0;
t(1)=0;

for i=2:1000
 T(i)=T(i-1)-3*h*(T(i-1)-Tr)/(R*ds*Cp);
 t(i)=t(i-1)+deltat;
end

plot (t,T)
xlabel('Time (s)')
ylabel('Temperature, °C')
```

And we obtain:

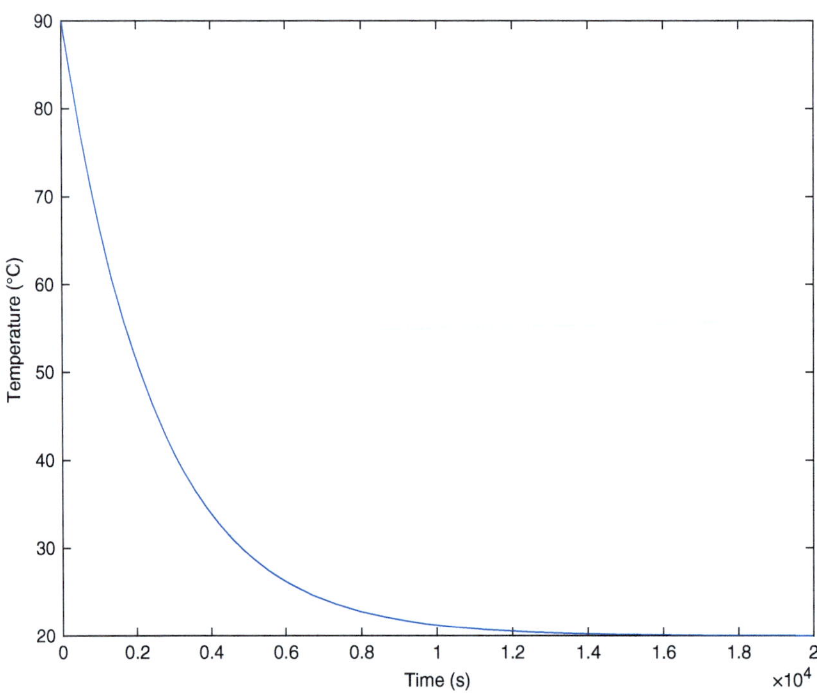

(b) Now we have that:

$$\frac{\delta T}{\delta t} = \frac{k_{\text{term}}}{r^2 \rho_S C_p} \frac{\delta}{\delta r}\left(r^2 \frac{\delta T}{\delta r}\right) = \frac{k_{\text{term}}}{r^2 \rho_S C_p}\left(2r\frac{\delta T}{\delta r} + r^2 \frac{\delta^2 T}{\delta r^2}\right)$$

So, applying FDM:

$$\frac{T_i^{t+1} - T_i^t}{\Delta t} = \frac{k_{\text{term}}}{r^2 \rho_S C_p}\left(2r\frac{T_{i-1}^t - T_i^t}{\Delta r} + r^2 \frac{T_{i-1}^t + T_{i+1}^t - 2T_i^t}{\Delta r^2}\right)$$

Rearranging:

$$T_i^{t+1} = \frac{k_{\text{term}}\Delta t}{r^2 \rho_S C_p}\left(2r\frac{T_{i-1}^t - T_i^t}{\Delta r} + r^2 \frac{T_{i-1}^t + T_{i+1}^t - 2T_i^t}{\Delta r^2}\right) + T_i^t$$

And:

$$T_i^{t+1} = T_{i-1}^t\left(\frac{2}{\Delta r}\frac{k_{\text{term}}\Delta t}{r\rho_S C_p} + \frac{1}{\Delta r^2}\frac{k_{\text{term}}\Delta t}{\rho_S C_p}\right)$$
$$+ T_i^t\left(1 - \frac{2}{\Delta r}\frac{k_{\text{term}}\Delta t}{r\rho_S C_p} - \frac{2}{\Delta r^2}\frac{k_{\text{term}}\Delta t}{\rho_S C_p}\right) + T_{i+1}^t\left(\frac{1}{\Delta r^2}\frac{k_{\text{term}}\Delta t}{\rho_S C_p}\right)$$

The stability conditions of this solution are:

$$\left(\frac{2}{\Delta r}\frac{k_{\text{term}}\Delta t}{r\rho_S C_p} + \frac{1}{\Delta r^2}\frac{k_{\text{term}}\Delta t}{\rho_S C_p}\right) > 0$$

$$\left(1 - \frac{2}{\Delta r}\frac{k_{\text{term}}\Delta t}{r\rho_S C_p} - \frac{2}{\Delta r^2}\frac{k_{\text{term}}\Delta t}{\rho_S C_p}\right) > 0$$

$$\left(\frac{1}{\Delta r^2}\frac{k_{\text{term}}\Delta t}{\rho_S C_p}\right) > 0$$

$$\left(1 - \frac{2}{\Delta r}\frac{k_{\text{term}}\Delta t}{r\rho_S C_p} - \frac{2}{\Delta r^2}\frac{k_{\text{term}}\Delta t}{\rho_S C_p} + \frac{2}{\Delta r}\frac{k_{\text{term}}\Delta t}{r\rho_S C_p} + \frac{1}{\Delta r^2}\frac{k_{\text{term}}\Delta t}{\rho_S C_p} + \frac{1}{\Delta r^2}\frac{k_{\text{term}}\Delta t}{\rho_S C_p}\right) \leq 1$$

The only condition that gives some information is the second one, so:

$$1 > \left(\frac{2}{\Delta r}\frac{k_{\text{term}}}{r\rho_S C_p} - \frac{2}{\Delta r^2}\frac{k_{\text{term}}}{\rho_S C_p}\right)\Delta t$$

And finally:

$$\Delta t < \frac{1}{\left(\frac{2}{\Delta r}\frac{k_{\text{term}}}{r\rho_S C_p} - \frac{2}{\Delta r^2}\frac{k_{\text{term}}}{\rho_S C_p}\right)}$$

Boundary condition at $r = 0$ (center of the particle) is:

$$-k_{\text{term}}\frac{T_{i+1} - T_i}{\delta r}\bigg]_{i=1} = 0$$

So:

$$T_1^{t+1} = T_2^t$$

And at the surface of the particle:

$$-k_{\text{term}} \frac{T_N^t - T_{N-1}^t}{\Delta r} = h\left(T_{N-1}^t - T_r\right)$$

And then:

$$T_N^t = T_{N-1}^t - h\left(T_{N-1}^t - T_r\right)\frac{\Delta r}{k_{\text{term}}}$$

Using a procedure similar to that shown in previous problems, we can use Matlab to solve this system.

# Part III

# Catalytic and Multiphase Reactor Design

# 6

# Reaction Rate in Catalytic Processes

## Summary of Equations for the Catalytic Reactor Design

### Rate in Heterogeneous Systems

The rate of a chemical process is defined by the following expression:

$$r_A = \frac{g_A}{(\text{extensive property})}$$

Different properties are used for defining reaction rate and corresponding valid units:

| Extensive property | Rate equation symbol (valid units) | Kinetic constant (valid units) assuming first-order kinetics | Relationship with generation term |
|---|---|---|---|
| External surface of catalyst | $(-r_A)$, $\text{kmol/s·m}^2_{\text{catalyst}}$ | $k$, $\text{m}^3_{\text{F}}/\text{m}^2_{\text{catalyst}}\text{·s}$ | $(-r_A) = g_A/S_{\text{ext}}$ |
| Weight of catalyst | $(-r'_A)$, $\text{kmol/s·kg}_{\text{catalyst}}$ | $k'$, $\text{m}^3_{\text{F}}/\text{kg}_{\text{catalyst}}\text{·s}$ | $(-r'_A) = g_A/W$ |
| Volume of reactor mix | $(-r''_A)$, $\text{kmol/s·m}^3_{\text{R}}$ | $k''$, $\text{m}^3_{\text{F}}/\text{m}^3_{\text{R}}\text{·s}$ | $(-r''_A) = g_A/V$ |
| Volume of catalyst | $(-r'''_A)$, $\text{kmol/s·m}^3_{\text{catalyst}}$ | $k'''$, $\text{m}^3_{\text{F}}/\text{m}^3_{\text{catalyst}}\text{·s}$ | $(-r'''_A) = g_A/V_{\text{p}}$ |

$$g_A = r_A S_{\text{ext}} = r'_A W = r''_A V = r'''_A V_{\text{p}}$$

$$r'_A = S_S \cdot r_A \qquad k' = S_S \cdot k$$
$$r''_A = r_A \cdot a \qquad k'' = k \cdot a$$
$$r'''_A = r_A \cdot a/\varepsilon_S \qquad k''' = k \cdot a/\varepsilon_S$$
$$r''_A = r'''_A \cdot (1 - \varepsilon_b) \qquad k'' = k''' \cdot (1 - \varepsilon_b)$$
$$r'''_A = r'_A \cdot \rho_s \qquad k''' = k' \cdot \rho_s$$

with (valid units):

$S_S$ = specific surface $(m^2_{surface}/kg_{solid})$
$a$ = ratio $m^2_{surface}/m^3_{reactor}$
$\varepsilon_s$ = solid load $(m^3_{solid}/m^3_{reactor})$
$\varepsilon_b$ = bed porosity $(m^3_{holes}/m^3_{reactor})$, i.e., $(1 - \varepsilon_b) = m^3_{solid}/m^3_{reactor} = \varepsilon_S$
$\rho_s$ = catalyst density $(kg_{solid}/m^3_{solid})$

Note that on many occasions, "$a$" refers to the surface area by weight of catalyst, for example, $m^2/kg$ of catalyst, so the relationships will differ from those shown.

### Rate of External Diffusion

$$(-r_A) = -\frac{g_A}{S_{ext}} = k_C \cdot (C_{Ag} - C_{As})$$

with $k_C$ in units of $(kmol/s \cdot m^2_C)/(kmol/m^3_F) = m^3_F/m^2_C \cdot s = m/s$.

Many times, the product $(k_C a)$ is used, with $k_C$ in $m^3_F/m^2_C \cdot s$ and the ratio "$a$" in $m^2/m^3_R$, so valid units for $(k_C a)$ would be $s^{-1}$.

### Dimensionless Numbers and Their Relationship

$$Sh = \frac{(k_C d_p)}{D_A}$$

In a packed bed, usually Sh is expressed using the porosity of the bed:

$$Sh = \frac{(k_C d_p)}{D_A} \frac{\varepsilon_b}{1 - \varepsilon_b}$$

$$Re = \frac{d_p \cdot \rho \cdot u}{\mu}$$

$$Sc = \frac{\mu}{\rho \cdot D_A}$$

$$j_D = \frac{Sh}{Sc^{\frac{1}{3}} \cdot Re}$$

$$j_D = \frac{0.725}{Re^{0.41} - 0.15}$$

$$We = \frac{\left(-r_A'''\right)_{observed} \cdot L^2}{D_e \cdot C_{As}}$$

$u$ = Superficial rate, calculated as the flow/total section of the reactor (m/s)
$k_C$ = Individual mass transfer coefficient $(kmol/s \cdot m^2/kmol/m^3)$
$S_{ex}$ = External surface of the particle $(m^2)$
$C_{Ag}$ = Concentration of reagent A in the fluid $(kmol/m^3)$
$C_{AS}$ = Concentration on the surface $(kmol/m^3)$
$D_A$ = Coefficient of molecular diffusion $(m^2/s)$
$d_p$ = Effective diameter of the particle, that is, the diameter of a sphere that has the same outer area

**Internal Diffusion Effect**

For a pseudo-first-order chemical reaction:

$$\frac{C_A}{C_{As}} = \frac{\cosh(m(L-x))}{\cosh(mL)}$$

$$\eta = \frac{\overline{(-r_A)}}{(-r_A)_{\text{without diffusion effects}}}$$

$$\eta = \frac{\tanh(mL)}{mL}$$

For a first-order kinetics: $(-r_A)_{\text{without diffusion effects}} = k \cdot C_{As}$ with concentration corresponding to the surface. In this case: $(-r_A) = k \cdot C_{As} \cdot \eta$.

In previous equations, $L$ is the generalized length:

$$L = \frac{\text{Volume of the particles}}{\text{External surface available for entry and difussion of the reagent}}$$

That for the usual geometries is:

| Flat particles (reactant diffusing by both sides) | Flat particles (reactant diffusing by one side) | Spheres | Cylinder (very long) |
|---|---|---|---|
| $L = \text{thickness}/2$ | $L = \text{thickness}$ | $L = R/3$ | $L = R/2$ |

$$m = \sqrt{\frac{k'''}{D_e}} \quad \text{(first order)}$$

$$m = \left[\frac{3k''' \cdot C_{As}}{2 \cdot D_e}\right]^{\frac{1}{2}} \quad \text{(second order)}$$

$$m = \left[\frac{k''' \cdot C_{As}^{n-1}(n+1)}{2 \cdot D_e}\right]^{\frac{1}{2}} \quad \text{($n$-th order)}$$

Note that valid units of the diffusion coefficient, by definition, are $m^3_F/m_c \cdot s$, which are usually indicated as $m^2/s$. Correspondingly, valid units for $k_{AL}$ are $m^3_F/m^2_c \cdot s$, usually m/s. In this way, units of $k'''$ should be $m^3_F/m^3_c \cdot s$. In the previous expressions, the kinetic constant, $k'''$, should be given in units of inverse of time, in such a way that "$m$" is in units of inverse of length and "mL" is dimensionless.

In the case of reversible reactions, the following expression is introduced for a first-order reaction:

$$m = \sqrt{\frac{k'''_{\text{forward}}}{X_{Aeq}D_e}}$$

being $X_{Aeq}$ the conversion of the reaction in the equilibrium.

For kinetics other than first-order, $k''_{ap}$ is the first-order apparent rate constant. In many systems, where chemical reaction is second-order, we will do:

$$(-r''_A) = k'' C_A C_B = (k'' C_B)C_A = k''_{ap} C_A$$

In general, for any kinetic law, we will group in $k''_{ap}$ all the variables except $C_A$:

$$(-r''_A) = k'' C_A^n C_B^m = \left(k'' C_A^{n-1} C_B^m\right) C_A = k''_{ap} C_A$$

**Combination of Resistances**

In the general case, all steps in the chemical reaction process can affect the reaction rate, so the flow of reactant through the external layer could be expressed using $(kmol/s \cdot m^2)$:

$$(-r_A) = -\frac{g_A}{S_{ext}} = k_C \cdot (C_{Ag} - C_{As})$$

In the common case, $C_{As}$ does not equal zero. At the same time, the diffusion through the pores and chemical reaction rate are given by:

$$\overline{(-r_A)} = k C_{As} \eta$$

Obviously, there is no accumulation of reactant species A on the external surface, so these two rates must be equal, so:

$$(-r_A) = k_C \cdot (C_{Ag} - C_{As}) = k C_{As} \eta$$

Note that the previous equality is absolutely always true, regardless of the most important resistance to the process. If one of the resistances is very small, for example, when $k_C$ is very large, the difference $(C_{Ag} - C_{As})$ will be very small so that the concentration profiles in the film will be very small. From the previous equation, doing the combination of resistances, we get:

$$(-r_A) = \frac{(C_{Ag} - C_{As})}{\frac{1}{k_C}} = \frac{C_{As}}{\frac{1}{k\eta}} = \frac{C_{Ag}}{\frac{1}{k_C} + \frac{1}{k\eta}}$$

We can define a global efficiency factor as:

$$(-r_A) = \frac{C_{Ag}}{\frac{1}{k_C} + \frac{1}{k\eta}} = k C_{Ag} \Omega$$

$$\Omega = \frac{1/k}{\frac{1}{k_C} + \frac{1}{k\eta}} = \frac{1}{\frac{k}{k_C} + \frac{1}{\eta}}$$

Note that $\Omega$ has no units and represents an efficiency for the whole process.

**Process of Absorption (No Reaction)**

$$(-r_A) = \frac{p_A}{\frac{1}{k_{Ag}} + \frac{H_A}{k_{AL}}}$$

$$(-r''_A) = (-r_A) \cdot a = k_{Ag} \cdot a \cdot (p_A - p_{Ai}) = k_{AL} \cdot a \cdot (C_{Ai} - C_A)$$

In the equation, valid units for $k_{Ag}$ are, for example, $(kmol/s \cdot m^2_{interface} \cdot atm)$, whereas $k_{AL}$ can be used in $m^3_L/m^2_{interface} \cdot s = m/s$.

Rate of the absorption process can also be calculated using global mass transfer coefficients, with driving force $\left(p_A - p^*_A\right)$ and $\left(C^*_A - C_A\right)$, using the equations:

$$\frac{1}{K_{Ag}} = \frac{1}{k_{Ag}} + \frac{H_A}{k_{AL}}$$

$$\frac{1}{K_{AL}} = \frac{1}{k_{AL}} + \frac{1}{H_A \cdot k_{Ag}}$$

where pressures and concentrations refer to equilibrium, that is:

$$p_A^* = H_A \cdot C_A$$

$$C_A^* = p_A/H_A$$

In this way:

$$\left(-r_A''\right) = (-r_A) \cdot a = K_{Ag} \cdot a \cdot \left(p_A - p_A^*\right) = K_{AL} \cdot a \cdot \left(C_A^* - C_A\right)$$

In the equation, valid units for $K_{Ag}$ are, for example, (kmol/s·m$^2_{\text{interface}}$·atm), whereas $K_{AL}$ can be used in m$^3_{\text{L}}$/m$^2_{\text{interface}}$·s = m/s.

For more details, please consult Conesa (2019).

**Problem 6.1** The reaction $A + B \rightarrow$ products takes place in an isothermal, constant-density fluid in a continuous fluidized bed reactor with a well-mixed solid catalyst. The rate law for the reaction is as follows:

$$-r_A = k\, C_A\, C_B \ (\text{mol/kg} - \text{cat/s})$$

where the reaction mechanism is unknown. Two catalyst varieties, Catalyst 1 and Catalyst 2, are available, with identical physical and chemical properties except that Catalyst 2 has double the density of active sites compared to Catalyst 1. Steady-state data from the experiments are provided:

| | $W$ (kg) | $Q_0$ (l/s) | $C_{A0}$ (mol/l) | $C_{B0}$ (mol/l) | $X_A$ |
|---|---|---|---|---|---|
| Catalyst 1 | 150 | 100 | 0.10 | 0.10 | 0.450 |
| Catalyst 2 | 100 | 200 | 0.15 | 0.20 | 0.667 |

(a) Determine the apparent rate constants $k$ for each run, using appropriate units.
(b) Based on your answer to part (a) and knowledge of the two catalysts, propose a plausible mechanism for how the active sites of the catalyst facilitate the reaction. If you cannot fully answer part (a), explain how the answer to part (a) would address this question.

**Solution to Problem 6.1**

For details refer the Wiley website at http://www.wiley-vch.de/ISBN9783527354115

(a) In a fluidized bed, the solid phase behaves as a CSTR, and the fluid phase has a particular bubble behavior that is between a plug flow and a perfectly mixed flow. Usually, one of these two extremes is assumed for the gas phase. Let us assume CSTR for both the solid and gas. A mass balance in the gas phase is:

$$n_{A0} - n_A = (-k'C_A C_B) \cdot W$$

With the kinetic constant $k'$ expressed in the units of mass of catalyst. From this:

$$Q_0 C_{A0} - Q_0 C_{A0}(1 - X_A) = (-k'C_{A0}(1 - X_A)(C_{B0} - C_{A0}X_A)) \cdot W$$

Solving for the constant:

$$\frac{Q_0 C_{A0} X_A}{(C_{A0}(1 - X_A)(C_{B0} - C_{A0}X_A)) \cdot W} = k'$$

For both catalysts:

Catalyst 1: $k' = 9.917 \, l^2/\text{mol·kg·s}$

Catalyst 2: $k' = 40.080 \, l^2/\text{mol·kg·s}$

(b) The difference between both constants could be due to the chemical reaction step: Catalyst 2 has double the number of active sites, but the $k'$ value is almost four times that of Catalyst 1. This could indicate that chemical reaction is the controlling step. External diffusion (from the gas to the catalyst surface) is probably not affecting the reaction.

**Problem 6.2**   The first-order reaction A → products is conducted in an isothermal liquid using a solid catalyst in a packed-bed, tubular reactor. The fluid flow past each particle is in the turbulent regime, characterized by a Reynolds number (Re) of $10^4$. Steady-state experiments are performed with different average particle diameters ($d_p$) and the following data are obtained:

| $d_p$ (cm) | $W$ (kg) | $Q_0$ (l/s) | $X_A$ |
| --- | --- | --- | --- |
| 0.10 | 100 | 100 | 0.92 |
| 0.15 | 80 | 100 | 0.73 |

(a) Based on the provided data, assess the apparent influence of intraparticle diffusion resistance on the observed kinetics.
(b) Is it necessary to consider the impact of external mass transfer resistance? Provide a quantitative explanation.

**Solution to Problem 6.2**

(a) We will calculate reaction rate in each run based on the weight of the catalyst. Doing a differential mass balance for the reacting species:

$$\text{Input} + \text{Generation} = \text{Exit} + \text{Accumulation}$$

$$(n_A) + (r'_A) \, dW_{\text{cat}} = (n_A + dn_A) + 0$$

$$\frac{dn_A}{dW_{cat}} = \frac{Q_0 dC_A}{dW_{cat}} = -\frac{Q_0 C_{A0} dX_A}{dW_{cat}} = r_A'$$

In this way, we can estimate $-r_A' = \frac{Q_0 C_{A0} \Delta X_A}{\Delta W_{cat}} = \frac{Q_0 C_{A0}(X_A - 0)}{(W_{cat} - 0)} = \frac{Q_0 C_{A0} X_A}{W_{cat}}$

For both cases, we found $\frac{(-r_A')_1}{(-r_A')_2} = \frac{0.92 C_{A0}}{0.9125 C_{A0}} = 1.008$

As we see, there are not many differences between these two cases, so probably the internal diffusion is not interfering with the reaction rate.

(b) In the turbulent regime ($Re > 4 \cdot 10^3$), mass transfer in the film is usually very fast and is not necessary to be considered in the calculation of the reaction rate.

**Problem 6.3**  The gas-phase reaction A → products is conducted at a constant temperature in a well-mixed, constant-volume batch reactor containing a solid catalyst. The reactor has a fluid volume of $V = 10\,l$ and a catalyst weight of $W = 5\,kg$, with a density of 2.0 kg/l. The catalyst particles have a slab geometry, with a mean thickness of 0.4 cm (l = 0.2 cm). For the encountered concentrations of A, the reaction follows zero-order kinetics.

It can be demonstrated that the slab geometry with zero-order kinetics is unique in the following manner: as the Thiele modulus (mL) increases, the effectiveness factor transitions directly from 1 to 1/mL, without an intermediate region of pore diffusion resistance.

The concentration of A is measured at 2-minute intervals, resulting in the following data:

| Time (min) | 0 | 2 | 4 | 6 | 8 | 10 | 12 | 14 |
|---|---|---|---|---|---|---|---|---|
| $C_A$ (mol/l) | 0.1000 | 0.0800 | 0.0600 | 0.0400 | 0.0210 | 0.0078 | 0.001 | 0.000 |

(a) Estimate the value of the rate constant $k$, considering the catalyst weight as the basis.

(b) Estimate the value of $D_e$, the effective diffusion coefficient of A, in $cm^2/s$.

**Solution to Problem 6.3**

For details refer the Wiley website at http://www.wiley-vch.de/ISBN9783527354115

(a) In the batch reactor:

$$(r_A') \cdot W = \frac{dC_A}{dt}$$

$$dC_A = k_{ap}' \cdot W \cdot dt$$

$$(C_{A0} - C_A) = k_{ap}' \cdot W \cdot t$$

We can calculate the value of the constant for each data point:

| Time (min) | 0 | 2 | 4 | 6 | 8 | 10 | 12 | 14 |
|---|---|---|---|---|---|---|---|---|
| $C_A$ (mol/l) | 0.1 | 0.08 | 0.06 | 0.04 | 0.021 | 0.0078 | 0.001 | 0 |
| $k_{ap}'$ (mol/kg/min) | — | 0.002 | 0.002 | 0.002 | 0.001975 | 0.001844 | 0.00165 | |

The average value of the constant is $k'_{ap} = 0.0072 \text{ mol/kg·min}$.

(b) In a slab where zero-order reaction is taking place:

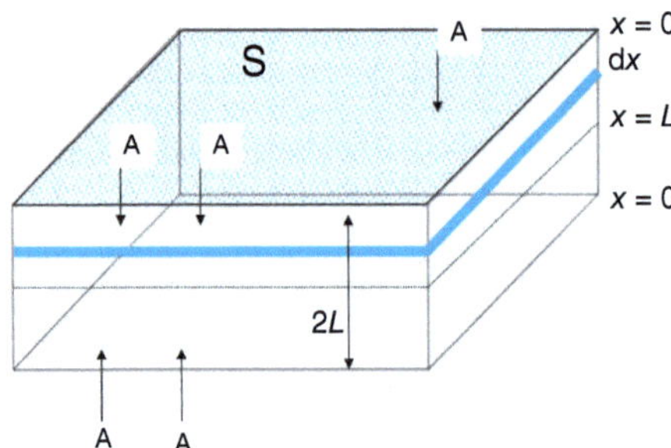


Boundary conditions:

$$C_A = C_{As} \quad \text{at } x = 0$$

$$\frac{dC_A}{dx} = 0 \quad \text{at } x = L$$

From a balance in the volume differential, we can check that:

$$\frac{dN_A}{dx} = r'''_A = k'''$$

With NA in mol/s·m$^2$c, $x$ in m, and $r_A$ in mol/s·m$^3$c
And using Fick's law:

$$\frac{dC_A^2}{dx^2} = \frac{k'''}{D_e}$$

Doing two integrations of the previous equation:

$$\frac{dC_A}{dx} = \frac{k'''}{D_e}x + M_1$$

$$C_A = \frac{k'''}{D_e}\frac{x^2}{2} + M_1 x + M_2$$

Where $M_1$ and $M_2$ are integration constants. Using the boundary conditions:

$$C_{As} = M_2$$

$$0 = \frac{k}{D_e}L + M_1$$

And then:

$$C_A = \frac{k'''}{D_e}\frac{x^2}{2} + \left(-\frac{k'''}{D_e}\right)Lx + C_{As}$$

Using the relations $C = C_A/C_{As}$ and $X = x/L$ we find that:

$$C = \left(\frac{k'''}{2C_{As}D_e}L^2\right)X^2 + 1 - \left(\frac{k'''}{2C_{As}D_e}L^2\right)$$

The product $\left(\sqrt{\frac{k'''}{2C_{As}D_e}}L\right)$ corresponds to the Thiele modulus for zero-order kinetics, (mL), and so:

$$C = (mL)^2 X^2 + 1 - (mL)^2$$

It is clear that for mL < 1, previous equation yields $C > 0$ at $X = 0$, and the effectiveness factor, defined as the ratio of observed rate to intrinsic rate, is equal to 1 because everywhere in the catalyst the observed rate is $k'''$, unaffected by reactant concentration, which is diminishing yet remaining positive toward the center. This trend would hold true until mL = 1, when the reactant concentration becomes zero at the center. In this sense, this equation holds as long as the concentration does not become zero in the slab. In the case that $C = 0$ before reaching the center of the particle, what corresponds to the case that mL > 1. As we observe in the calculation in (a), the apparent kinetic constant ($k'_{ap} = k' \cdot \eta$) changes with time, in such a way that we can estimate the effectiveness factor as the ratio $k'$ (at time $t$)/$k'$ (at time zero). The values are shown:

| Time (min) | 0 | 2 | 4 | 6 | 8 | 10 | 12 | 14 |
|---|---|---|---|---|---|---|---|---|
| $k'_{ap}$ (mol/kg/min) | — | 0.002 | 0.002 | 0.002 | 0.001975 | 0.001844 | 0.00165 | |
| Efectiveness factor | — | 1 | 1 | 1 | 0.9875 | 0.922 | 0.825 | |

As we can check, $k'$ equals 0.002 mol/kg/min. Multiplying by the density of the solid catalyst:

$$k''' = 0.002 \, \frac{\text{mol}}{\text{kg} \cdot \text{min}} \cdot 2 \cdot 10^{-3} \, \frac{\text{kg}}{\text{cm}^3} = 4 \cdot 10^{-6} \, \frac{\text{mol}}{\text{cm}^3 \cdot \text{min}}$$

From these data, we can calculate the diffusivity bearing in mind $\eta = 1/\text{mL}$. From the data at times between 8 and 12, we have that:

$$\eta = \frac{1}{\sqrt{\frac{k'''}{2C_{As}D_e}}\,L}$$

and:

$$D_e = k''' \cdot \eta^2 \cdot \frac{L^2}{2C_{As}}$$

giving an average value of $D_e = 6.68 \cdot 10^{-4}$ cm$^2$/min $= 1.11 \cdot 10^{-5}$ cm$^2$/s.

**Problem 6.4** At a point inside a packed-bed reactor, the partial pressures of A and B are 0.96 and 1.44 atm, respectively. The temperature is 300 °C and the total absolute pressure is 8 atm both conditions remain constant inside the catalyst. The reaction and its intrinsic kinetics are:

$$A + 2B \rightarrow C \quad r' = k' p_A \sqrt{p_B} = 1.2 \cdot 10^{-6} \, \frac{\text{mol}}{\text{g} \cdot \text{s} \cdot \text{atm}^{1.5}} p_A \sqrt{p_B}$$

The catalyst is spherical with a radius of 0.635 cm, its porosity is 0.45, its density is 1.12 g/cm$^3$, and it only has micropores with an average radius of 100 Å. The molecular weights of A and B are 120 and 2, respectively. Assuming that there are no external resistances to mass transfer, calculate the point reaction rate.

**Solution to Problem 6.4**

For details refer the Wiley website at http://www.wiley-vch.de/ISBN9783527354115

In different books (e.g. Levenspiel (1999)), we can find the relation between effective diffusivity $(D_e)$ and Knudsen diffusivity $(D_{KA})$, as well as the equation to calculate the later.

$$D_{eA} = D_{KA} \cdot \varepsilon^2$$

$$D_{KA} = \frac{4}{3} r \left( \frac{2 R_g T}{P \cdot M_A} \right)^{0.5}$$

where $r$ is the radius of the pore and $M_A$ is the molecular weight of the species being diffused. In our case, $D_{eA} = 4.3 \cdot 10^{-3}$ cm$^2$/s and $D_{eB} = 0.0333$ cm$^2$/s.

Before continuing, let us express the rate equation as a function of the concentrations:

$$r' = k' p_A \sqrt{p_B} = k' \left( \frac{C_A}{R_g T} \right) \left( \frac{C_B}{R_g T} \right)^{0.5} = \cdots = 12.2 \frac{\text{mol}}{\text{g} \cdot \text{s} \cdot \left( \frac{\text{mol}}{\text{cm}^3} \right)^{1.5}} \cdot C_A \cdot C_B^{0.5}$$

In terms of volume of catalyst:

$$k''' = k' \cdot \rho_S = 12.2 \frac{\text{mol}}{\text{g} \cdot \text{s} \cdot \left( \frac{\text{mol}}{\text{cm}^3} \right)^{1.5}} \cdot 1.12 \frac{\text{g}}{\text{cm}_C^3} = 13.7 \frac{\text{mol}}{\text{cm}_C^3 \cdot \text{s} \cdot \left( \frac{\text{mol}}{\text{cm}^3} \right)^{1.5}}$$

The pseudo-first-order for "A" will be given by:

$$r_A''' = k_{ap\,A}''' \cdot C_A = \left( k''' \cdot C_B^{0.5} \right) \cdot C_A$$

So:

$$k_{ap\,A}''' = \left( k''' \cdot C_B^{0.5} \right)$$

$$= \left( 13.7 \frac{\text{mol}}{\text{cm}_C^3 \cdot \text{s} \cdot \left( \frac{\text{mol}}{\text{cm}^3} \right)^{1.5}} \cdot \left( \frac{1.44}{82 \cdot 573} \right)^{0.5} \right) = 7.58 \cdot 10^{-2}\ \text{s}^{-1}$$

We can now estimate effectiveness factors for A:

$$(\text{mL})_A = \left( \frac{k_{ap\,A}'''}{D_{eA}} \right)^{\frac{1}{2}} \cdot \frac{R}{3} = 0.511 \rightarrow \eta_A = 0.921$$

Similarly, for B:

$$r_B''' = k_{ap\,B}''' \cdot C_B = \left( k''' \cdot C_A \cdot C_B^{-0.5} \right) \cdot C_B$$

In such a way that the apparent constant for B is:

$$k'''_{\text{ap B}} = \left(k''' \cdot C_A \cdot C_B^{-0.5}\right) = \dots = 5.05 \cdot 10^{-2}\ \text{s}^{-1}$$

And the efficiency:

$$(\text{mL})_B = \left(\frac{k'''_{\text{ap B}}}{D_{\text{eB}}}\right)^{\frac{1}{2}} \cdot \frac{R}{3} = 0.15 \rightarrow \eta_B = 0.993$$

In this way, the reaction rate would be:

$$r'''_A = k'''_{\text{ap A}} \cdot C_A \cdot \eta_A = 1.43 \cdot 10^{-3} \frac{\text{mol}}{\text{cm}^3 \cdot \text{s}}$$

$$r'''_B = k'''_{\text{ap B}} \cdot C_B \cdot \eta_B = 1.54 \cdot 10^{-3} \frac{\text{mol}}{\text{cm}^3 \cdot \text{s}}$$

Estequiometrically, we have the relationship $r_B = 2 \cdot r_A$, so $r_A$ and $r_B/2$ both should be equal; the minor one is the actual rate $r' = 7.68 \cdot 10^{-4}\ \frac{\text{mol}}{\text{cm}^3 \cdot \text{s}}$.

**Problem 6.5** A cylindrical catalytic converter measures 1/8 in. in diameter, 1/4 in. in length, and weighs 0.0551 g. It has a surface area of 12.9 m$^2$, and its average pore radius is 32 Å. Estimate the effective diffusivity, in cm$^2$/s, at 250 °C for a component whose molecular weight is 78.

**Solution to Problem 6.5**

To calculate the effective diffusivity, one must first know which mechanism is the predominant. In this case, where there is a pore size of 3.2 nm (between 2 and 100 nm), the Knudsen diffusivity will predominate. Therefore, the effective diffusivity of the component whose molecular weight is 78 can be calculated as:

$$D_e = \varepsilon D_K / \tau$$

where:

      $\varepsilon$ is the porosity of the catalyst

      $D_K$ is the Knudsen diffusivity

      $\tau$ is the tortuosity of the pore

Since there are no values to calculate the tortuosity, a good approximation is to assume that $\tau = 1/\varepsilon$ so that the expression for $D_e$ becomes:

$$D_e = \varepsilon^2 D_K$$

The porosity of the catalyst is calculated assuming that the pores are cylindrical, we have:

$$\bar{r} = \left(\frac{2\varepsilon}{\rho_s \cdot S_S}\right)$$

where $r$ is the average pore radius, $\varepsilon$ is the porosity of the catalyst, $\rho_s$ is the bulk density of the catalyst (mass of solid/total volume including pores), and $S_S$ is the internal specific surface.

$$\text{Volume of particles} = \pi R^2 \cdot \text{length}$$

$$= \pi \cdot \left(\frac{0.3175}{2}\right)^2 \cdot 0.635 = 0.05\,\text{cm}^3 = 5 \cdot 10^{-8}\,\text{m}^3$$

$$\rho_s\left(\frac{\text{g}}{\text{m}^3}\right) = \frac{0.0551\,\text{g}}{5 \cdot 10^{-8}\,\text{m}^3} = 1\,095\,974.71\,\frac{\text{g}}{\text{m}^3}$$

$$S_S\left(\frac{\text{m}^2}{\text{g}}\right) = 234.11$$

Clearing, the porosity is calculated as $\varepsilon = 0.4105$

On the other hand, the Knudsen diffusivity can be calculated as:

$$D_K = Kr\left(\frac{T}{M_A}\right)^{0.5}$$

where $K$ is a constant that is $9.7 \cdot 10^{-5}$, $r$ is the average pore radius in angstroms, $T$ is the temperature in K, and $M_A$ is the molecular weight of the compound diffusing through the catalyst in g/mol.

$$D_K\left(\frac{\text{cm}^2}{\text{s}}\right) = 0.008\,038$$

And therefore, the effective diffusivity is:

$$D_e\left(\frac{\text{cm}^2}{\text{s}}\right) = 0.001\,354\,892$$

**Problem 6.6**  A first-order heterogeneous irreversible reaction takes place within a spherical granule of catalyst that is uniformly impregnated with platinum: A → B. For a position just midway along the radius of the granule ($r = 0.5R$), the concentration of the reactant is equal to one-tenth of its concentration on the outer surface of the catalyst. The concentration of A at the outer surface of the catalyst is 0.001 M. The particle diameter is 0.002 cm, and the effective diffusion coefficient is 0.1 cm²/s.

(a) What is the concentration of A at a distance of 0.0003 cm from the outer surface of the catalyst?
(b) If the effectiveness factor is desired to be equal to or greater than 0.8, to what size should the diameter of the catalyst be reduced?

**Solution to Problem 6.6**

(a) During the diffusion of reactant "A" through a particle, the following is fulfilled:

$$\frac{C_A}{C_{As}} = \frac{\cosh(m(L - x))}{\cosh(mL)}$$

For spheres, $L = R/3$. At the point where we have data:

$$\frac{\left(\frac{C_{As}}{10}\right)}{C_{As}} = \frac{\cosh\left(m\left(\frac{R}{3} - \frac{R}{2}\right)\right)}{\cosh\left(\frac{mR}{3}\right)}$$

From that expression, we can calculate $(mR/3) = 6$. As we know $R$, we obtain $m = 6 \cdot 3/0.001 = 9000\,\text{cm}^{-1}$

At the point where we should calculate $C_A$:

$$\frac{C_A}{C_{As}} = \frac{\cosh\left(m\left(\frac{R}{3} - 0.000\,03\right)\right)}{\cosh\left(\frac{mR}{3}\right)}$$

Obtaining $C_A = 4.1 \cdot 10^{-4}$ M.

(b) Using the expression:

$$\eta = \frac{\tanh(\text{mL})}{\text{mL}}$$

We can check that for $\eta > 0.8$ we obtain mL $< 1.25$ and so $L < 1.39 \cdot 10^{-4}$ cm, and this corresponds to $R < 4.17 \cdot 10^{-4}$ cm.

**Problem 6.7**  A tubular reactor is completely covered inside by a film of a porous catalyst with a thickness of 1 mm. Its thickness can be considered insignificant in relation to the radius of the reactor. The density of the catalytic layer is $1.3\,\text{g/cm}^3$, and its effective diffusivity is $0.006\,\text{cm}^2/\text{s}$. At a point inside this reactor, we have 0.015 for the mole fraction of A at 1.1 atm and 250 °C. At the operating conditions: $(k_C a) = 0.023\,\text{l/s} \cdot \text{g}_{cat}$. The catalytic intrinsic rate for the overall reaction 2A → Products is given by:

$$r' = k'C_A^2 = 6.07\,\frac{1^2}{\text{g}_{cat} \cdot \text{s} \cdot \text{mol}}C_A^2$$

Determine the reaction rate and the ratio of the concentration on the surface to the global.

**Solution to Problem 6.7**

For details refer the Wiley website at http://www.wiley-vch.de/ISBN9783527354115

At the selected point, we have:

$$k_C a \cdot (C_{Ag} - C_{As}) = k'C_{As}^2\eta_A$$

With:

$$p_A = 0.015 \cdot 1.1 = 0.0165\,\text{atm}$$

$$C_{Ag} = \frac{p_A}{R_g T} = 3.84 \cdot 10^{-4}\,\frac{\text{mol}}{\text{l}}$$

$$0.023\,\frac{1}{\text{s} \cdot \text{g}_{cat}} \cdot \left(3.84 \cdot 10^{-4}\,\frac{\text{mol}}{\text{l}} - C_{As}\right) = 6.07\,\frac{1^2}{\text{g}_{cat} \cdot \text{s} \cdot \text{mol}}C_{As}^2\eta$$

And, for second-order reaction:

$$mL = \left[ \frac{3k'''C_{As}}{2D_e} \right]^{\frac{1}{2}} L$$

Remember that:

$$k''' = k' \cdot \rho_s = 6.07 \frac{L^2}{g_{cat} \cdot s \cdot mol} \cdot 1.3 \frac{g_{cat}}{cm^3_{cat}} = 7.891 \frac{L^2}{cm^3_{cat} \cdot s \cdot mol}$$

For a layer of catalyst with reactant diffusing only by one side, $L$ = thickness = 0.1 cm. We will find all the values by a trial-and-error strategy:

We finally get $\eta = 0.997$ and $C_{AS} = 0.000\,351$ mol/l. We can now calculate the reaction rate:

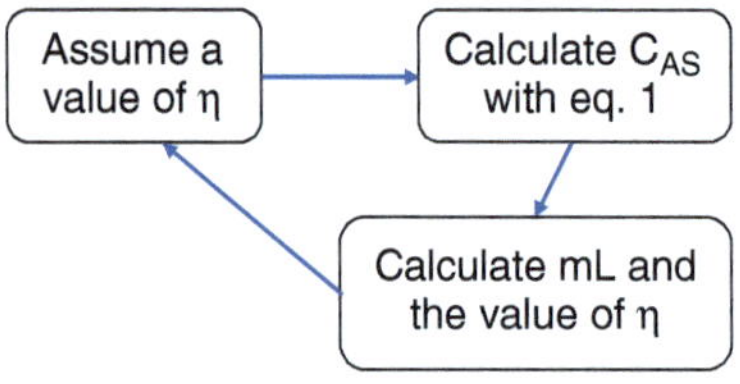



$$(-r'_A) = k'C^2_{As}\eta_A = 7.48 \cdot 10^{-7} \frac{mol}{s \cdot g_{cat}}$$

**Problem 6.8**    For the catalytic reaction of A to produce B and C, the internal concentration profiles are known in mol/l:

$$C_A = 0.2 + 0.18 \cdot \left( \frac{r}{R} \right)^2$$

$$C_B = 0.18 - 0.10 \cdot \left( \frac{r}{R} \right)^2$$

$$C_C = 0.1 - 0.025 \cdot \left( \frac{r}{R} \right)^2$$

where $R$ is the radius of the spherical catalyst grain. It has been determined that there are no significant differences in internal temperature. The intrinsic rate constants, reaction orders, and stoichiometry of the reactions involved are unknown. However, the effective diffusivities of A, B, and C have been determined to be 0.010, 0.014, and 0.016 cm²/s, respectively.

(a) Calculate the global selectivity (defined as the ratio between the rate of production of B and the rate of production of C).
(b) Determine the local selectivities in the center of the catalyst and in the limit near the outer surface of the catalyst.
(c) Determine the effectiveness factor for the consumption of reagent A.

NOTE: If any item cannot be calculated with the data provided, justify it.

**Solution to Problem 6.8**

For details refer the Wiley website at http://www.wiley-vch.de/ISBN9783527354115

(a) and (b) With the proportioned profiles, we can calculate:

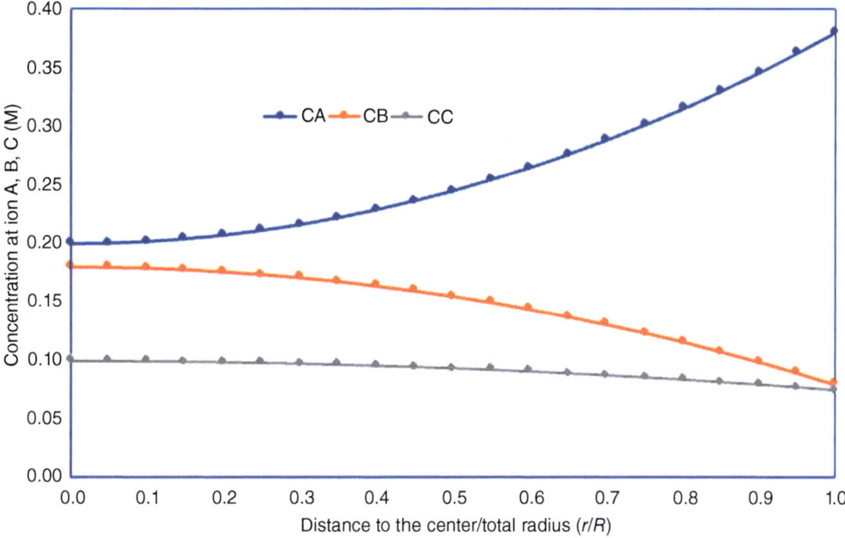


And also the average concentration of each species, being:

$$\overline{C_A} = 0.2615 \text{ M}$$

$$\overline{C_B} = 0.1458 \text{ M}$$

$$\overline{C_C} = 0.0914 \text{ M}$$

Defining the selectivity as the ratio of reaction rates:

$$S_{\frac{B}{C}} = \frac{r_B}{r_C} = \frac{kC_A^n b\eta_A}{kC_A^n c\eta_A} \approx \frac{\overline{C_B}}{\overline{C_C}} = \frac{0.1458}{0.0914} = 1.6$$

For the local selectivities, we can do similar reasoning and calculate the ratios among $C_B$ and $C_C$, obtaining:

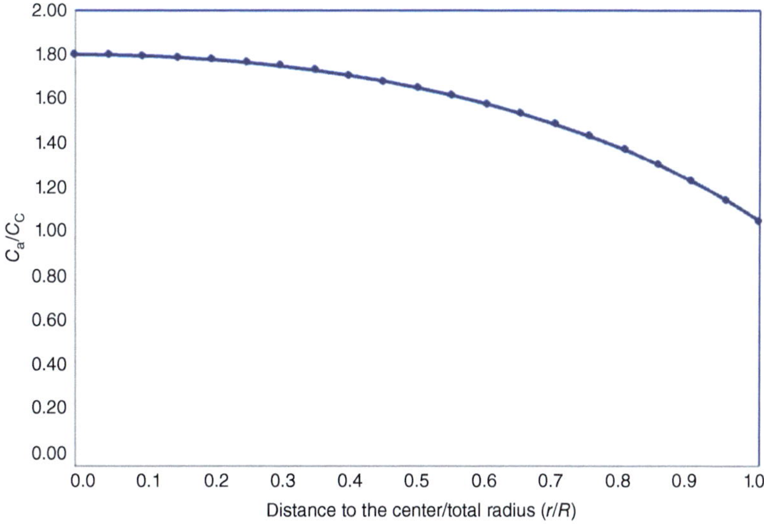

This indicates that the selectivity is higher in the center.

(c) For obtaining the effectiveness factor of "A," we have:

$$\frac{C_A}{C_{As}} = \frac{\cosh(m(L-x))}{\cosh(mL)}$$

By giving values to "mL," we can obtain a good coincidence for mL = 1.23, obtaining:

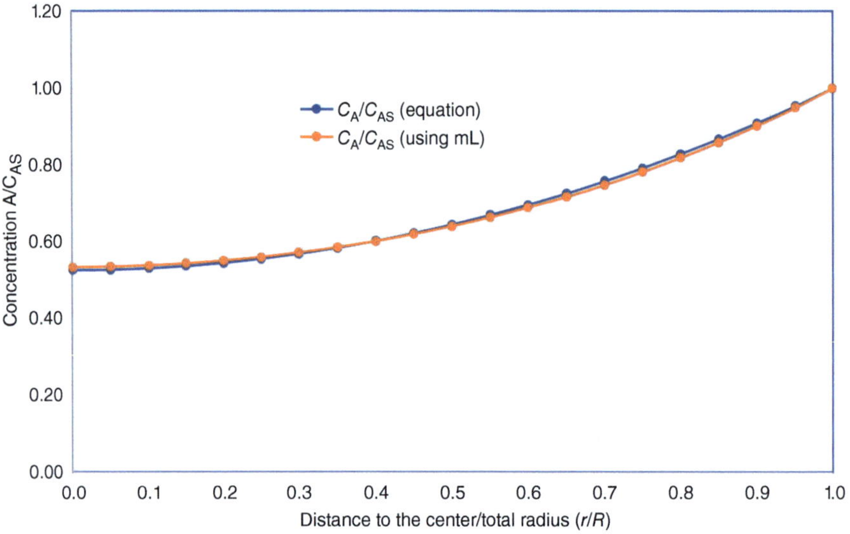


The effectiveness factor for "A" is: $\eta = \tanh(mL)/mL = 0.682$.

**Problem 6.9**  The following two reactions are carried out simultaneously on a spherical catalyst:

$$2A \rightarrow B \ (\text{rate constant} = k_1)$$

$$\tfrac{1}{2}A + B \rightarrow C \ (\text{rate constant} = k_2)$$

Both are irreversible, and the order of reaction with respect to each reactant corresponds to its stoichiometric number. The intrinsic reaction constants, $k_1$ and $k_2$, are 2.0 and 0.5, respectively. If concentrations in (M) are used. The resulting reaction rates have units of moles of A consumed per minute per cm$^3$ of granule. For given operating conditions (intragranular isothermal), we have the following profiles:

| $r/R$ | 0.0 | 0.1 | 0.2 | 0.3 | 0.4 | 0.5 | 0.6 | 0.7 | 0.8 | 0.9 | 1.0 |
| --- | --- | --- | --- | --- | --- | --- | --- | --- | --- | --- | --- |
| $C_A$, M | 2 | 2.07 | 2.14 | 2.28 | 2.45 | 2.79 | 3.21 | 3.80 | 4.61 | 5.76 | 7.42 |
| $C_B$, M | 10 | 9.97 | 9.92 | 9.83 | 9.68 | 9.44 | 9.06 | 8.46 | 7.48 | 5.85 | 2.97 |

(a) Use the finite differences method to calculate $r_A$ and $r_B$. Note that the effective diffusivities are unknown.

(b) At this point in the reactor. Is B reactant or product?

**Solution to Problem 6.9**

For details refer the Wiley website at http://www.wiley-vch.de/ISBN9783527354115

(a)  In the system, we can write:

$$\left(-r_A'''\right) = k_1''' \cdot C_A^2 + k_2''' \cdot C_A^{\frac{1}{2}} \cdot C_B$$

$$r_B''' = k_1''' \cdot C_A^2 - k_2''' \cdot C_A^{\frac{1}{2}} \cdot C_B$$

As well, we can also distinguish:

$$r_B''' \text{ (product)} = k_1''' \cdot C_A^2$$

$$\left(-r_B'''\right) \text{ (reactant)} = k_2''' \cdot C_A^{\frac{1}{2}} \cdot C_B$$

in order to know if the species B is reactant or product. Doing the corresponding calculations, we get:

| $r/R$ | $C_A$ (mol/l) | $C_B$ (mol/l) | $(-r_A)$ (mol/min·cm$^3$) | $r_B$ (mol/min·cm$^3$) | $(r_B)$ (product) | $(-r_B)$ reactant |
|---|---|---|---|---|---|---|
| 0 | 2 | 10 | 15.07 | 0.93 | 8.00 | 7.07 |
| 0.1 | 2.07 | 9.97 | 15.74 | 1.40 | 8.57 | 7.17 |
| 0.2 | 2.14 | 9.92 | 16.42 | 1.90 | 9.16 | 7.26 |
| 0.3 | 2.28 | 9.83 | 17.82 | 2.98 | 10.40 | 7.42 |
| 0.4 | 2.45 | 9.68 | 19.58 | 4.43 | 12.01 | 7.58 |
| 0.5 | 2.79 | 9.44 | 23.45 | 7.68 | 15.57 | 7.88 |
| 0.6 | 3.21 | 9.06 | 28.72 | 12.49 | 20.61 | 8.12 |
| 0.7 | 3.8 | 8.46 | 37.13 | 20.63 | 28.88 | 8.25 |
| 0.8 | 4.61 | 7.48 | 50.53 | 34.47 | 42.50 | 8.03 |
| 0.9 | 5.76 | 5.85 | 73.38 | 59.34 | 66.36 | 7.02 |
| 1 | 7.42 | 2.97 | 114.16 | 106.07 | 110.11 | 4.05 |

Average reaction rate for A: $\left(\overline{-r_A'''}\right) = 37.45 \, \dfrac{\text{mol}}{\text{min} \cdot \text{cm}^3}$ and $\left(\overline{-r_B'''}\right) = 22.94 \, \dfrac{\text{mol}}{\text{min} \cdot \text{cm}^3}$
The average rate of reaction in the particle can be used to calculate the efficiency factor by dividing it by the rate of reaction in the case of no diffusion, i.e., the rate of reaction at the end of the particle (outside of the catalyst), according to the definition:

$$\eta_A = \frac{\left(\overline{-r_A'''}\right)}{\left(-r_A'''\right)_{\text{no dif problems}}} = \frac{37.45}{114.16} = 0.328$$

In the same way:

$$\eta_B = \frac{\left(\overline{-r_B'''}\right)}{\left(-r_B'''\right)_{\text{no dif problems}}} = \frac{22.94}{106.07} = 0.216$$

(b) The rate of the reaction in which C is produced has a lower kinetic constant in addition to having a lower reaction order, so the B component does not disappear to a large extent. Therefore, in the particle, component B behaves as product.

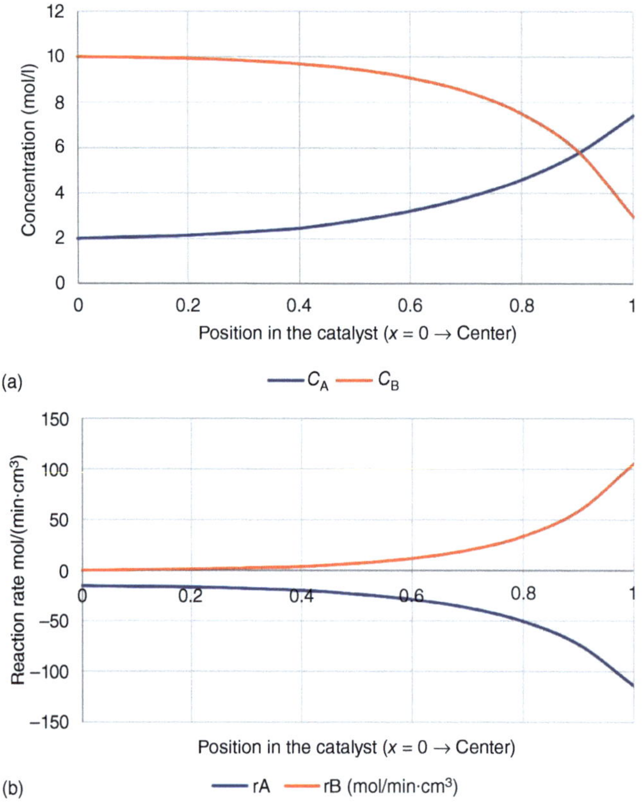

(a)

(b)

**Problem 6.10**  In a continuous reactor, the following reaction system is carried out under isothermal conditions:

$$A + 2B \underset{k_2}{\overset{\overrightarrow{k_1}}{\rightleftarrows}} C$$

$$A + 2B \overset{\overrightarrow{k_3}}{\phantom{=}} D$$

The rate constants, referred to mole of reaction, $k_1$ and $k_3$ are 712 and 213 $\frac{l^3}{s \cdot g_{cat} \cdot mol^2}$. The reaction orders coincide with the stoichiometric coefficients. The first reaction is reversible, and its equilibrium constant is 89 000 $l^2/mol^2$. The mass transfer coefficients, $(k_C \cdot a)$ for A, B, C, and D are 0.10, 0.16, 0.095, and 0.092, $l/s \cdot g_{cat}$, respectively. Assume that internal effects are negligible.

(a) Usually, it is desired to know $r_1$ and $r_2$ starting from the fact that the global concentrations are known, with the concentrations on the surface being an intermediate result. State the minimum number of necessary equations that must be solved simultaneously and describe very briefly the procedure to evaluate both rates if the global concentrations are known at a given point in the reactor.

(b) To simplify the mathematical solution, assume that the surface concentrations are known: 0.0051, 0.0063, 0.0047, and 0.00050 M for A, B, C, and D, respectively. Calculate the four global concentrations and the point selectivity defined as production of C divided by local production of D at that point in the reactor.

**Solution to Problem 6.10**

For details refer the Wiley website at http://www.wiley-vch.de/ISBN9783527354115

(a) During the reaction, we have:

$$r_1' = k_1' C_A C_B^2$$

$$r_2' = k_2' C_C$$

$$r_3' = k_3' C_A C_B^2$$

For the different species we have (subindex "s" refers to the surface, "g" to the gas phase or global concentrations):

$$r_A' = k_{cA} a(C_{Ag} - C_{As}) = k_1' C_{As} C_{Bs}^2 - k_2' C_{cs} + k_3' C_{As} C_{Bs}^2$$

$$r_B' = k_{cB} a(C_{Bg} - C_{Bs}) = \left[ k_1' C_{As} C_{Bs}^2 - k_2' C_{cs} + k_3' C_{As} C_{Bs}^2 \right] \cdot 2$$

$$r_C' = k_{cC} a(C_{Cs} - C_{Cg}) = k_1' C_{As} C_{Bs}^2 - k_2' C_{cs}$$

$$r_D' = k_{cD} a(C_{Ds} - C_{Dg}) = k_3' C_{As} C_{Bs}^2$$

Note that the flow of reactants "A" and "B" occurs in the opposite direction to that of the products "C" and "D." The reaction kinetics is evaluated at the surface because there are no internal effects.

We know the values of $k_1'$ and $k_3'$, and:

$$K_{\text{equilibrium}} = \frac{k_1'}{k_2'}$$

So, we can calculate $k_2' = 8 \cdot 10^{-3}$ 1/s·g$_{\text{cat}}$. As we also know the values of the mass transfer coefficients, with the three equations, we can find the desired concentrations. This can be done by numerical methods preferably. Once we have the values of the concentrations at the surface, we can simply calculate $r_1$ and $r_2$.

(b) Given that: CAS = 0.0051 M, CBS = 0.0063 M, CCS = 0.0047 M, and CDS = 0.00050 M;

$$0.1(C_{Ag} - 0.0051) = 712 \cdot 0.051 \cdot 0.0063^2 - 8 \cdot 0.0047 + 213 \cdot 0.0051 \cdot 0.0063^2$$

And so: $C_{As} = 0.006\,59$ M

Also:

$$0.16(C_{Bg} - 0.0063)$$
$$= [712 \cdot 0.051 \cdot 0.0063^2 - 8 \cdot 0.0047 + 213 \cdot 0.0051 \cdot 0.0063^2] \cdot 2$$

And then: $C_{BS} = 0.008\,12$ M

In the same way: $C_{CS} = 0.003\,58$ M and $C_{DS} = 0.000\,031$ M
Finally, the selectivity can be calculated as:

$$S_{\frac{C}{D}} = \frac{r_C}{r_D} = 2.471$$

**Problem 6.11**  The solid density of an alumina particle is $3.8\,\text{g/cm}^3$, the pellet density is $15\,\text{g/cm}^3$, and the internal surface is $200\,\text{m}^2/\text{g}$. Compute the pore volume per gram, the porosity, and the mean pore radius.

**Solution to Problem 6.11**

The different characteristics can be calculated by assuming $1\,\text{cm}^3$ of bed. In that volume, we have $3.8\,\text{g}$ of particles, which, according to their density, corresponds to a volume of $3.8/15 = 0.253\,\text{cm}^3$ of particles.

$$\text{Porosity} = \varepsilon = \frac{V_{\text{inter particle}}}{V_{\text{total}}} = \frac{(1 - 0.253)}{1} = 0.746$$

$$\frac{\text{Pore volume}}{\text{gram}} = \frac{0.746\,\text{cm}^3_{\text{pores}}/\text{cm}^3_{\text{bed}}}{3.8\,\text{g/cm}^3_{\text{bed}}} = 0.1963\,\text{cm}^3_{\text{pores}}/\text{g}$$

And the mean pore radius is:

$$\bar{r} = 2 \cdot \frac{V}{S} = 2 \cdot \frac{0.1963\,\text{cm}^3_{\text{pores}}/\text{g}}{200\,\frac{\text{m}^2}{\text{g}} \cdot 10\,000\,\frac{\text{cm}^2}{\text{m}^2}} = 0.000\,000\,196\,3\,\text{m} = 1963\,\text{Å}$$

**Problem 6.12**  A series of experiments were performed using various sizes of crushed catalysts to determine the importance of pore diffusion. The reaction may be assumed to be first-order and irreversible. The surface concentration of reactant was $2 \times 10^{-4}\,\text{mol/cm}^3$.

Data:

| Diameter of sphere (cm): | 0.25 | 0.075 | 0.025 | 0.0075 |
|---|---|---|---|---|
| $r$ observed (mol/h cm$^3$): | 0.22 | 0.70 | 1.60 | 2.40 |

(a) Determine the "true" rate constant $k$ and the effective diffusivity $D_e$ from the above data.
(b) Predict the effectiveness factor and the expected rate of reaction for a commercial cylindrical catalyst pellet if dimensions are $0.5\,\text{cm} \times 0.5\,\text{cm}$.

**Solution to Problem 6.12**

(a) In order to find the rate constant, we need to find the efficiency factor. It can initially be assumed that for the smallest diameter of the spheres, there is no resistance to diffusion in the pores, so $\eta = 1$. The assumption that is made can be verified with the Thiele module, since if it is less than 0.4, the internal diffusion is considered negligible. You have to:

$$\eta = \frac{\overline{\left(-r'''_A\right)}}{\left(-r'''_A\right)_{\text{without diffusion effects}}}$$

$$\overline{(-r'''_A)} = k''' C_{As} \eta$$

$$k''' = \frac{\overline{(-r_A)}}{C_{As}\eta} = \frac{2.40}{(1 \cdot 2 \cdot 10^{-4})} = 1200\,\text{h}^{-1}$$

The Thiele modulus is used to calculate the effective diffusivity. Values of $D_e$ have to be assumed until the values of $\eta$ match. We calculate the $\eta$ with the data in the table.

Is obtained:

| $d_p$ (cm) | $(-r'''_A)$ (mol/cm$^3$·h) | Actual $\eta$ |
|---|---|---|
| 0.25 | 0.22 | 0.0916 |
| 0.075 | 0.7 | 0.2916 |
| 0.025 | 1.6 | 0.6666 |
| 0.0075 | 2.4 | 1 |

Now:

$$m = \left[\frac{k'''}{D_e}\right]^{\frac{1}{2}}$$

$$\eta = \frac{\tanh(mL)}{mL}$$

By iterations, $D_e$ values are obtained for the first three particle sizes; for the smallest one, it cannot be solved, so an average of the largest sizes is used. Is obtained:

| $d_p$ (cm) | $(-r'''_A)_{actual}$ (mol/cm$^3$·h) | $\eta$ | $D_e$ (cm$^2$/h) | mL | $L = d_p/6$ (cm) | $\eta$ |
|---|---|---|---|---|---|---|
| 0.25 | 0.22 | 0.0917 | 0.1787 | 10.7986 | 0.0417 | 0.0926 |
| 0.075 | 0.7 | 0.2917 | 0.1600 | 3.4229 | 0.0125 | 0.2915 |
| 0.025 | 1.6 | 0.6667 | 0.1256 | 1.2877 | 0.0042 | 0.6667 |
| 0.0075 | 2.4 | 1.0000 | 0.1548 | 0.3481 | 0.0013 | 0.9615 |

When obtaining a value of mL < 0.4 for the smallest particle, the initial assumption is considered valid.

(b) For the cylindrical pellet, bearing in mind the whole surface of the pellets:

$$L = \frac{\text{Volume of the particles}}{\text{External surface available for entry and difussion of the reagent}}$$

$$= \frac{\text{Volume of the particles}}{\text{Surface of the bases} + \text{Surface of the cylinder}}$$

$$= \frac{\pi R^2 \text{Length}}{2\pi R\text{Length} + 2\pi R^2} = \frac{R \cdot \text{Length}}{2 \cdot \text{Length} + 2R}$$

With cylinders of dimensions $0.5\,\text{cm} \times 0.5\,\text{cm}$:

$$L = \frac{0.5 \cdot 0.5}{2 \cdot 0.5 + 2 \cdot 0.5} = 0.5\,\text{cm}$$

$$\text{mL} = \left[\frac{k'''}{D_e}\right]^{\frac{1}{2}} \cdot L = \left[\frac{1200\,\text{h}^{-1}}{0.16\,\frac{\text{cm}^2}{\text{h}}}\right]^{\frac{1}{2}} \cdot (0.5\,\text{cm}) = 43.30$$

As diffusion control, $\text{mL} > 4$ is required, considering the diffusion in the important pores. So, the following is approximated:

$$\eta = 1/\text{mL} = 0.023$$

Therefore, the reaction rate $(-r_A''')_{\text{actual}} = k''' C_{As} \eta = 0.068\,76\,\text{mol/cm}^3 \cdot \text{h}$.

**Problem 6.13**   The following rates were observed for a first-order irreversible reaction, carried out on a spherical catalyst:

For $d_p = 0.6\,\text{cm}$, $r_{\text{obs}}' = 0.090\,\text{mol/g}_{\text{cat}} \cdot \text{h}$
For $d_p = 0.3\,\text{cm}$, $r_{\text{obs}}' = 0.162\,\text{mol/g}_{\text{cat}} \cdot \text{h}$

In both cases, strong diffusional limitations were observed. Determine the true rate of reaction. Is diffusional resistance still important with $d_p = 0.1\,\text{cm}$?

**Solution to Problem 6.13**
For details refer the Wiley website at http://www.wiley-vch.de/ISBN9783527354115

Comparing both particle sizes:

$$\text{mL} = \left[\frac{k'''}{D_e}\right]^{\frac{1}{2}} L$$

$$\frac{\text{mL}_1}{\text{mL}_2} = \frac{L_1}{L_2} = \frac{d_{p1}}{d_{p2}} = \frac{0.6}{0.3} = 2$$

$$\frac{(-r_A')_{\text{Obs1}}}{(-r_A')_{\text{Obs2}}} = \frac{k' C_{As} \eta_1}{k' C_{As} \eta_2} = \frac{\eta_1}{\eta_2}$$

As we know that:

$$\eta = \frac{\tanh(\text{mL})}{\text{mL}}$$

$$\frac{(-r_A)_{\text{Obs1}}}{(-r_A)_{\text{Obs2}}} = \frac{0.09}{0.162} = \frac{\dfrac{\tanh(\text{mL}_1)}{\text{mL}_1}}{\dfrac{\tanh(\text{mL}_2)}{\text{mL}_2}} = \frac{\dfrac{\tanh(2\text{mL}_2)}{2\text{mL}_2}}{\dfrac{\tanh(\text{mL}_2)}{\text{mL}_2}} = \frac{\tanh(2\text{mL}_2)}{2\tanh(\text{mL}_2)}$$

By using trial-and-error method, we can estimate the values of:

$$\eta_1 = 0.344 \qquad \eta_2 = 0.619 \qquad \text{mL}_1 = 2.88 \qquad \text{mL}_2 = 1.11$$

Also, we can calculate $(-r_A')_{\text{true}} = k' C_{As} = 0.2615\,\frac{\text{mol}}{\text{g}_{\text{cat}} \cdot \text{h}}$

For $d_{\text{p}} = 0.1\,\text{cm}$, $\text{mL}_3 = \text{mL}_1/6 = 2.88/6 = 0.48$ and:

$$\eta_3 = \frac{\tanh(0.48)}{0.48} = 0.9296$$

So, the internal diffusion is not very important in the case where the particle diameter is 0.1 cm.

**Problem 6.14**  A second-order gas-phase reaction, A → R, occurs in a catalyst pellet and has a rate coefficient $k = 3.86\,\text{m}^3/\text{kmol·s}$. The reactant pressure is 1 bar, the temperature is 600 K, the molecular diffusivity is $D_{\text{A}} = 0.10\,\text{cm}^2/\text{s}$, and the reactant molecular weight is 60 g/mol. The catalyst pellets have the following properties:

Radius of sphere $R = 9\,\text{mm}$
Pellet density $= 12\,\text{g/cm}^3$
Internal surface area $= 100\,\text{m}^2/\text{g}$
Internal void fraction $= 0.60$

(a)  Estimate the effective diffusivity.
(b)  Determine if there may be pore diffusion limitations and what might be done to eliminate them, if present.

**Solution to Problem 6.14**

(a)  In pellets:

$$D_{\text{e}} = \varepsilon^2 \cdot D_{\text{A}} = 3.6 \cdot 10^{-6}\,\frac{\text{m}^2}{\text{s}}$$

(b)  Let us calculate the effectiveness factor:

$$C_{\text{As}} = \frac{p_{\text{A}}}{R_{\text{g}}T} = 20.056\,\frac{\text{mol}}{\text{m}^3}$$

For second-order reactions:

$$\text{mL} = \left[\frac{3k''' \cdot C_{\text{As}}}{2 \cdot D_{\text{e}}}\right]^{\frac{1}{2}} \left(\frac{R}{3}\right) = 0.5388$$

And then:

$$\eta = \frac{\tanh(\text{mL})}{\text{mL}} = 0.9133$$

We can conclude that some limitations are present, but the effectiveness factor is quite close to unity. If we want to eliminate the possible effects and increase reaction rate, we can increase the pressure (increasing $C_{\text{As}}$), decrease the temperature (also increasing $C_{\text{As}}$), and, obviously, decrease the diameter of the particles.

**Problem 6.15**  The production rate of a heterogeneously catalyzed, first-order reaction in a 0.75 cm diameter spherical pellet is $3.25 \times 10^{-3}\,\text{mol/cm}^3\text{·s}$ when the catalyst is exposed to pure, gaseous A at a pressure of 1 atm and a temperature of 525 K. This reaction's activation energy is $E = 18.600\,\text{cal/mol}$. Further, the effective diffusivity of A in the pellet is $D_{\text{e}} = 0.009\,\text{cm}^2/\text{s}$ at 525 K and that diffusion is in the regime of Knudsen flow. The bulk fluid and the external surface concentrations can be assumed to be the same. Find the production rate if the catalyst is changed to a cylindrical pellet that is 0.5 cm in diameter and 1.0 cm in length, and the temperature is increased to 600 K.

**Solution to Problem 6.15**

For details refer the Wiley website at http://www.wiley-vch.de/ISBN9783527354115

For the bed with no effects of external diffusion, the reaction rate is:

$$\left(-r_A'''\right) = k''' C_A \eta$$

And we have that:

$$\eta = \frac{\tanh(mL)}{mL}$$

$$L = \frac{R}{3} \text{(spheres)}$$

$$mL = \left[\frac{k'''}{D_e}\right]^{\frac{1}{2}} L = \sqrt{\frac{k'''}{0.009}} \left(\frac{0.75}{3}\right)$$

Also, we know that the reaction rate is:

$$\left(-r_A'''\right) = 3.25 \cdot 10^{-3} \frac{\text{mol}}{\text{cm}^3 \cdot \text{s}} = k''' \left(\frac{P_A}{R_g T}\right) \eta$$

From these equations, the only parameter unknown is $k'''$ that can be estimated. The only thing to bear in mind is that $C_A$ should be used in mol/cm$^3$, so the value of $R_g$ to be used is 82 atm·cm$^3$/mol·K. We get a value of $k''' = 135\,880.94\ \text{s}^{-1}$ at 525 K. The efficiency factor in these conditions is very low, in the order of $10^{-3}$.

We will find all the values by a trial-and-error strategy:

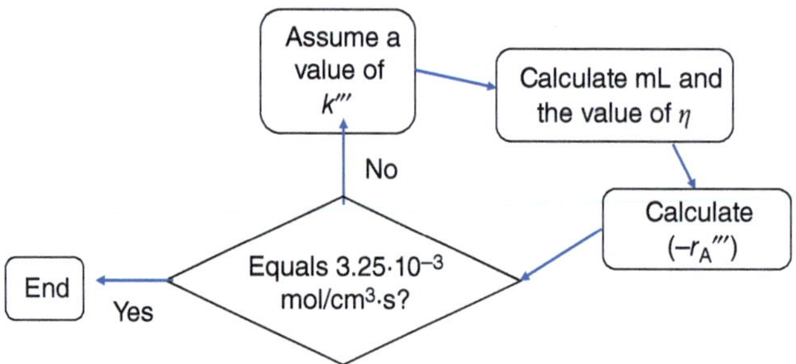


For the reaction at 600 K, the true chemical constant would be:

$$k_{T1}''' = k_{T_2}''' \exp\left(-\frac{E}{R}\left(\frac{1}{T_1} - \frac{1}{T_2}\right)\right)$$

$$k_{600}''' = 135\,880.94 \exp\left(-\frac{18600}{1.987}\left(\frac{1}{600} - \frac{1}{525}\right)\right) = 1\,286\,634\ \text{s}^{-1}$$

For the cylindrical pellet, bearing in mind the whole surface of the pellets:

$$L = \frac{\text{Volume of the particles}}{\text{External surface available for entry and difussion of the reagent}}$$

$$= \frac{\text{Volume of the particles}}{\text{Surface of the bases} + \text{Surface of the cylinder}}$$

$$= \frac{\pi R^2 \text{Length}}{2\pi R \text{Length} + 2\pi R^2} = \frac{R \cdot \text{Length}}{2 \cdot \text{Length} + 2R}$$

In our case, $L = 0.1677$ cm, and the efficiency factor is:

$$\mathrm{mL} = \sqrt{\frac{1\,286\,634}{0.009}}\,(0.1677) = 1992$$

$$\eta = \frac{\tanh(\mathrm{mL})}{\mathrm{mL}} = 0.0005$$

And the reaction rate in the new conditions is:

$$\left(-r_A'''\right) = \frac{k''' p_A}{RT}\eta = 0.015 \frac{\mathrm{mol}}{\mathrm{cm}^3 \cdot \mathrm{s}}$$

**Problem 6.16** In a laboratory, research is being conducted on the heterogeneously catalyzed decomposition of A. The reaction kinetics of this decomposition can be described by $(-r_A) = k \cdot C_A$, where "$k$" is dependent on the temperature following the Arrhenius equation. The catalyst used consists of spherical particles with an internal surface area of $62\,\mathrm{m}^2/\mathrm{g}$. All experiments are conducted isothermally and in the same reactor, with an inlet concentration of A always equal to $10\,\mathrm{mol/m}^3$. The diffusion coefficient $D_e$ remains constant and does not vary with temperature. Your task is to calculate the effectiveness factor for each experiment (1–8) based on the provided results below.

| Experiment (−) | Temperature (°C) | $n_{A0}$ (mol/h) | Weight catalyst (g) | Diameter catalyst (mm) | Conversion (−) |
| --- | --- | --- | --- | --- | --- |
| 1 | 127 | 1 | 400 | 8 | $4.62\ 10^{-4}$ |
| 2 | 127 | 1 | 400 | 1.5 | $4.62\ 10^{-4}$ |
| 3 | 127 | 2 | 800 | 8 | $4.62\ 10^{-4}$ |
| 4 | 177 | 10 | 400 | 8 | $2.98\ 10^{-3}$ |
| 5 | 177 | 10 | 400 | 1.5 | $2.98\ 10^{-3}$ |
| 6 | 177 | 20 | 800 | 8 | $2.98\ 10^{-3}$ |
| 7 | 350 | 1000 | 400 | 8 | $3.44\ 10^{-2}$ |
| 8 | 350 | 2000 | 800 | 8 | $3.44\ 10^{-2}$ |

**Solution to Problem 6.16**

For details refer the Wiley website at http://www.wiley-vch.de/ISBN9783527354115

Assuming the CSTR behavior of the reactor:

$$n_{A0} - n_A = \left(r_A'''\right) V_{cat} = -k''' C_A \eta V_{cat}$$

$$-n_{A0} X_A = -k''' C_{A0}(1 - X_A)\eta \left(\frac{W_{cat}}{\rho_{cat}}\right)$$

$$\frac{n_{A0} X_A}{C_{A0}(1 - X_A)W_{cat}} = \frac{k''' \cdot \eta}{\rho_{cat}}$$

In the different runs:

| Exp. | $n_{A0}$ (mol/h) | $W_{cat}$ (g) | $d_{cat}$ (mm) | $L_{cat}$ (m) | Conversion $X_A$ | $k''' \cdot \eta / \rho_{cat}$ (m³/g·h) |
|---|---|---|---|---|---|---|
| 1 | 1 | 400 | 8 | $1.33 \cdot 10^{-3}$ | $4.62 \cdot 10^{-4}$ | $1.16 \cdot 10^{-7}$ |
| 2 | 1 | 400 | 1.5 | $2.50 \cdot 10^{-4}$ | $4.62 \cdot 10^{-4}$ | $1.16 \cdot 10^{-7}$ |
| 3 | 2 | 800 | 8 | $1.33 \cdot 10^{-3}$ | $4.62 \cdot 10^{-4}$ | $1.16 \cdot 10^{-7}$ |
| 4 | 10 | 400 | 8 | $1.33 \cdot 10^{-3}$ | $2.98 \cdot 10^{-3}$ | $7.47 \cdot 10^{-6}$ |
| 5 | 10 | 400 | 1.5 | $2.50 \cdot 10^{-4}$ | $2.98 \cdot 10^{-3}$ | $7.47 \cdot 10^{-6}$ |
| 6 | 20 | 800 | 8 | $1.33 \cdot 10^{-3}$ | $2.98 \cdot 10^{-3}$ | $7.47 \cdot 10^{-6}$ |
| 7 | 1000 | 400 | 8 | $1.33 \cdot 10^{-3}$ | $3.44 \cdot 10^{-2}$ | $8.91 \cdot 10^{-3}$ |
| 8 | 2000 | 800 | 8 | $1.33 \cdot 10^{-3}$ | $3.44 \cdot 10^{-2}$ | $8.91 \cdot 10^{-3}$ |

Assuming that the efficiency is unity for the run with smaller particles, the ratio of mass transfer coefficient to catalyst density is:

$$\frac{k'''}{\rho_{cat}} = 1.16 \cdot 10^{-7} \frac{m^3}{g \cdot h} = 3.21 \cdot 10^{-11} \frac{m^3}{g \cdot s}$$

From this run, we also assume that mL = 0.4 (the lower limit for assuming unity efficiency), giving us:

$$mL = \left(\frac{k'''}{D_e}\right)^{0.5} \left(\frac{d_p}{6}\right) = 0.4$$

This leads to the equation:

$$0.4 = \left(\frac{3.21 \cdot 10^{-11} \cdot \rho_{cat}}{D_e}\right)^{0.5} \cdot \frac{1.5 \cdot 10^{-3}}{6}$$

We can solve for the value of $D_e/\rho_{cat}$:

$$\frac{D_e}{\rho_{cat}} = 1.25 \cdot 10^{-17} \frac{m^5}{g \cdot s}$$

Note that effective diffusivity cannot be calculated if density is unknown, but these parameters are assumed to be constant across runs. Thus, for subsequent runs, mL can be estimated using this derived value:

$$mL = \left(\frac{k'''/\rho_{cat}}{D_e/\rho_{cat}}\right)^{0.5} \left(\frac{d_p}{6}\right) = \left(\frac{k'''/\rho_{cat}}{1.25 \cdot 10^{-17}}\right)^{0.5} \left(\frac{d_p}{6}\right)$$

In the different runs:

| Exp. | mL | $\eta$ |
| --- | --- | --- |
| 1 | 2.133 | 0.456 |
| 2 | 0.400 | 0.950 |
| 3 | 2.133 | 0.456 |
| 4 | 17.155 | 0.058 |
| 5 | 3.217 | 0.310 |
| 6 | 17.155 | 0.058 |
| 7 | 592.268 | 0.002 |
| 8 | 592.268 | 0.002 |

Here, the efficiency is calculated with the usual expression as a function of mL. Note that efficiency for the run no. 2 is not unity because arbitrarily we have assigned a value of mL equal to 0.4.

**Problem 6.17** The heterogeneously catalyzed first-order surface reaction $A \rightarrow P$ is being studied in an ideal plug flow reactor (PFR) with an empty volume of 250 ml. The reactor has been completely filled with 200 g of porous, spherical catalyst. The catalyst data are as follows:

Internal surface area: $200\,\text{m}^2/\text{g}$
Apparent density of a catalyst particle: $1600\,\text{kg/m}^3$
Porosity of the catalyst bed: 0.5
Diameter of a catalyst particle: 4 mm
Effective diffusion coefficient in catalyst particle: $0.3 \times 10^{-11}\,\text{m}^2/\text{s}$
Kinetics: $(-r_\text{A}) = kC_\text{A}$, with $k = k_0 \cdot \exp(-E/R_\text{g}T)$, and $T$ in K.

The reactor is fed with a stream of pure A at a flow rate of $Q_0 = 500\,\text{ml/min}$ and an initial concentration of $C_{\text{A}0} = 0.01\,\text{mol/l}$. In the first experiment, the temperature in the reactor was 400 K, and a degree of conversion of 0.6321 was achieved. In the second experiment, the temperature is 500 K, and a degree of conversion of 0.8647 is achieved.

Your task is to determine the value of $E/R_\text{g}$ for the following two extremes:

(a) In both experiments, the concentration of A in the porous catalyst hardly depends on the distance to the external surface of the particle.
(b) In both experiments, the concentration of A in the porous catalyst strongly depends on the distance to the external surface of the particle.

Assume that the densities of A, P, and their mixtures, as well as the diffusion coefficient, are independent of temperature and composition. Additionally, neglect concentration gradients in the film surrounding the catalyst particles.

**Solution to Problem 6.17**

(a) With a kinetic constant following Arrhenius law, we have that:

$$\frac{E}{R_g} = \frac{\ln\left(\frac{k'_{T2}}{k'_{T1}}\right)}{\frac{1}{T_1} - \frac{1}{T_2}}$$

In the reactor:

$$Q_0 dC_A = -r'_A dW$$

$$Q_0 C_{A0}(-dX_A) = -k' C_{A0}(1 - X_A)dW$$

$$k'' = -\frac{Q_0 \ln(1 - X_A)}{W}$$

For the two runs, we find $k'$ (1st run) $= 2500\,\text{ml/min·kg}_{cat}$ and $k'$ (2nd run) $= 5000\,\text{ml/min·kg}_{cat}$. Using the relationship mentioned, we can finally say that $E/R_g = 1386\,\text{K}$.

(b) Now, effectiveness factor should be calculated.

$$(\text{mL})_1 = \sqrt{\frac{k'''}{D_e}}\left(\frac{d_p}{6}\right) = \sqrt{\frac{k'\frac{\text{ml}}{\text{min·kg}} \cdot 10^{-6}\frac{\text{ml}}{\text{m}^3} \cdot \frac{1}{60}\frac{\text{min}}{\text{s}} \rho_{cat}\frac{\text{kg}}{\text{m}^3}}{D_e\frac{\text{m}^2}{\text{s}}}}\left(\frac{R}{6}\right) \quad m = 9.94$$

$$\eta_1 = \frac{\tanh(\text{mL}_1)}{\text{mL}_1} = 0.1$$

Similarly, $(\text{mL})_2 = 14.06$ and $\eta_2 = 0.0711$. Now, the mass balance gives:

$$k'\eta = -\frac{Q_0 \ln(1 - X_A)}{W}$$

For the two runs, we find $E/R_g = 2080\,\text{K}$.

**Problem 6.18**  The batch saponification of ethyl acetate,

$$CH_3COOC_2H_5 + NaOH \rightarrow CH_3COONa + C_2H_5OH$$

was carried out in a 200 ml reactor at 26 °C. The initial concentrations of both reactants were 0.051 M.

(a) From the following time versus concentration data, determine the specific rate and tabulate it as a function of composition of the reacting mixture.

| Time (s) | NaOH (mol/l) |
| --- | --- |
| 30 | 0.0429 |
| 90 | 0.0340 |
| 150 | 0.0282 |
| 210 | 0.0240 |
| 270 | 0.0209 |
| 390 | 0.0164 |
| 630 | 0.0118 |
| 1110 | 0.0067 |

(b) Determine a suitable reaction rate model for this system.

**Solution to Problem 6.18**

For details refer the Wiley website at http://www.wiley-vch.de/ISBN9783527354115

(a) Let us first plot the given data:

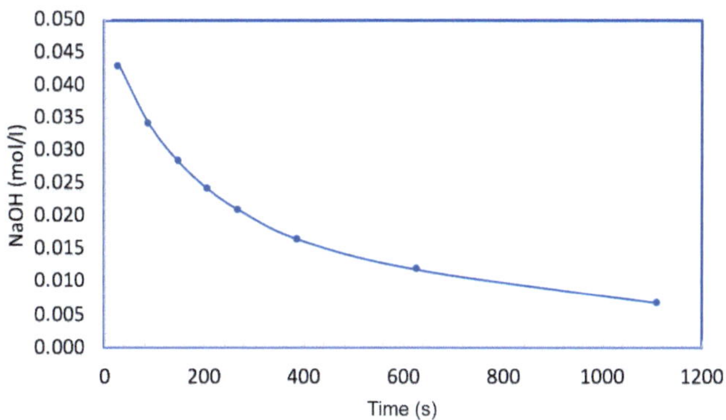

We will calculate the reaction rate using the mass balance for a batch reactor:

$$\frac{dC_A}{dt} = r_A$$

We will be using the finite differences approximation:

$$r_A = \frac{\left(C_A^{t+1} - C_A^t\right)}{\Delta t}$$

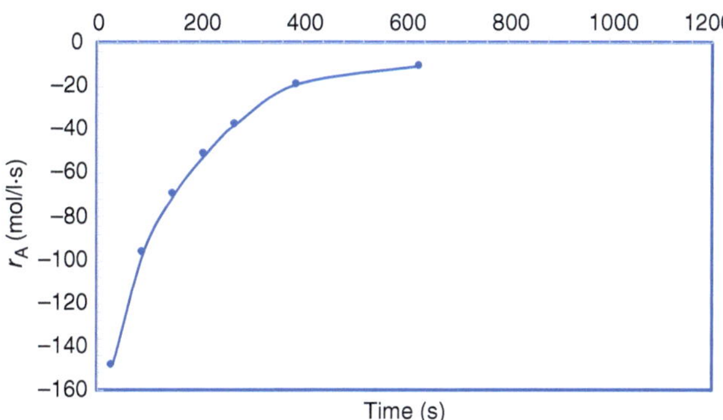

If the reaction is:

| Zero-order | $\frac{dC_A}{dt} = r_A = -k$ | $C_A = C_{A0} - kt$ |
|---|---|---|
| First-order | $\frac{dC_A}{dt} = r_A = -kC_A$ | $C_A = C_{A0}\exp(-kt)$ |
| Second-order | $\frac{dC_A}{dt} = r_A = -kC_A^2$ | $1/C_A = 1/C_{A0} + kt$ |

(b) We can check which of the above is more accurate in representing the data given, resulting in a second-order reaction:

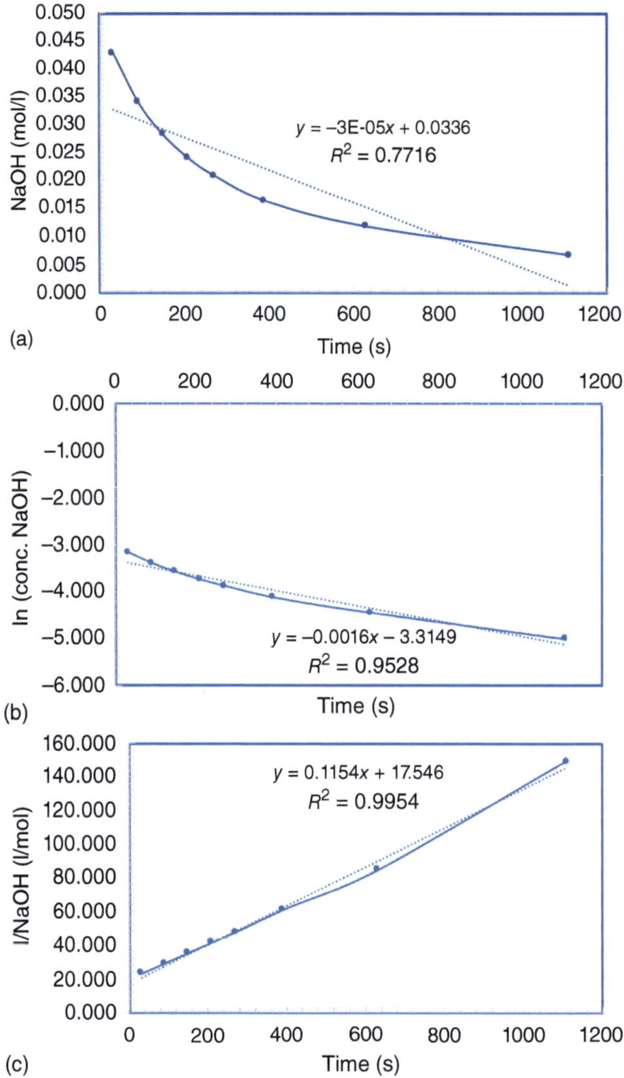


**Problem 6.19**  In a fixed-bed reactor, the oxidation reaction of $SO_2$ to $SO_3$ with air has been studied. The purpose of the study is to determine if the effect of external mass transfer is significant under operating conditions. The catalyst consists of 0.3 cm diameter spherical $Pt/Al_2O_3$ particles, with the Pt deposited only on its outer surface. The specific external area of the particles is 1.05 m²/kg, and the void fraction of the bed is 0.43. The pressure in the reactor is 790 mmHg, and the gases circulate with a velocity of 0.41 m/s. The temperature of the pellets is 480 °C,

and the feed stream contains 6.42% $SO_2$ and 93.58% air (molar percent). Using the correlation:

$$j_D = \frac{0.725}{(Re)^{0.41} - 0.15}$$

Estimate the mass transfer coefficient between the fluid and the surface of the solid and the difference in partial pressure found in the film.

*Data:* Assume that the properties of the mixture are those of air. For air, at 480 °C, we have:

Viscosity $= 3.72 \cdot 10^{-5}$ kg/m·s
Average molecular weight $= 28.9$ g/mol
$SO_2$ diffusivity in air $= 6.3\ 10^{-5}$ m²/s
$SO_3$ diffusivity in air $= 5.6\ 10^{-5}$ m²/s
Diffusivity of $O_2$ in $N_2 = 9.6\ 10^{-5}$ m²/s
Equilibrium constant of the reaction $= 73$ atm$^{-0.5}$

**Solution to Problem 6.19**

In the reactor, the mass transfer is given by:

$$\left(-r'_{SO_2}\right) = k_C \cdot a\,(C_{SO_2,g} - C_{SO_2,s})$$

where the difference is between the concentration in the gas and on the surface of the catalyst. The density of the mix can be assumed to be that of the air (very low concentration), that is:

$$\rho_{air} = P_{total} \cdot \frac{M_{air}}{R_g T} = 0.488\ \frac{kg}{m^3}$$

And so:

$$Sc = \frac{\mu}{\rho \cdot D_A} = 1.21$$

$$Re = \frac{d_p \cdot \rho \cdot u}{\mu} = 16.13$$

$$j_D = \frac{0.725}{(Re)^{0.41} - 0.15} = 0.2435$$

$$j_D = \frac{Sh}{Sc^{\frac{1}{3}} \cdot Re}$$

$$Sh = j_D \cdot Sc^{\frac{1}{3}} \cdot Re = 4.187$$

$$Sh = \frac{(k_C d_P)}{D_A}$$

$$Sh \cdot \frac{D_A}{d_p} = k_C = 0.087\,92\ \frac{m}{s}$$

Now:

$$\left(-r'_{SO_2}\right) = k_C \cdot a\,(C_{SO_2,g} - C_{SO_2,s})$$

So:

$$(C_{SO_2,g} - C_{SO_2,s}) = \frac{\left(-r'_{SO_2}\right)}{k_C \cdot a} = \frac{\left(0.0189 \frac{mol}{g \cdot h} \cdot \frac{1}{3600} \frac{s}{h}\right)}{\left(0.087\,92 \frac{m}{s}\left(1.05 \frac{m^2}{kg} \cdot \frac{1}{1000} \frac{g}{kg}\right)\right)}$$

$$= 0.057 \frac{mol}{m^3} = 5.7 \cdot 10^{-5} \frac{mol}{l}$$

Expressed as difference of pressure:

$$\Delta p = \Delta C \cdot RT = 0.003\,51 \text{ atm}$$

**Problem 6.20**   A reaction of the type $A + B \rightarrow P$ takes place in a catalytic reactor. The kinetics of the reaction is first-order with respect to each of the reactants, with a kinetic constant at 250 °C of $7.8 \times 10^{-3}$ m³/mol·s. The catalyst has an indeterminate shape, but by means of nitrogen porosimetry, it is specified that the area of the catalyst is 800 m²/g and its density is 1220 g/cm³. The reactor will be a fixed bed, and the catalyst will be arranged so that it occupies 0.3 m³ of catalyst/m³ bed. The diffusivities of A and B are, respectively, $4 \cdot 10^{-4}$ and $8 \cdot 10^{-3}$ m²/s. At the entrance to the bed, $C_A = 100$ mol/m³ and $C_B = 10$ mol/m³. Calculate:

(a)  The efficiency factor for "A"
(b)  The efficiency factor for "B"
(c)  The reaction rate.

*Data:* Consider that the external diffusion step is very fast.

**Solution to Problem 6.20**

(a)  and (b) The chemical step is represented by:

$$(-r_A) = kC_A C_B$$

To calculate the effectiveness factors, we need to know the $L = $ generalized length. We have that:

$$800 \frac{m^2}{g} \cdot 1220 \frac{g}{cm^3} \cdot 10^{-6} \frac{cm^3}{m^3} = 0.916 \text{ m}^{-1} = \frac{S_{external}}{V_{particles}}$$

And then $L = 1/0.916 = 1.024$ m.

When working with Thiele modulus, the reaction rate should be $n$th order with respect to the reactant being considered, so for the reactant "A," the kinetic law should be expressed as:

$$(-r_A) = (kC_B)C_A \eta_A = k_{psA} C_A \eta_A$$

where $k_{psA}$ represents a pseudo-first-order rate constant. Using the Thiele modulus:

$$m_A L = \left[\frac{k_{psA}}{D_A}\right]^{\frac{1}{2}} L = 10.11$$

$$\eta_A = \frac{\tanh(m_A L)}{m_A L} = 0.098$$

The reaction rate, for A, is:

$$\left(-r''_A\right) = (k''C_B)C_A\eta_A = (7.8 \cdot 10^{-3} \cdot 100) \cdot 10 \cdot 0.0098 = 0.7644 \frac{\text{mol}}{\text{m}^3 \cdot \text{s}}$$

On the hand of B, we should do:

$$\left(-r''_B\right) = (k''C_A)C_B\eta_B = k''_{psB}C_B\eta_B$$

where $k''_{psB}$ represents a pseudo-first-order rate constant. Using the Thiele modulus:

$$m_B L = \left[\frac{k''_{psB}}{D_B}\right]^{\frac{1}{2}} L = 3.19$$

$$\eta_B = \frac{\tanh(m_B L)}{m_B L} = 0.31$$

The reaction rate, for B, is:

$$\left(-r''_B\right) = (k''C_A)C_B\eta_B = (7.8 \cdot 10^{-3} \cdot 10) \cdot 100 \cdot 0.31 = 2.418 \frac{\text{mol}}{\text{m}^3 \cdot \text{s}}$$

(c) Obviously, the system cannot react different amounts of A and B, being the global rate the lower one, i.e., 0.7644 mol/m³·s.


**Problem 6.21**   The elimination of $SO_2$ by reduction is a significant process in industrial systems. It has been observed that the reaction follows first-order kinetics, given as $r = k \cdot C_{SO_2}$, using a 5% $V_2O_5/TiO_2$ catalyst. At a temperature of 500 K, the kinetic constant ($k$) is determined to be 0.566 cm³/g$_{cat}$·s, and the activation energy is found to be 110 kJ/mol.

Now, we need to calculate the effectiveness factor and reaction rate at the inlet of a commercial unit that contains an extruded cellular monolith made of 5% $V_2O_5/TiO_2$. This monolith has a channel wall thickness of 1.35 mm and an open frontal area of 64%. We will consider typical commercial operating inlet conditions of 400 ppm of $SO_2$, 350 °C, and 1 atm. For our calculations, we can assume that the catalyst in the square monolith walls can be treated as a flat plate. Furthermore, we will neglect the films mass transfer resistance.

Additional data provided include the catalyst density of 1.48 g/cm³ and an effective diffusion coefficient ($D_e$) of 0.07 cm²/s.

**Solution to Problem 6.21**
For calculating the effectiveness factor, both the diffusion and the chemical reaction rate are needed at the system temperature, then:

$$k'_{623K} = k'_{500K} \exp\left\{\frac{E}{R_g}\left(\frac{1}{623} - \frac{1}{500}\right)\right\} = 105 \frac{\text{cm}^3}{\text{g}_{cat} \cdot \text{s}}$$

The equivalent length can be calculated:

$$\frac{V_{cat}}{S_{cat}} = \frac{\text{length} \cdot \text{width} \cdot \text{thickness}}{2 \cdot (\text{length} \cdot \text{width})} = \frac{\text{thickness}}{2}$$

where the factor of 2 accounts for each wall having two sides. The Thiele modulus is then:

$$\Phi = m \cdot L = \sqrt{\frac{k'''}{D_e}} \cdot \left(\frac{\text{thickness}}{2}\right)$$

$$= \sqrt{\frac{105 \,\frac{\text{cm}^3}{\text{g}_{\text{cat}}\cdot\text{s}} \cdot 1.48 \,\frac{\text{g}_{\text{cat}}}{\text{cm}^3}}{0.07 \,\frac{\text{cm}^2}{\text{s}}}} \cdot \left(\frac{1.35 \cdot 10^{-1} \,\text{cm}}{2}\right) = 3.18$$

And the effectiveness factor is $\eta = \frac{\tanh(mL)}{mL} = \frac{\tanh(3.18)}{3.18} = 0.31$

For calculating the reaction rate:

$$C_{SO_2} = 0.400 \,\frac{\text{g}}{\text{m}^3} \cdot \frac{1}{64} \,\frac{\text{mol}}{\text{g}} \cdot 10^{-3} \,\frac{\text{m}^3}{\text{l}} \cdot 10^{-3} \,\frac{1}{\text{cm}^3} = 6.25 \cdot 10^{-9} \,\frac{\text{mol}}{\text{cm}^3}$$

$$\left(-r'_{SO_2}\right) = k' \cdot C_{SO_2} \cdot \eta = 105 \cdot 6.25 \cdot 10^{-9} \cdot 0.31 = 2.03 \cdot 10^{-7} \,\frac{\text{mol}}{\text{g}_{\text{cat}} \cdot \text{s}}$$

**Problem 6.22**   Mordenite zeolite is being applied to an isothermal catalytic process. This one-dimensional zeolite is available in three different sizes, all cylindrical, with the channels arranged in the axial direction. The data on the dimensions are shown below, along with other data on the irreversible reaction.

| Zeolite | Length (mm) | Radius (mm) |
| --- | --- | --- |
| A | 100 | 10 |
| B | 30 | 15 |
| C | 16 | 20 |

$k' = 0.217 \; 10^{-6} \,\text{m}^3/\text{s}\cdot\text{kg zeolite}$

$D_e = 1 \; 10^{-12} \,\text{m}^2/\text{s}$

Zeolite density $= 1800 \,\text{kg/m}^3$

(a) Calculate the Thiele modulus and the efficiency factor for these three catalysts. What is the reaction order?

(b) Which catalyst would you choose for maximum performance?

**Solution to Problem 6.22**

(a) One of the important points in this problem is to calculate a valid characteristic length "*L*." In a cylinder, if it is not considered infinitely long:

$$\text{Volume} = \pi R^2 \cdot \text{Length}$$

$$\text{External surface} = 2\pi R \text{Length} + 2\pi R^2$$

In this way:

$$L = \frac{\text{Volume}}{\text{External surface}} = \frac{\pi R^2 \text{Length}}{2\pi R \text{Length} + 2\pi R^2} = \frac{R \cdot \text{Length}}{2 \cdot \text{Length} + 2R}$$

With the data given:

Catalyst A: $L_A = 4.53 \cdot 10^{-3}$ m

Catalyst B: $L_B = 5.00 \cdot 10^{-3}$ m

Catalyst C: $L_C = 4.40 \cdot 10^{-3}$ m

For all three catalysts, the value of "m" is:

$$m = \sqrt{\frac{k'''}{D_e}} = \sqrt{\frac{2.17 \cdot 10^{-7}\,\frac{m^3}{s \cdot kg} \cdot 1800\,\frac{kg}{m^3}}{10^{-12}\,\frac{m^2}{s}}} = 12\,763.1\ \mathrm{m}^{-1}$$

We can calculate Thiele modulus and effectiveness for each catalyst:

| Zeolite | mL | Effectiveness |
| --- | --- | --- |
| A | 57.8 | 0.017 |
| B | 63.8 | 0.015 |
| C | 56.1 | 0.018 |

(b) All three catalysts have bad effectiveness, and the best one is "C."

**Problem 6.23**  A commercial process for the dehydrogenation of ethylbenzene uses 3-mm spherical catalyst particles. The rate constant is $15\,\mathrm{s}^{-1}$, and the diffusivity of ethylbenzene in steam is $4 \cdot 10^{-5}$ m$^2$/s under reaction conditions.

(a) Assume that the pore diameter is large enough that this bulk diffusivity applies. Determine a likely lower bound for the isothermal effectiveness factor.

(b) Repeat the problem assuming a pore diameter of $20\,\mathrm{nm} = 2 \cdot 10^{-8}$ m. The reaction temperature is $625\,^{\circ}$C.

(c) How fine would you have to grind the ethylbenzene catalyst for laboratory kinetic studies to give the intrinsic kinetics?

**Solution to Problem 6.23**

(a) The lower efficiency is obtained for the highest value of "$L$." Assume that $L = R_p = 1.5\,\mathrm{mm}$ and $D_e = D_A$:

$$mL = \sqrt{\frac{k'''}{D_e}}\,L = 0.919$$

and:

$$\eta = \frac{\tanh(mL)}{mL} = 0.79$$

(b) For the Knudsen diffusivity (Conesa and Font Montesinos 2002):

$$D_K = \frac{d_{pore}}{3} \cdot \left(8 \cdot R_g \cdot \frac{T}{\pi \cdot M_A}\right)^{\frac{1}{2}}$$

$$D_K = \frac{2 \cdot 10^{-8}}{3} \cdot \left(8 \cdot 8.314\,\frac{J}{mol \cdot K} \cdot \frac{898\ \mathrm{K}}{3.1416 \cdot 0.106\,\frac{kg}{mol}}\right)^{\frac{1}{2}} = 2.82 \cdot 10^{-6}\,\frac{m^2}{s}$$

is an order of magnitude less than the ordinary diffusivity. We have that:

$$\mathrm{mL} = \sqrt{\frac{k'''}{D_{\mathrm{K}}}}\, L = 3.46$$

and:

$$\eta = \frac{\tanh(\mathrm{mL})}{\mathrm{mL}} = 0.29$$

(c) For efficiency equals unity $\mathrm{mL} < 0.4$, so:

$$\sqrt{\frac{k'''}{D_{\mathrm{K}}}}\, L < 0.4$$

We can obtain $L < 1.74 \cdot 10^{-4}$ m.

**Problem 6.24**  The cumene cracking reaction has been extensively researched by various investigators due to its clean nature and its utility in comparing the activities of different cracking catalysts. Integral flow reactors have been employed to gather data on this reaction. In such reactors, the conversion variation with reactor space time remains relatively unaffected by changes in the form of the reaction rate expression, particularly at high degrees of conversion that are far from equilibrium. Therefore, it is common to report data for these systems using a simplified first-order reaction rate constant, assuming negligible volume changes during the reaction. A commonly used expression to calculate this constant is:

$$-n_{\mathrm{A0}}X_{\mathrm{A}} = -k'C_{\mathrm{A0}}(1 - X_{\mathrm{A}})\eta W_{\mathrm{cat}}$$

$$\frac{n_{\mathrm{A0}}X_{\mathrm{A}}}{C_{\mathrm{A0}}(1 - X_{\mathrm{A}})\eta W_{\mathrm{cat}}} = k'$$

where $W_{\mathrm{cat}}$ is the weight of catalyst employed and the reaction rate constant is expressed per unit weight of catalyst. If data published are expressed in these terms, the apparent rate constant at 510 °C is approximately 0.716 cm$^3$/g$_{\mathrm{catalyst}}$·s. If the catalyst employed in this study has the properties enumerated below, determine the effectiveness factor for the catalyst. The value of the combined diffusivity can be estimated at $6.19 \cdot 10^{-3}$ cm$^2$/s.

Data on catalyst properties (silica–alumina) follow.

| | |
|---|---|
| Equivalent diameter | 0.43 cm |
| Particle density | 1.14 g/cm$^3$ |
| Specific surface area | 342 cm$^2$/g |
| Porosity | 0.51 |
| Void volume per gram | 0.447 cm$^3$/g |

**Solution to Problem 6.24**

Because we are given the apparent rate constant rather than the true rate constant, a trial-and-error solution will be required. Either of the approaches described earlier may be used, but we can employ:

$$mL = \left[\frac{k'''}{D_e}\right]^{\frac{1}{2}}\left(\frac{D}{6}\right) = \left[\frac{\frac{k'''_{ap}}{\eta}}{D_e}\right]^{\frac{1}{2}}\left(\frac{D}{6}\right) = \left[\frac{\frac{0.716\,\frac{cm^3}{g\cdot s}}{\eta}\cdot 1.14\,\frac{g}{cm^3}}{6.19\cdot10^{-3}\,\frac{cm^2}{s}}\right]^{\frac{1}{2}}\left(\frac{0.43}{6}\right) = \frac{0.8227}{\eta^{0.5}}$$

$$\eta = \frac{\tanh(mL)}{mL}$$

Solving the nonlinear relationship, we can find $mL = 1.15$ and $\eta = 0.714$.

# 7

# Catalytic Reactor Design

**Problem 7.1**   Consider the decomposition of A: A → R, carried out in a tubular, packed-bed catalytic reactor. The reaction follows first-order kinetics:

$$-r'_A \; (\mathrm{mol/kg_{cat}/s}) = k' C_A$$

The reactor/catalyst system has the following characteristics:

| Reactor setup: | Catalyst properties: |
|---|---|
| Reactor volume: 50 l | Catalyst density: 2.0 kg/l |
| Catalyst weight: 80 kg | Particle geometry: spherical |
| Fluid flow rate: 1.0 l/s | Particle diameter: 0.1 cm |
| Feed concentration of A: 100 mmol/l | Effective diffusivity of A: $5 \times 10^{-6}$ cm$^2$/s |
| Exit concentration of A: 55 mmol/l | |

(a) Calculate the apparent rate constant, $k'$, in appropriate units.
(b) Determine whether the reaction is operating with minimal, intermediate, or strong pore diffusion resistance.

**Solution to Problem 7.1**

(a) Let us do a mass balance with a first-order reaction:

$$\mathrm{Input} + \mathrm{Generation} = \mathrm{Exit} + \mathrm{Accumulation}$$

$$(n_A) + \left(r'_A\right) \mathrm{d}W_{\mathrm{cat}} = (n_A + \mathrm{d}n_A) + 0$$

$$\frac{\mathrm{d}n_A}{\mathrm{d}W_{\mathrm{cat}}} = \frac{Q_0 \mathrm{d}C_A}{\mathrm{d}W_{\mathrm{cat}}} = -\frac{Q_0 C_{A0} \mathrm{d}X_A}{\mathrm{d}W_{\mathrm{cat}}} = r'_A = -k'_{\mathrm{ap}} C_{A0}(1 - X_A)$$

$$\frac{Q_0 \mathrm{d}X_A}{(1 - X_A)} = -k'_{\mathrm{ap}} \mathrm{d}W_{\mathrm{cat}}$$

*Problem Solving in Chemical Reactor Design*, First Edition. Juan A. Conesa.
© 2025 WILEY-VCH GmbH. Published 2025 by WILEY-VCH GmbH.

Defining the space time: $\tau = \frac{W_{cat}}{Q_0}$ (having units of weight-time/volume), we find that:

$$1 - X_A = \exp\left(-k'_{ap}\tau\right)$$

With the data given:

$$\left(\frac{55}{100}\right) = \exp\left(-k'_{ap}\frac{80}{1}\right)$$

And we get $k'_{ap} = 0.007\,47\ \mathrm{l/kg_{cat}\cdot s}$. Note that this is an apparent constant, i.e., it would include the effect of the internal diffusion (effectiveness factor, $k'_{ap} = k'\cdot\eta$).

(b) For estimating in which regime the reactor is running, we calculate the Weisz modulus:

$$We = \frac{\left(-r'''_A\right)_{observed} \cdot L^2}{D_e \cdot C_{As}}$$

$$\left(-r'''_A\right)_{Observed} = \left(-r'_A\right)_{Observed} \cdot \rho_{cat} = \left(k'_{ap}C_{A0}(1 - X_A)\right) \cdot \rho_{cat}$$

$$= 0.007\,47\frac{1}{\mathrm{kg}\cdot\mathrm{s}} \cdot 55\,\frac{\mathrm{mmol}}{1} \cdot 2\,\frac{\mathrm{kg}}{1} = 0.822\,\frac{\mathrm{mmol}}{1\cdot\mathrm{s}}$$

$$We = \frac{0.822\,\frac{\mathrm{mmol}}{1\cdot\mathrm{s}} \cdot \left(\frac{0.1}{6}\right)^2\mathrm{cm}^2}{5\cdot10^{-6}\,\frac{\mathrm{cm}^2}{\mathrm{s}} \cdot 55\,\frac{\mathrm{mmol}}{1}} = 0.8303$$

In the expression, $L = R/3 = d_p/6$ for spheres. As the We value is between 0.15 and 4, the influence of the internal diffusion is intermediate.

We can also try to calculate the effectiveness factor:

$$mL = \left[\frac{k''}{D_e}\right]^{\frac{1}{2}}\left(\frac{D}{6}\right) = \left[\frac{k''_{ap}}{\eta}{D_e}\right]^{\frac{1}{2}}\left(\frac{D}{6}\right) = \left[\frac{\frac{0.007\,47\,\frac{1}{\mathrm{kg}\cdot\mathrm{s}} \cdot 2\,\frac{\mathrm{kg}}{1}}{\eta}}{5\cdot10^{-6}\,\frac{\mathrm{cm}^2}{\mathrm{s}}}\right]^{\frac{1}{2}}\left(\frac{0.1}{6}\right) = \frac{0.9134}{\eta^{0.5}}$$

$$\eta = \frac{\tanh(mL)}{mL}$$

Solving the nonlinear relationship, we can find $\eta = 0.740$.

**Problem 7.2**  A reaction is being conducted in a continuous, well-mixed fluidized bed reactor with a catalyst load of 10 kg. The feed stream, which has a constant density, contains 1.0 M reactant (A) and is supplied at a rate of 0.10 l/s. Under these operating conditions, the concentration of A leaving the reactor at steady state is measured to be 0.20 M, corresponding to an 80% conversion.

The catalyst particles in the reactor are well characterized and have a density of 2.0 kg/l. The effective diffusion coefficient of A within the catalyst particles is determined to be $1\cdot10^{-6}$ cm$^2$/s, and the mean $V_p/A_p$ ratio is 0.015 cm.

Note: The reaction order ($n$) is uncertain, but you are confident that it falls between 0 and 2.

(a) Determine whether the reaction is operating with minimal, intermediate, or strong pore diffusion resistance.

When the flow rate is increased to 0.25 l/s, the steady-state concentration of A in the exit stream is found to be 0.33 M, corresponding to a 67% conversion.

(b) Determine the rate law for the reaction in the form of $-r_A = kC_A^n$ (specify $n$ and $k$). This rate law represents the behavior that would be observed if there were no pore diffusion resistance.

**Solution to Problem 7.2**

For details refer the Wiley website at http://www.wiley-vch.de/ISBN9783527354115

(a) The balance in the CSTR is:

$$Q_0 C_{A0} + r_A' \cdot W = Q_0 C_A$$

$$r_A' = Q_0(C_{A0} - C_A)/W$$

With the data: $r_A' = -0.008$ mol/kg·s. Now we can calculate Weisz modulus:

$$We = \frac{\left(-r_A'''\right)_{observed} \cdot L^2}{D_e \cdot C_{As}}$$

With the data:

$$We = \frac{0.008\,\frac{mol}{kg \cdot s}.2\,\frac{kg}{l} \cdot 0.015^2\,cm^2}{10^{-6}\,\frac{cm^2}{s} \cdot 0.2\,\frac{mol}{l}} = 18$$

As this modulus is higher than 4, the resistance in the pores is very strong.

(b) Without knowing the rate law, $(-r_A') = k' C_A^n$, we can do the calculations and deduce the value of "$n$" that makes constant the value of "$k$." We have that:

$$r_A' = Q_0(C_{A0} - C_A)/W = k' C_A^n$$

And so:

$$k' = Q_0(C_{A0} - C_A)/\left(C_A^n \cdot W\right)$$

We have that the constant is equal in the two situations mentioned for a value of $n = 1.4755$, i.e., approximately 1.5 and:

$$k' = \frac{0.086 L^{1.5}}{(mol^{0.5} \cdot kg \cdot s)}$$

**Problem 7.3**  A small reactor that can be considered "differential" and contains eight granules of a spherical catalyst of 0.32 cm in diameter, each granule weighs approximately 0.020 g. The reactor is 1 cm long and has an internal diameter of 0.58 cm. At a given total temperature and pressure, the concentrations of A at the inlet and outlet are 0.0325 and 0.0308 mol/l, respectively, while the volumetric flow rate of the feed is 250 cm³/s. The reaction is first-order and occurs in the gas phase. The porosity of the catalyst particle is 0.35, the tortuosity of its pores is 5, and the combined diffusivity of A at operating conditions is 0.022 cm²/s.

Calculate the intrinsic rate constant for the catalytic consumption of A if there are no temperature profiles within the catalyst. In addition, consider that the effects of changing the number of moles are negligible and that there is no external resistance to mass transfer because the rate of linearity is high in the packed bed.

**Solution to Problem 7.3**

First, let us calculate the porosity of the bed. We can do:

$$\varepsilon = 1 - \frac{V_{\text{bed}}}{V_{\text{catal}}} = 1 - \frac{\rho_{\text{bed}}}{\rho_{\text{catal}}}$$

The densities are:

$$\frac{\rho_{\text{bed}}}{\rho_{\text{catal}}} = \frac{\dfrac{\text{Total weight of catalyst}}{\text{volume of bed}}}{\dfrac{\text{weight of granule}}{\text{volume of granule}}} = \frac{\dfrac{0.02 \cdot 8}{\pi \cdot \left(\frac{0.58}{2}\right)^2 \cdot 1}}{\dfrac{0.02}{\frac{4}{3}\pi(0.159)^3}} = \frac{0.606}{1.193}$$

With this, $\varepsilon = 0.493$

Considering the differential reactor as a CSTR:

$$r_{\text{A}}'' = \frac{C_{\text{A0}} - C_{\text{A}}}{\bar{t}}$$

With the residence time being:

$$\bar{t} = \frac{V_{\text{bed}}}{Q_0} \cdot \varepsilon = 5.2 \cdot 10^{-4}\ \text{s}$$

giving a value of $r_{\text{A}}'' = 3.265\ \frac{\text{mol}}{\text{l·s}}$. The reaction rate can also be expressed in terms of weight of catalyst:

$$r_{\text{A}}' = 3.265\ \frac{\text{mol}}{\text{l} \cdot \text{s}} \cdot \frac{\varepsilon}{\rho_{\text{bed}}} = 2.65 \cdot 10^{-3}\ \frac{\text{mol}}{\text{g} \cdot \text{s}}$$

Using the first-order kinetics expression:

$$r_{\text{A}}' = k' \cdot \eta \cdot C_{\text{A}}$$

As the bed is differential, an average value of $C_{\text{A}}$ can be used, giving finally a value of $k' \cdot \eta = 0.084\,92\ \text{l/g·s}$.

As in Problem 7.1, we know the relationship between effectiveness factor and Thiele modulus (mL), and using the same procedure, we get:

$$k' = 0.0936\ \frac{1}{\text{g} \cdot \text{s}}$$

And

$$\eta = 0.896$$

**Problem 7.4**   The process of cracking cumene into benzene and propylene took place in a fixed bed of zeolite particles under atmospheric pressure and at a temperature of $262\,^\circ\text{C}$. Nitrogen was present in large excess. At a specific location within the reactor, where the partial pressure of cumene was measured to be 0.0689 atm, an observed reaction rate of 0.153 $\text{kmol/kg}_{\text{cat}}\cdot\text{h}$ was recorded.

We need to demonstrate that, given these operating conditions, the partial pressure drop and temperature drop across the external film surrounding the particles can be considered negligible.

Further data available:

Molecular weight $= 34.37\,$kg/kmol
Flowrate $= 0.094\,$kg/m·h
$C_\mathrm{p} = 0.33\,$kcal/kg·K
$\mathrm{Pr} = 0.846$
$a = 45\,\mathrm{m^2_{cat}}/\mathrm{kg_{cat}}$
$(-\Delta H) = 41\,816\,$kcal/kmol
gas density $= 0.66\,$kg/m$^3$
thermal conduct gas $= 0.037\,$kcal/m·h·K
$\mathrm{Re} = 0.052$
$D_{\mathrm{Am}} = 0.096\,\mathrm{m^2}/\mathrm{h}$
$G = 5647\,$kg/m$^2$·h

**Solution to Problem 7.4**
External temperature gradient between the medium and the particle surface occurs when there is a temperature difference between the catalyst surface and the external fluid (Conesa and Font Montesinos 2002). The heat transferred to the fluid must be equal to that generated at the catalyst surface:

$$\text{Heat transferred} = h \cdot S_{\mathrm{cat}} \cdot (T_\mathrm{g} - T_\mathrm{s})$$

$$\text{Heat generated} = (-r_\mathrm{A}) \cdot \Delta H_\mathrm{r} \cdot V_{\mathrm{cat}}$$

where heat is given in J/s, $(-r_\mathrm{A})$ in kmol A/s m$^3$ cat, $(\Delta H_\mathrm{r})$ is the enthalpy of reaction in J/kmol, $V_{\mathrm{cat}}$ is the volume of the catalyst pellets, and $S_{\mathrm{cat}}$ is the outer surface of the pellet. $T_\mathrm{s}$ is the temperature of the surface of the particle, $T_\mathrm{g}$ is the temperature of the fluid surrounding the particle, and $h$ is the individual heat transport coefficient (J/s m$^2$ °C). Equating these two expressions, we can arrive at:

$$(T_\mathrm{g} - T_\mathrm{s}) = \frac{(-r_\mathrm{A}) \cdot \Delta H_\mathrm{r} \cdot V_{\mathrm{cat}}}{h \cdot S_{\mathrm{cat}}} = \frac{(-r_\mathrm{A}) \cdot \Delta H_\mathrm{r} \cdot L}{h}$$

where $L$ is the characteristic length.

The value of $h$ can be calculated in some cases by introducing the parameter $j_\mathrm{H}$:

$$j_\mathrm{H} = \left(\frac{h}{C_\mathrm{p} u \rho}\right)(\mathrm{Pr})^{\frac{2}{3}}$$

where $(u \cdot \rho)$ is the mass flow rate, Pr is the Prandt number ($\mathrm{Pr} = C_\mathrm{p}\mu/K_{\mathrm{term}}$), $C_\mathrm{p}$ is the specific heat of the fluid, and $K_{\mathrm{term}}$ is the thermal conductivity of the fluid.

In different books (e.g. Conesa and Font Montesinos (2002); Levenspiel (1999)). We can find correlations of this parameter with Reynolds number. In the present case, $\mathrm{Re} = 0.052$ and $j_\mathrm{H} = 0.052$. With this, the value of "$h$" (heat transfer coefficient) can be estimated.

On the other hand, the Thodos equation related $j_D$ with Re:

$$j_D = \frac{0.725}{(Re^{0.41} - 0.15)}$$

Being:

$$j_D = \frac{Sh}{(Sc)^{\frac{1}{3}} Re}$$

And:

$$Sc = \frac{\mu}{\rho D}$$

$$Re = (d_p \rho u)/\mu$$

The parameter $j_D$ is useful to calculate the mass transfer coefficient from the Sherwood number:

$$Sh = \frac{k_C d_p}{D}$$

The mass transfer coefficient, $k_C$, will be used to calculate the external partial pressure increment:

$$(p_g - p_s) = \frac{r_A \cdot M_A \cdot \rho_p}{a \cdot u \cdot \rho_g \cdot j_D}$$

Doing the corresponding calculations, we find with the present data:

$$j_D = 7.498$$

$$j_H = 0.052$$

$$Sc = 1.484$$

$$(p_g - p_s) = 0.000\,384\,\text{bar}$$

$$(T_g - T_s) = -0.060\,44\,\text{K}$$

**Problem 7.5**

(a) The kinetics of the catalytic reaction $A \leftrightarrow R + S$ are given by:

$$r' = \frac{dX_A}{d\left(\frac{W}{n_{A0}}\right)} = \frac{k' K_A (p_A - p_R \cdot p_S / K)}{(1 + p_A \cdot K_A + K_R \cdot p_R + K_S \cdot p_S)^2}$$

The reaction is taking place isothermally in a packed-bed reactor with plug flow at a temperature of 275 °C. The feed consists of a mixture containing 0.155 mol of water per mole of reactant, where water acts as an inert diluent and is not adsorbed on the catalyst. The following data are given:

Total flow rate = 42 kmol/h
Bed density = 1500 kg/m³
Bed diameter = 0.05 m

Catalyst rate constant, $k' = 4.3593\,\text{kmol/kg}_{\text{cat}}\cdot\text{h}$
Total pressure, $P_{\text{total}} = 3\,\text{atm}$
Reaction equilibrium constant, $K_A = 0.430\,39\,\text{atm}^{-1}$
Overall reaction rate constant, $K = 0.589\,\text{atm}$
Sum of rate constants, $(K_R + K_S) = 2.8951\,\text{atm}^{-1}$

We need to calculate the length of the reactor required to achieve exit conversions of:

(a1) 40%

(a2) 70%

(b) Now, let us consider the same reaction conditions but in a multitubular reactor. The length of each tube is 3 m. The total feed per tube is 4 kmol/h. The desired annual production is 20 000 metric tons, and the product's molecular weight is 44. We assume that one year of operation is equivalent to 8000 hours. Determine the number of tubes required to meet the production target.

**Solution to Problem 7.5**

For details refer the Wiley website at http://www.wiley-vch.de/ISBN9783527354115

(a) From the rate law, as $p_R = p_S$:

$$r' = \frac{dX_A}{d\left(\frac{W}{n_{A0}}\right)} = \frac{k'K_A(p_A - p_R \cdot p_S/K)}{(1 + p_A \cdot K_A + (K_R + K_S) \cdot p_S)^2}$$

$$\frac{(1 + p_A \cdot K_A + (K_R + K_S) \cdot p_S)^2 \, n_{A0}dX_A}{k'K_A(p_A - p_R \cdot p_S/K)} = dW$$

To integrate the previous expression, we need to know the relationship between conversion and partial pressures:

$$n_A = n_{A0}(1 - X_A)$$

$$n_R = n_{R0} + n_{A0}X_A = n_{A0}X_A$$

$$n_S = n_R$$

$$p_A = P_{\text{total}} \cdot n_A/n_0$$

$$p_R = p_S = P_{\text{total}} \cdot n_R/n_0$$

being $n_0$ the total flow of gas entering the system. In this way:

$$W = \int_0^{X_A} \frac{\left(1 + \frac{n_{A0}}{n_0}(1 - X_A) \cdot K_A + (K_R + K_S) \cdot \frac{n_{A0}}{n_0}X_A\right)^2 n_{A0}dX_A}{k'K_A\left(\frac{n_{A0}}{n_0}(1 - X_A) - \frac{\left(\frac{n_{A0}}{n_0}X_A\right)^2}{K}\right)}$$

Using a spreadsheet:

| $X_A$ | $n_A$ (kmol/s) | $n_R$ (kmol/s) | $n_S$ (kmol/s) | $P_A$ (atm) | $P_R$ (atm) | $P_S$ (atm) | $r_A = n_{A0} \cdot dX_A/dW$ (kmol/h)/(kg cat) | $n_{A0}/r_A$ | $dW$ (kg cat) |
|---|---|---|---|---|---|---|---|---|---|
| 0 | 3.636 | 0.000 | 0.000 | 2.597 | 0.000 | 0.000 | 1.086 | 3.347 | 0.000 |
| 0.02 | 3.564 | 0.073 | 0.073 | 2.545 | 0.052 | 0.052 | 0.945 | 3.848 | 0.072 |
| 0.04 | 3.491 | 0.145 | 0.145 | 2.494 | 0.104 | 0.104 | 0.824 | 4.413 | 0.083 |
| ... | | | | | | | | | |
| 0.36 | 2.327 | 1.309 | 1.309 | 1.662 | 0.935 | 0.935 | 0.017 | 213.114 | 3.047 |
| 0.38 | 2.255 | 1.382 | 1.382 | 1.610 | 0.987 | 0.987 | −0.004 | −920.719 | −7.076 |
| 0.4 | 2.182 | 1.455 | 1.455 | 1.558 | 1.039 | 1.039 | −0.024 | −154.711 | −10.754 |

We can check that the maximum conversion that can be obtained in this reactor is 0.36. In that situation, a longer reactor will produce a lower conversion due to the chemical equilibrium among the species.

Taking $X_A = 0.36$, the amount of catalyst needed is 3.047 kg. The length of the reactor can be calculated using:

$$L\,(\text{m}) = W(\text{kg}_{\text{cat}}) \cdot \frac{1}{\rho_{\text{cat}}} \left( \frac{\text{m}^3_{\text{cat}}}{\text{kg}_{\text{cat}}} \right) \cdot \frac{1}{\varepsilon} \left( \frac{\text{m}^3_{\text{react}}}{\text{m}^3_{\text{cat}}} \right) \cdot \frac{1}{S} \left( \frac{1}{\text{m}^2_{\text{react}}} \right)$$

Using the properties of the bed, the total length is 10.6 m. It is not possible to reach a conversion higher than 0.36.

(b) Suppose the reaction is carried out under the same conditions in a multitubular reactor. The tubes length is 3 m. The total feed per tube is 4 kmol/h. An annual production of 20 000 metric tons of product is required. The molecular weight of the product is 44. One year on stream is equivalent to 8000 hours. Determine the number of tubes required to meet the production.

For an annual production of 2000 ton:

$$2000 \frac{\text{ton}}{\text{year}} \cdot 1000 \frac{\text{kg}}{\text{ton}} \cdot \frac{1}{44} \frac{\text{kmol}}{\text{kg}} \cdot \frac{1}{8000} \frac{\text{year}}{\text{h}} = 56.82 \frac{\text{kmol}}{\text{h}}$$

We have $L = 3$ m and $n_{A0} = 4$ kmol/h·tube, so the number of tubes is:

$$N_{\text{tubes}} = \frac{\text{Total production}}{n_{A0} \cdot (1 - X_A)} = \frac{56.82}{4 \cdot (1 - 0.36)} = 22.19 \text{ tubes} \approx 23$$

**Problem 7.6** Let us suppose that we are setting out front first principles to investigate the dehydrogenation of ethylbenzene, which is a well-established process for manufacturing styrene:

$$C_6 H_5 \cdot CH_2 \cdot CH_3 \longleftrightarrow C_6 H_5 \cdot CH = CH_2 + H_2$$

There is available a catalyst that will give a suitable rate of reaction at 560 °C. At this temperature, the equilibrium constant for the reaction above is $K_P = 100\,\text{mbar} = 10^4\,\text{N/m}^2$.

(a) If a feed of pure ethylbenzene is used at 1 bar pressure, determine the fractional conversion at equilibrium.
(b) If the feed to the process consists of ethylbenzene diluted with steam in the ratio 15 mol steam: 1 mol ethylbenzene, determine the new fractional conversion at equilibrium.

**Solution to Problem 7.6**

For details refer the Wiley website at http://www.wiley-vch.de/ISBN9783527354115

(a) $P$ (bar) $= 1$         pressure at which the reaction occurs, in bar.
     $P$ (Pa) $= 100\,000$     pressure at which the reaction occurs, in Pa.

Subscript EB will be used to refer to ethylbenzene, W to refer to water vapor, S to refer to styrene, and H to refer to hydrogen.
No inlet flow data are given, so a calculation basis is assumed.

$$n_{EB0} \text{ (mol/s)} = 100$$

$$n_{W0} \text{ (mol/s)} = 0$$

$$n_{S0} \text{ (mol/s)} = 0$$

$$n_{H0} \text{ (mol/s)} = 0$$

$$n_0 = n_{EB0} + n_{W0} + n_{S0} + n_{H0}$$

$$n_0 \text{ (mol/s)} = 100$$

The moles of each component at the outlet of the reactor are calculated as follows:

$$n_{EB} = n_{EB0}(1 - X_{EB})$$

$$n_W = n_{W0}$$

$$n_S = n_{S0} + n_{EB0}X_{EB}$$

$$n_H = n_{H0} + n_{EB0}X_{EB}$$

The total moles that will exist once equilibrium occurs are therefore:

$$n = n_{EB} + n_W + n_S + n_H = n_{EB0} - n_{EB0}X_{EB} + n_{W0} + n_{S0} + n_{EB0}X_{EB} + n_{H0} + n_{EB0}X_{EB}$$

$$n = n_0 + n_{EB0}X_{EB}$$

So the mole fractions of each component at equilibrium will be

$$y_{EB} = n_{EB}/n = (n_{EB0}(1 - X_{EB}))/(n_0 + n_{EB0}X_{EB})$$

$$y_W = n_W/n = n_{W0}/(n_0 + n_{EB0}X_{EB})$$

$$y_S = n_S/n = (n_{S0} + n_{EB0}X_{EB})/(n_0 + n_{EB0}X_{EB})$$

$$y_H = n_H/n = (n_{H0} + n_{EB0}X_{EB})/(n_0 + n_{EB0}X_{EB})$$

The partial pressures can be calculated using Dalton's law as follows:

$$P_{EB} = P y_{EB}$$

$$P_W = P y_W$$

$$P_S = P y_S$$

$$P_H = P y_H$$

Then, in equilibrium, it holds that:

$$K_P = P_{S,out} P_{H,out} / P_{EB,out}$$

Therefore, substituting in the values of the partial pressures and clearing the value of $X_{EB}$, it is obtained that:

$$(n_{EB02} \cdot (K_P + P)) \cdot X_{EB2} + (n_{EB0} \cdot (K_P \cdot (n_0 - n_{EB0}) + P \cdot (n_{S0} + n_{H0}))) \cdot X_{EB}$$
$$+ (P \cdot n_{S0} \cdot n_{H0} - K_P \cdot n_{EB0} \cdot n_0) = 0$$

That can be solved and gives: $X_{EB} = 0.301$.

(b) The above process is repeated with the same conditions, except that at the inlet there is a steam flow rate of 15 mol steam for each mole of ethylbenzene fed:

$$n_{EB0} \ (\mathrm{mol/s}) = 100$$

$$n_{W0} \ (\mathrm{mol/s}) = 1500$$

$$n_{S0} \ (\mathrm{mol/s}) = 0$$

$$n_{H0} \ (\mathrm{mol/s}) = 0$$

$$n_0 = n_{EB0} + n_{W0} + n_{S0} + n_{H0}$$

and gives: $X_{EB} = 0.735$.


**Problem 7.7**  The following second-order liquid-phase catalytic reaction is conducted in an isothermal fixed-bed reactor:

$$A \rightarrow B \quad r = k \cdot C_A^2$$

The intrinsic rate constant is 161 l/mol·s. The feed is 0.8 l/s of a solution of A with the concentration of A at 4 mol/l. Spherical catalyst pellets of radius 0.12 cm are used to pack the reactor. The catalyst has a pellet density of 0.88 g/cm$^3$, and the reactor bed density is 0.5 g/cm$^3$. The effective diffusivity of A inside the catalyst pellet is $3.14 \cdot 10^{-3}$ cm$^2$/s. You wish to achieve 99.9% conversion of A.

(a) First, neglect both diffusional resistance and mass transfer resistance. What is the mass of catalyst required to achieve the desired conversion?
(b) Next, consider diffusional resistance, but neglect mass transfer resistance. What are the Thiele modulus and effectiveness factor at the entrance of the reactor?

What are the Thiele modulus and effectiveness factor at the exit of the reactor? Given these two effectiveness factors, compute an upper bound and a lower bound on the mass of catalyst required to achieve the desired conversion.

(c) Next, consider both diffusional resistance and mass transfer resistance. Assume that we have estimated the mass transfer coefficient to be $k_C = 0.35\,\text{cm/s}$. Find the effectiveness factor at the entrance of the reactor. Find the effectiveness factor at the exit of the reactor.

Given these two effectiveness factors, compute an upper bound and a lower bound on the mass of catalyst required to achieve the desired conversion.

**Solution to Problem 7.7**

For details refer the Wiley website at http://www.wiley-vch.de/ISBN9783527354115

(a) For a fixed-bed reactor with a second-order reaction, the mass balance would be:

$$dn_A = -n_{A0} \cdot dX_A = r_A'' dV = -k'' C_A^2 \cdot \eta \cdot dV = -k'' C_{A0}^2 (1 - X_A)^2 \cdot \eta \cdot dV$$

$$\int dV = \int_0^{X_A} \frac{n_{A0} \cdot dX_A}{k'' C_{A0}^2 (1 - X_A)^2 \cdot \eta}$$

But at this point, efficiency is unity. So:

$$V = \int_0^{0.999} \frac{(4 \cdot 0.8) \cdot dX_A}{161 \cdot 4^2 (1 - X_A)^2 \cdot 1} = 12.48\,\text{l}$$

The weight of catalyst is:

$$W = 12.48\,\text{l} \cdot 0.5\,\frac{\text{g}}{\text{cm}^3} \cdot 1000\,\frac{\text{cm}^3}{\text{l}} = 6243\,\text{g}$$

(b) If efficiency is not unity, at the input:

$$mL = \left[\frac{3k'' \cdot C_{As}}{2 \cdot D_e}\right]^{\frac{1}{2}} L = \left[\frac{3 \cdot 16 \cdot 4}{2 \cdot 3.14 \cdot 10^{-3}}\right]^{\frac{1}{2}} \left(\frac{0.12}{3}\right) = 6.994$$

$$\eta = \frac{\tanh(mL)}{mL} = 0.143$$

And, at the exit:

$$C_A = C_{A0}(1 - X_A) = 0.004\,\frac{\text{mol}}{\text{l}}$$

$$mL = \left[\frac{3k'' \cdot C_{As}}{2 \cdot D_e}\right]^{\frac{1}{2}} L = \left[\frac{3 \cdot 16 \cdot 0.004}{2 \cdot 3.14 \cdot 10^{-3}}\right]^{\frac{1}{2}} \left(\frac{0.12}{3}\right) = 0.2212$$

$$\eta = \frac{\tanh(mL)}{mL} = 0.984$$

Using the expression in (a), at the input:

$$V = \int_0^{0.999} \frac{(4 \cdot 0.8) \cdot dX_A}{161 \cdot 4^2 (1 - X_A)^2 \cdot 0.143} = 87.33\,\text{l}$$

$$W = 87.33\,\text{l} \cdot 0.5\,\frac{\text{g}}{\text{cm}^3} \cdot 1000\,\frac{\text{cm}^3}{\text{l}} = 43\,662\,\text{g}$$

And, at the exit:

$$V = \int_0^{0.999} \frac{(4 \cdot 0.8) \cdot dX_A}{161 \cdot 4^2(1-X_A)^2 \cdot 0.984} = 12.69\,\mathrm{l}$$

$$W = 12.69\,\mathrm{l} \cdot 0.5\,\frac{\mathrm{g}}{\mathrm{cm}^3} \cdot 1000\,\frac{\mathrm{cm}^3}{\mathrm{l}} = 6345\,\mathrm{g}$$

So, the lower and upper limits for the weight of catalyst needed are 6345 and 43 662 g, respectively.

(c) Now we have that:

$$(-r_A) = kC_{As}^2\eta = k_C(C_{Ag} - C_{As})$$

We know the values of $C_{Ag}$ (in the bulk gas), $k$, and $k_C$. We will assume a value for the concentration in the surface ($C_{As}$), calculate the Thiele modulus (mL), and then the efficiency factor to check for the previous equality. The value of $k_C$ is in cm/s, and we need it in units of $(\text{time})^{-1}$.

$$k_C''' = 0.35\,\frac{\mathrm{cm}}{\mathrm{s}} \cdot \frac{6}{d_p} = 17.5\,\mathrm{s}^{-1}$$

where $6/d_p$ is the ratio between the surface and the volume of a sphere. Using the data, we get:

| Input | Exit |
| --- | --- |
| $C_A$ (mol/l) = 4 | $C_A$ (mol/l) = 0.004 |
| $k$ (l/mol/s) = 16 | $k$ (l/mol/s) = 16 |
| $k_m$ (cm/s) = 0.35 | $k_m$ (cm/s) = 0.35 |
| $a_s$ (cm$^2$/cm$^3$) = 50 | $a_s$ (cm$^2$/cm$^3$) = 50 |
| $C_{A,S}$ (mol/l) = 2.785 79 | $C_{A,S}$ (mol/l) = 0.005 47 |
| $D_e$ (cm$^2$/s) = 0.003 14 | $D_e$ (cm$^2$/s) = 0.003 14 |
| $L$ (cm) = 0.04 | $L$ (cm) = 0.04 |
| mL = 5.837 | mL = 0.259 |
| Efficiency = 0.171 | Efficiency = 0.978 |
| $\eta k C_{A,S}^2 = 21.273$ | $\eta k C_{A,S}^2 = 0.000$ |
| $k_m \cdot a_s \cdot (C_A - C_{A,S}) = 21.249$ | $k_m \cdot a_s \cdot (C_A - C_{A,S}) = -0.026$ |

Using the conditions at the input: $V = 75.03\,\mathrm{l}$, $W = 36\,513\,\mathrm{g}$
Using the conditions at the exit: $V = 12.76\,\mathrm{l}$, $W = 6582\,\mathrm{g}$.

**Problem 7.8**  You have been asked to design a packed-bed reactor to carry out the following reaction in the gas phase: 2A → B. It is required to achieve a conversion fraction of 0.9 for 5 m$^3$/s of feed containing 50% (molar basis) A and 50% of an inert I. The operating conditions are 28 °C and 1.1 atm; the pressure drop is considered negligible. The diameter of the catalyst is 1.28 cm, and its particle and packed-bed densities are 1.2 and 0.66 g/cm$^3$, respectively. Assume negligible external resistance to mass transfer.

The intrinsic rate, per mole of reaction, is given by:

$$r' = 0.069 \frac{L^2}{\text{mol} \cdot \text{g} \cdot \text{s}} C_A^2$$

The effectiveness factor can be calculated by:

$$\eta = 0.943 - 0.781 \log(mL)$$

The effective diffusivity of A is $0.005 \, \text{cm}^2/\text{s}$. Calculate:

(a) The weight of catalyst needed.
(b) The volume of the packed bed, in $\text{m}^3$.

**Solution to Problem 7.8**

For details refer the Wiley website at http://www.wiley-vch.de/ISBN9783527354115

(a) and (b) The total amount of gas entering the reactor is:

$$n_0 = \frac{PQ}{RT} = \frac{1.1 \cdot 5 \cdot 1000}{0.082 \cdot 301} = 222.83 \, \frac{\text{mol}}{\text{s}}$$

For the reactant:

$$n_{A0} = n_0 \cdot 0.5 = 111.41 \, \frac{\text{mol}}{\text{s}}$$

$$C_{A0} = \frac{111.41}{5000} = 0.022\,83 \, \frac{\text{mol}}{\text{l}}$$

At the exit:

$$n_A = n_{A0}(1 - X_A) = n_{A0} \cdot 0.1 = 111.14 \, \text{mol/s}$$

Note that the reaction rate given in the statement refers to mole reacting, but the stoichiometry indicated should be accounted for, in such a way that:

$$r'_A = -2 \cdot r' = -0.138 \frac{L^2}{\text{mol} \cdot \text{g} \cdot \text{s}} C_A^2$$

In the FPR:

$$dn_A = -n_{A0} \cdot dX_A = r'_A dW = -2k' C_A^2 \cdot \eta \cdot dW = -2k' C_{A0}^2 (1 - X_A)^2 \cdot \eta \cdot dW$$

$$W = \int_0^{X_{Af}} \frac{Q \cdot dX_A}{2k' C_{A0}(1 - X_A)^2 \cdot \eta} = \frac{Q}{2k' C_{A0}} \int_0^{X_{Af}} \frac{dX_A}{(1 - X_A)^2 \cdot \eta}$$

Note that the effectivity depends on the concentration, i.e., the conversion, so the analytical integration cannot be done and will be integrated numerically.

For using the concentration in $\text{mol/cm}^3$, the units of the constant should be modified:

$$k' = 0.069 \, \frac{L^2}{\text{g} \cdot \text{mol} \cdot \text{s}} = 0.069 \, \frac{\text{dm}^6}{\text{g} \cdot \text{mol} \cdot \text{s}} \cdot (10^3)^2 \left( \frac{\text{cm}^3}{\text{dm}^3} \right)^2 = 690\,00 \, \frac{\text{cm}^6}{\text{g} \cdot \text{mol} \cdot \text{s}}$$

Here:

$$\frac{Q}{2k' C_{A0}} = 1.625 \cdot 10^6 \, \text{g}$$

And then:

$$W = 3.25 \cdot 10^6 \sum_0^{0.9} \frac{\Delta X_A}{(1 - X_A)^2 \cdot \eta}$$

For second-order reaction:

$$\mathrm{mL} = \left[ \frac{3k' \cdot C_A}{2 \cdot D_e} \right]^{\frac{1}{2}} L$$

As this module depends on $C_A$, it will be changing along the reactor as well as the efficiency of the reaction. At the input conditions:

$$\mathrm{mL} = \left[ \frac{3 \cdot 0.069 \cdot 0.0232}{2 \cdot 0.005} \right]^{\frac{1}{2}} \left( \frac{0.64}{3} \right) = 12.3$$

$$\eta = 0.943 - 0.781 \log(\mathrm{mL}) = 0.09$$

For doing the integration, we will use a spreadsheet in the form:

| $X_A$ | $C_A$ (mol/cm$^3$) | $r_A$ (mol/g·s) | mL | $\eta$ | Weight catalyst (g) | Volume (cm$^3$) |
|---|---|---|---|---|---|---|
| 0.00 | $2.2 \cdot 10^{-5}$ | $3.4 \cdot 10^{-5}$ | 12.3 | 0.09 | 0.0 | 0.0 |
| 0.01 | $2.2 \cdot 10^{-5}$ | $3.4 \cdot 10^{-5}$ | 12.2 | 0.09 | 176 996.6 | 0.5 |
| 0.02 | $2.2 \cdot 10^{-5}$ | $3.3 \cdot 10^{-5}$ | 12.2 | 0.10 | 354 331.5 | 1.1 |
| ... | ... | | | | | |
| 0.88 | $2.7 \cdot 10^{-6}$ | $4.9 \cdot 10^{-7}$ | 4.3 | 0.45 | 40 810 875.4 | 123.7 |
| 0.89 | $2.5 \cdot 10^{-6}$ | $4.1 \cdot 10^{-7}$ | 4.1 | 0.47 | 43 502 182.2 | 131.8 |
| 0.90 | $2.2 \cdot 10^{-6}$ | $3.4 \cdot 10^{-7}$ | 3.9 | 0.48 | 46 628 251.5 | 141.3 |

Resulting in a total weight of almost 47 tons of catalyst and a volume of 70.6 m$^3$ (accounting for the bed density mentioned in the statement). This would indicate that it is a very slow process in the mentioned conditions.

Graphically, the problem gives:

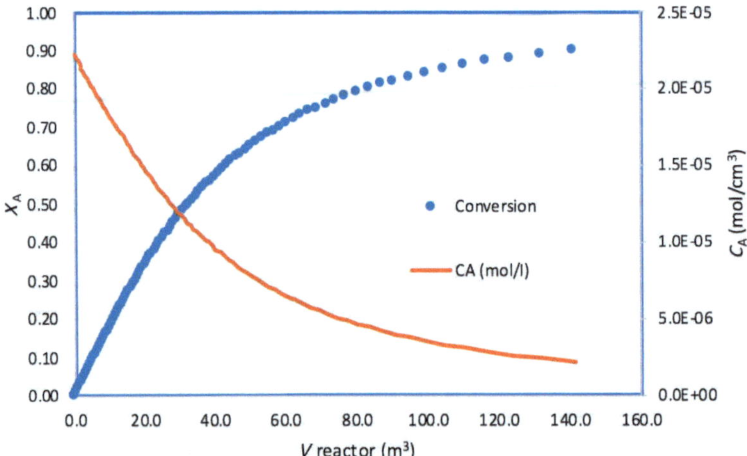

**Problem 7.9**  The gas-phase catalytic reaction $A + \tfrac{1}{2}B \rightarrow C$ is conducted in an isothermal tubular reactor. The feed consists of $1\,\text{m}^3/\text{min}$ with 98% A and 2% B at $1\,\text{atm}$ and $373\,\text{K}$. The mass transfer coefficient of B $(k_C a)_B$, is $0.15\,\text{l/s·g}$. Suppose that for A, we can neglect its external resistance to mass transfer. Also, ignore the effect of the change in number of moles on the mole fraction because the associated error would be less than 1%. The catalytic rate (including internal effects) is given by:

$$r' = 5.714 \cdot 10^{-4} \frac{\text{mol reacted}}{\text{s} \cdot \text{g} \cdot \text{atm}^{1.8}} p_A p_B^{0.8}$$

(a) Estimate the amount of catalyst needed to react half of the reactant B fed.

(b) Compare the amount of catalyst needed in case of neglecting the resistance to mass transfer of B.

**Solution to Problem 7.9**

For details refer the Wiley website at http://www.wiley-vch.de/ISBN9783527354115

(a) For the reactant B, we can write the following, considering the units of the different constants and equating mass transfer rate to the reaction rate:

$$r_B' = \frac{5.714 \cdot 10^{-4}}{2}\frac{\text{mol B}}{\text{s} \cdot \text{g} \cdot \text{atm}^{1.8}}\, p_A p_{Bs}^{0.8}\,\text{atm}^{1.8} = (k_C a)_B \frac{\text{mol}}{1 \cdot \text{g}}(C_{Bg} - C_{Bs})\frac{\text{mol}}{1}$$

Note that $p_B$ in the intrinsic reaction rate is estimated at the surface of the catalyst $(p_{Bs})$ that is related to its concentration by:

$$p_{Bs} = C_{Bs} \cdot R_g T$$

At the input, we have:

$$n_0 = \left(1000\,\frac{1}{\text{min}} \cdot 1\,\text{atm}\right)/(0.082\,\text{atm} \cdot \text{l/mol/K} \cdot 373\,\text{K}) = 32.69\,\frac{\text{mol}}{\text{min}}$$

$$n_{B0} = 0.02 n_0 = 0.654\,\frac{\text{mol}}{\text{min}}$$

$$n_{A0} = 0.98 n_0 = 32.036\,\frac{\text{mol}}{\text{min}}$$

$$C_{B0} = 0.654/1 = 0.654\,\frac{\text{mol}}{\text{m}^3} = 6.54 \cdot 10^{-4}\,\frac{\text{mol}}{1}$$

$$p_{B0} = 0.02\,\text{atm}$$

At the exit, $n_B = 0.5 n_{B0} = 0.327\,\text{mol/min}$, $C_B = 3.27 \cdot 10^{-4}\,\text{mol/l}$, $n_A = n_{A0} - 2 n_{B0} X_B = 31.382\,\text{mol/min}$ and $p_A = \frac{31.382}{31.382 + 0.327} \cdot P_{total} = 0.9896\,\text{atm}$ and $p_B = 0.0103\,\text{atm}$

Using the first equation, we can calculate $C_{Bs}$, being $C_{Bs} = 5.80 \cdot 10^{-4}\,\text{mol/l}$ at the input and $C_{Bs} = 2.85 \cdot 10^{-4}\,\text{mol/l}$.

These values give reaction rates of $1.11 \cdot 10^{-5}\,\text{mol B/s/g}$ at the input and $0.636 \cdot 10^{-5}\,\text{mol B/s/g}$ at the exit. We can estimate that the reaction rate is more or less constant with a value of $8.71 \cdot 10^{-6}\,\text{mol B/s·g}$.

As the total amount of B converted is $0.327\,\text{mol/min} = 0.005\,45\,\text{mol/s}$, we have:

$$W = \frac{0.005\,45}{8.71 \cdot 10^{-6}} = 626\,\text{g of catalyst}$$

(b) If mass transfer of B is ignored, $C_{Bs} = C_B$, $p_B = p_{Bs}$, and the reaction rate would be given by:

$$r'_B = \frac{5.714 \cdot 10^{-4}}{2} \frac{\text{mol B}}{\text{s} \cdot \text{g} \cdot \text{atm}^{1.8}} p_A p_{Bs}^{0.8} \text{ atm}^{1.8}$$

At the input, $r_B = 1.22 \cdot 10^{-5}$ mol B/s/g and $0.703 \cdot 10^{-5}$ mol B/s/g at the exit. With an average of $9.64 \cdot 10^{-6}$ mol/s/g, the amount of catalyst needed is 565 g.

**Problem 7.10**  Two materials, mole sieve 5A and $\gamma$-Al$_2$O$_3$, can be used for drying a gas stream flowing at a rate of 100 ml/min at a temperature of 27 °C and a pressure of 1 bar. The gas stream contains 1 vol% water. The adsorption of water on these materials can be described as a surface reaction with the kinetics given by $(-r_A) = k \cdot C_A$.

(a) Your goal is to determine which drying material will result in the lowest water content when the gas stream is fed to an ideal plug flow reactor (PFR) with an empty volume of 250 ml, filled with one of the drying materials.
(b) Additionally, you need to calculate the water content that can be achieved using the chosen drying material.

The following data are provided for the two materials:

| | Units | Material 1 | Material 2 |
|---|---|---|---|
| Average pore diameter | A | 5 | 1040 |
| Internal surface area | m$^2$/g | 200 | 200 |
| Particle diameter | mm | 8 | 2 |
| Particle density | kg/m$^3$ | 1200 | 1500 |
| Bed porosity | — | 0.6 | 0.6 |
| Effective diffusion coeff. water in the particles | m$^2$/s | 2.4 10$^{-8}$ | 3.6 10$^{-6}$ |
| Adsorption rate constant | m$^3$/m$^2 \cdot$ s | 1.0 10$^{-8}$ | 3.0 10$^{-11}$ |

**Solution to Problem 7.10**

(a) The key point in this problem is to calculate the efficiency factor for both materials. For the first one, we have:

$$(\text{mL})_1 = \sqrt{\frac{k'''}{D_e}} \left(\frac{d_p}{6}\right) = \sqrt{\frac{1 \cdot 10^{-8} \frac{\text{m}^3}{\text{m}^2 \cdot \text{s}} \cdot 200 \frac{\text{m}^2}{\text{g}} \cdot 1200 \frac{\text{kg}}{\text{m}^3} 1000 \frac{\text{g}}{\text{kg}}}{2.4 \cdot 10^{-8} \frac{\text{m}^2}{\text{s}}}} (1.3 \cdot 10^{-3}) \text{ m} = 13$$

$$\eta_1 = \frac{\tanh(\text{mL}_1)}{\text{mL}_1} = 0.076\,92$$

Similarly, $(\text{mL})_2 = 0.065$ and $\eta_2 = 0.999$. With this data, we cannot already say that the second material is better because the kinetic constant is much lower. Let us calculate the water content at the exit.

In a PFR with a first-order reaction:

$$X_A = 1 - \exp\left(-k''' \cdot \eta \cdot \frac{V}{Q_0}\right)$$

Doing the calculations, $X_{A1} = 0.763$ and $X_{A2} = 0.047$, so the first material is much better.

(b) The water content will be reduced by 76.3% with the first material ($X_{A1} = 0.763$) and by 4.7% with the second one.

**Problem 7.11**   A small vessel with a volume of 50 ml is partially filled with a packed bed of solid, spherical catalyst KX, behaving like an ideal PFR. The decomposition of a gas stream pure of A, flowing through this reactor, is determined by first-order reaction kinetics: $(-r_A) = kC_A$, where:

$$(-r_A) = \text{decomposition rate of A per unit catalyst surface}$$
$$k = \text{reaction rate constant} = 2 \cdot 10^{-6}\ \text{m}^3/\text{m}^2 \cdot \text{s}$$
$$C_A = \text{concentration of A at the catalyst surface (mol/m}^3\text{)}$$

Data for catalyst KX:

$$p = \text{density of the catalyst particle} = 1000\ \text{kg/m}^3$$
$$Sv = \text{specific internal surface area} = 4.0 \cdot 10^8\ \text{m}^2/\text{m}^3$$
$$Sg = \text{specific internal surface area} = 400\ \text{m}^2/\text{g}$$
$$D_e = \text{effective diffusion coefficient of A in catalyst} = 2.0 \cdot 10^{-6}\ \text{m}^2/\text{s}$$

In the first experiment, the reactor is filled with particles with a diameter of 3 mm, which are loosely packed, resulting in a bed porosity of 0.60. In this situation, a conversion of A of 38.1% is obtained.

Determine the following:

(a) Flow rate $Q_0$ through the reactor in the first experiment.
(b) The catalyst bed is then "tapped" resulting in a close-packed bed with a decreased porosity of 0.52. No additional catalyst has been added to the bed. Determine the conversion of A obtained in the new steady state.
(c) After this experiment, the catalyst particles of 3 mm are replaced by an equal mass of particles with a diameter of 0.03 mm, which are also tapped. The porosity of the bed remains at 0.52. Determine the steady-state conversion of A in this situation.

**Solution to Problem 7.11**

(a) In the reactor:

$$Q_0 C_{A0}(-dX_A) = -k' C_{A0}(1 - X_A)\eta\, dW$$

$$W = -\frac{Q_0 \ln(1 - X_A)}{k'\eta}$$

$$(mL) = \sqrt{\frac{k'''}{D_e}}\left(\frac{d_p}{6}\right) = \sqrt{\frac{2\cdot10^{-6}\,\frac{m^3}{m^2_{cat}\cdot s}\cdot400\,\frac{m^2}{m^3}\cdot10^6\,\frac{g}{m^3_{cat}}}{2\cdot10^{-6}\,\frac{m^2}{s}}}\left(\frac{1.5\cdot10^{-3}}{6}\right)\quad m = 10$$

$$\eta = \frac{\tanh(mL)}{mL} = 0.1$$

The amount of catalyst is given by:

$$W = 50\,\text{mL}_{bed}\cdot\frac{10^{-6}\,m^3_{bed}}{\text{mL}_{bed}}\frac{(1-0.6)\,m^3_{cat}}{m^3_{bed}}1000\,\frac{kg_{cat}}{m^3_{cat}} = 0.02\,kg_{cat} = 20\,g_{cat}$$

Using the relation with the conversion: $X_A = 0.381$ and $Q_0 = 0.0033\,m^3/s$.

(b) Now:

$$W = 50\,\text{mL}_{bed}\cdot10^{-6}\frac{m^3_{bed}}{\text{mL}_{bed}}\frac{(1-0.56)\,m^3_{cat}}{m^3_{bed}}1000\,\frac{kg_{cat}}{m^3_{cat}} = 0.024\,kg_{cat} = 24\,g_{cat}$$

Using the previous value of the flowrate, the new conversion is $X_A = 0.390$.

(c) In this situation, mL = 0.1 and $\eta = 0.996$. Using the data above, the conversion now is $X_A = 0.990$.

**Problem 7.12**   A particular porous catalyst is available as spheres of seven different diameters, ranging from 0.1 µm up to 100 mm, with a factor of 10 increase in diameter. Unfortunately, the data for the conversion of a flow of pure A in an ideal PFR with a fixed mass of catalyst have been shuffled. Only the seven conversion levels are known, without their corresponding particle diameters. The seven conversion levels are 0.8647, 0.8647, 0.0326, 0.2701, 0.5432, 0.8115, and 0.8647. It is known that one of these conversion levels is incorrect, while the other six are correct.

(a) Determine which conversion level is incorrect and what its correct value should be.
    Using the same catalyst, two similar series of spherical particles with different diameters are available. However, these catalysts have been used in an industrial reactor for years, resulting in 50% poisoning of the active surface. In the first series, homogeneous poisoning has occurred, affecting the exterior and the center of the catalyst to the same extent.
(b) Determine the conversion in the smallest and the largest spheres of this series under the standard conditions mentioned in part (a).

In the catalyst particles of the second series, poisoning progressed from the exterior, resulting in two distinct zones within the spheres:

- A poisoned outer shell of the catalyst particle that is completely inactive.
- A non-poisoned center with the original activity.

Here are some additional notes to consider:

- The reaction is first-order in A.

- All parameters not mentioned may be assumed to be the same within a series and between series.
- Use the porous-sphere model in your calculations.

**Solution to Problem 7.12**

For details refer the Wiley website at http://www.wiley-vch.de/ISBN9783527354115

(a) In the PFR with a first-order reaction:

$$W = -\frac{Q_0 \ln(1 - X_A)}{k'\eta}$$

and the effectiveness factor can be calculated from $D_e$ and $k$, besides the values of particle diameter.

The higher conversion would correspond to the smallest particles, as the diffusion into them is much more favorable and the effectiveness factor will be higher.

Let us assume values for the constants that will make the problem much easier. For example, we will take values of $10^{-6}$ m$^2$/s for $D_e$, 0.02 mol/kg·s for $k'$, density of the catalyst 1600 kg/m$^3$, and $W = 1$ kg. With this, for the smallest particles:

$$Q_0 = -\frac{k'W\eta}{\ln(1 - X_A)} = -\frac{0.02 \cdot 1 \cdot 1}{\ln(1 - 0.8647)} = 0.009\,987 \; \frac{m^3}{s}$$

With all these values, we will calculate the effectiveness factors and conversions of all other particle diameters:

| $D$ (m) | $R$ (m) | mL | $\eta$ | $\ln(1 - X_A)$ | $X_A$ |
| --- | --- | --- | --- | --- | --- |
| 0.000 000 1 | 0.000 000 05 | 0.000 094 3 | 1.000 00 | −2.000 260 762 | 0.864 70 |
| 0.000 001 0 | 0.000 000 50 | 0.000 942 8 | 1.000 00 | −2.000 260 175 | 0.864 70 |
| 0.000 010 0 | 0.000 005 00 | 0.009 428 1 | 0.999 97 | −2.000 201 503 | 0.864 69 |
| 0.000 100 0 | 0.000 050 00 | 0.094 280 9 | 0.997 05 | −1.994 355 066 | 0.863 90 |
| 0.001 000 0 | 0.000 500 00 | 0.942 809 | 0.781 19 | −1.562 577 776 | 0.790 40 |
| 0.010 000 0 | 0.005 000 00 | 9.428 090 4 | 0.106 07 | −0.212 159 69 | 0.191 16 |
| 0.100 000 0 | 0.050 000 00 | 94.280 904 | 0.010 61 | −0.021 215 969 | 0.020 99 |

As we can see, the value of conversion that was wrong is 0.8115, and the previous table shows the ordered values. If we had assumed other values of the constants, the resulting $Q_0$ would have been different, but not the values of effectiveness factor and conversions.

(b) If the spheres are 50% poisoned, the conversion will change, since there will be half the active surface. In the definition of "$L$":

$$L = \frac{\text{Volume of the particles}}{\text{External surface available for entry and difussion of the reagent}}$$

the external surface will be half, so $L$ is double. With this:

| D (m) | R | mL | $\eta$ | $\ln(1 - X_A)$ | $X_A$ |
|---|---|---|---|---|---|
| 0.000 000 1 | 0.000 000 1 | 0.000 188 6 | 1.000 00 | −2.000 26 | 0.864 70 |
| 0.000 001 0 | 0.000 001 0 | 0.001 885 6 | 1.000 00 | −2.000 26 | 0.864 70 |
| 0.000 010 0 | 0.000 010 0 | 0.018 856 2 | 0.999 88 | −2.000 02 | 0.864 67 |
| 0.000 100 0 | 0.000 100 0 | 0.188 561 8 | 0.988 31 | −1.976 89 | 0.861 50 |
| 0.001 000 0 | 0.001 000 0 | 1.885 618 1 | 0.506 46 | −1.013 05 | 0.636 89 |
| 0.010 000 0 | 0.010 000 0 | 18.856 180 8 | 0.053 03 | −0.106 08 | 0.100 65 |
| 0.100 000 0 | 0.100 000 0 | 188.561 808 3 | 0.005 30 | −0.010 61 | 0.010 55 |

With the smallest particles, the effectiveness factor would be the same and so the conversion, but not with bigger particles.

**Problem 7.13** Air containing 2 vol% ammonia is to be passed through a column at a rate of $5000\,\text{m}^3/\text{h}$. The column is packed with 25 mm Raschig rings (interfacial area $74\,\text{m}^2/\text{m}^3$) and is operating at 20 °C and 1 atm. The aim is to remove 98% of the ammonia by absorption into water. Assuming a superficial air velocity of 1 m/s and a water flow rate of approximately twice the minimum required, calculate the required column diameter and packed height. The solubility of ammonia in water at 20 °C is given by:

$$H_A = \frac{Cg\left(\frac{\text{kg}}{\text{m}^3}\right)}{C_L\left(\frac{\text{kg}}{\text{m}^3}\right)} = 0.002\,02$$

The absorption of ammonia into water is a typical case where gas-phase resistance controls the mass transfer rates. The value of $k_G$ can be assumed to be 164 m/h.

**Solution to Problem 7.13**
For details refer the Wiley website at http://www.wiley-vch.de/ISBN9783527354115

The concentrations of ammonia in the gas at the bottom and top of column are:

$$C_{Ag0} = \frac{p_{A0}}{RT} = \frac{0.01}{0.082 \cdot 298} = 8.18 \cdot 10^{-4}\,\frac{\text{mol}}{\text{l}}$$

$$C_{Ag\,\text{exit}} = (1 - 0.98) \cdot C_{Ag0} = 1.63 \cdot 10^{-5}\,\frac{\text{mol}}{\text{l}}$$

The amount of ammonia to be removed is $8.18 \cdot 10^{-4} \cdot 5000 \cdot 0.98 = 4010\,\text{mol/h}$.
The minimum amount of water required is then:

$$L = G \cdot \frac{C_{Ag}}{C_{Al}} = 5000 \cdot H_A = 5000 \cdot 0.002\,02 = 10.1\,\frac{\text{m}^3}{\text{h}}$$

If water is used at 2 l $= 20.2\,\mathrm{m^3/h}$, then the $NH_3$ concentration in water leaving the column is:

$$C_{Al\ exit} = \frac{4010}{20.2} = 198.53\,\frac{mol}{m^3} = 0.198\,53\,\frac{mol}{l}$$

For a superficial gas velocity of $u_g = 1\,\mathrm{m/s}$, the required sectional area of the column is $5000/3600 = 1.388\,\mathrm{m^2}$, and the column is 1.33 m in diameter. In this situation, the superficial water rate is:

$$u_1 = \frac{20.2}{1.388} = 16.31\,\frac{m}{h}$$

In the vessel, a plug type of flow should be used for modeling, and then:

$$GdC_{Ag} = \left(-r_A'''\right) \cdot dV$$

$$\left(-r_A'''\right) = k_{Ag}a \cdot \left(C_{Ag} - C_{Ag}^*\right)$$

where $\left(C_{Ag} - C_{Ag}^*\right)$ is the difference between the concentration of ammonia in the gas and the concentration of ammonia in equilibrium with the liquid in contact with that gas. In this way:

$$GdC_A = k_{Ag}a \cdot \left(C_{Ag} - C_{Ag}^*\right) \cdot dV$$

$$\frac{dC_A}{\left(C_{Ag} - C_{Ag}^*\right)} = k_{Ag}a \cdot \frac{dV}{G}$$

$$\int_{C_{A0}}^{C_{A\ exit}} \frac{dC_A}{\left(C_{Ag} - C_{Ag}^*\right)} = k_{Ag}a \cdot \frac{V}{G} = k_{Ag}a \cdot \frac{h}{u_g}$$

The concentration of ammonia in the liquid, at the exit is:

$$C_{Al\ exit}^* = 0.1983\,\frac{mol}{l} \cdot 0.002\,02\,\frac{\left(\frac{mol}{l}\right)_g}{\left(\frac{mol}{l}\right)_l} = 4.01 \cdot 10^{-4}\,\frac{mol}{l}$$

So, with the data:

$$\int_{C_{A0}}^{C_{A\ exit}} \frac{dC_A}{\left(C_{Ag} - C_{Ag}^*\right)} = \ln\left(\frac{\left(C_{A0} - C_{Ag}^*\right)}{\left(C_{A\ exit} - C_{Ag}^*\right)}\right) = 164 \cdot 74 \cdot \frac{h}{3600}$$

$$\ln\left(\frac{4.17 \cdot 10^{-4}}{3.66 \cdot 10^{-3}}\right) = 164 \cdot 74 \cdot \frac{h}{3600}$$

and then: $h = 0.6438\,\mathrm{m}$.

In the following figure and table, we can find a scheme of the results of this system:

| $C_{Ag0}$ | $8.18 \cdot 10^{-4}$ | mol/l | 0.818 | mol/m³ | 0.013 91 | kg/m³ |
| --- | --- | --- | --- | --- | --- | --- |
| $C_{Ag\,exit}$ | $1.64 \cdot 10^{-5}$ | mol/l | 0.016 | mol/m³ | 0.000 28 | kg/m³ |
| $C^*_{Ag\,exit}$ | $4.06 \cdot 10^{-3}$ | kmol/m³ | 4.059 | mol/m³ | 0.069 00 | kg/m³ |
| | | | | | | |
| $C_{AL0}$ | 0 | mol/l | 0 | mol/m³ | 0 | kg/m³ |
| A absorbed | | | 4010.476 | mol/h | 68.178 10 | kg/h |
| $C_{AL\,exit}$ | $1.99 \cdot 10^{-1}$ | mol/l | 198.538 | mol/m³ | 3.375 15 | kg/m³ |
| $C^*_{Ag\,exit}$ | $4.01 \cdot 10^{-4}$ | mol/l | 0.401 | mol/m³ | 0.006 82 | kg/m³ |
| | | | | | | |
| $(C_{Ag0} - C^*_{Ag\,exit})$ | $4.17 \cdot 10^{-4}$ | mol/l | 0.417 | mol/m³ | 0.007 10 | kg/m³ |
| $(C_{Ag\,exit} - C^*_{Ag\,exit})$ | $3.66 \cdot 10^{-3}$ | mol/l | 3.658 | mol/m³ | 0.062 18 | kg/m³ |

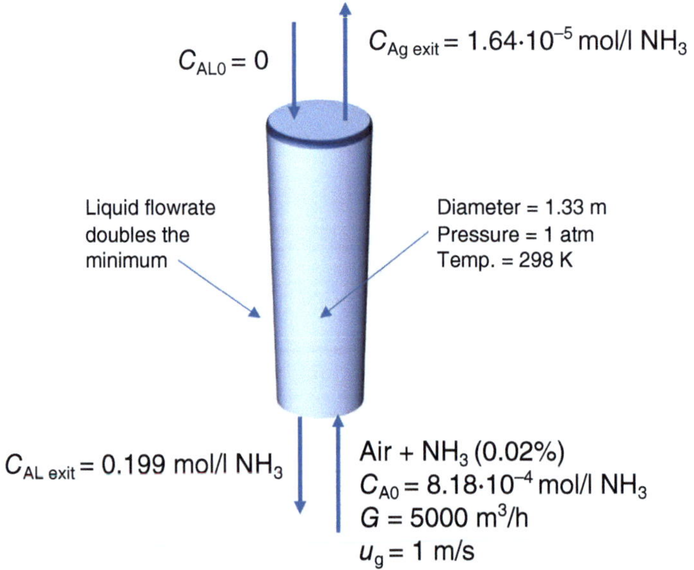


**Problem 7.14** An air–SO$_2$ mixture containing 10 vol% of SO$_2$ is flowing at 340 m³/h (20 °C, 1 atm). If 95% of the SO$_2$ is to be removed by absorption into water in a countercurrent-packed column operated at 20 °C, 1 atm, how much water (kg/h) is required? The absorption equilibria are given in table:

| $p_{SO_2}$ (mmHg) | 26 | 59 | 123 |
| --- | --- | --- | --- |
| $X$ (kg SO$_2$/100 kg H$_2$O) | 0.5 | 1.0 | 2.0 |

**Solution to Problem 7.14**

For details refer the Wiley website at http://www.wiley-vch.de/ISBN9783527354115

In order to solve the problem, first we will plot the equilibrium data in units of $Y$ (molar ratio in the gas) versus $X$ (molar ratio in the liquid):

| $p_{SO_2}$ (atm) | $y$ (mol $SO_2$/mol total) | $Y$ (mol $SO_2$/mol inert) | $X$ (kg $SO_2$/ 100 kg $H_2O$) | $X$ (molar) |
|---|---|---|---|---|
| 0.0342 | 0.0342 | 0.0354 | 0.5 | 0.0014 |
| 0.0776 | 0.0776 | 0.0842 | 1 | 0.0028 |
| 0.1618 | 0.1618 | 0.1931 | 2 | 0.0056 |

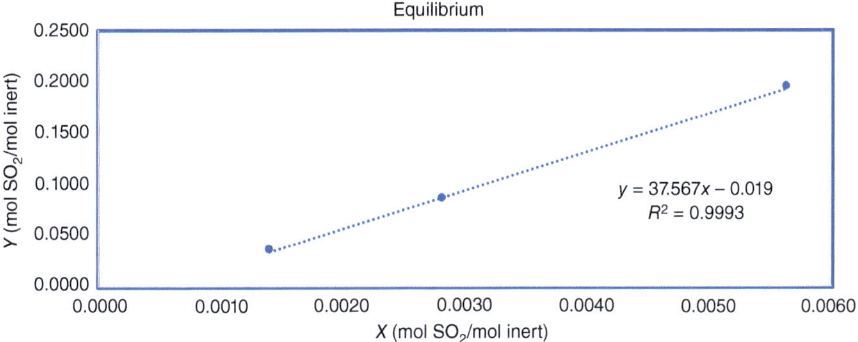


As we can see, a good approximation for the equilibrium data is $Y = 37.567X - 0.019$.

The input molar ratio in the gas is:

$$Y_0 = \frac{y_0}{1 - y_0} = \frac{0.1}{1 - 0.1} = 0.111$$

And the liquid enters with no $SO_2$, so $X_0 = 0$. As we know that enters $340\,\text{m}^3/\text{h}$ of gas, we can do the following transformation:

$$G' = G(1 - y_0) = 340\,\frac{\text{m}^3}{\text{h}} \cdot \rho_{\text{mix}}\,\frac{\text{kg}}{\text{m}^3} \cdot \left(M_{\text{mix}}\,\frac{\text{kg}}{\text{kmol}}\right)^{-1}$$

where the density of the mix and its molecular weight can be calculated by doing a proportional average:

$$\rho_{\text{mix}} = (1 - y_0)\rho_{\text{air}} + y_0\rho_{SO_2}$$

$$M_{\text{mix}} = (1 - y_0)M_{\text{air}} + y_0 M_{SO_2}$$

The exit molar ratio will be:

$$Y_{\text{exit}} = (1 - 0.95)Y_0 = 0.0056$$

In equilibrium with that gas, we will have:

$$Y_{\text{exit}} = 37.567 X_{\text{exit}} - 0.019$$

and then $X_{\text{exit}} = 0.035$ and the global mass balance in the column is:

$$L' \left( \frac{\text{kmol}}{\text{h}} \right) = \frac{G'(Y_0 - Y_{\text{exit}})}{X_{\text{exit}} - X_0} = 387.81 \, \frac{\text{kmol}}{\text{h}}$$

$$L' \left( \frac{\text{kg}}{\text{h}} \right) = L' \left( \frac{\text{kmol}}{\text{h}} \right) \cdot M_{\text{mix}} \left( \frac{\text{kg}}{\text{kmol}} \right) = 6980.67 \, \frac{\text{kg}}{\text{h}}$$

**Problem 7.15**   The absorption of ammonia in air into water at 20 °C is being investigated. The process involves a liquid film mass transfer coefficient of $2.70 \cdot 10^{-5}$ m/s and an overall coefficient of $1.44 \cdot 10^{-4}$ m/s. The solubility of ammonia in water at 20 °C is described by the equation:

$$H_{\text{A}} = \frac{Cg \left( \frac{\text{kg}}{\text{m}^3} \right)}{C_{\text{L}} \left( \frac{\text{kg}}{\text{m}^3} \right)} = 0.002\,02$$

Let us determine the gas film coefficient and the percentage resistance to mass transfer in the gas phase.

**Solution to Problem 7.15**

The mass transfer rate is usually calculated using individual mass transfer coefficients:

$$\left( -r_{\text{A}}''' \right) = (-r_{\text{A}}) \cdot a = k_{\text{Ag}} \cdot a \cdot (p_{\text{A}} - p_{\text{Ai}}) = k_{\text{AL}} \cdot a \cdot (C_{\text{Ai}} - C_{\text{A}})$$

but can also be calculated using global mass transfer coefficients, with driving force $\left( p_{\text{A}} - p_{\text{A}}^* \right)$ and $\left( C_{\text{A}}^* - C_{\text{A}} \right)$, using the equations:

$$\frac{1}{K_{\text{Ag}}} = \frac{1}{k_{\text{Ag}}} + \frac{H_{\text{A}}}{k_{\text{AL}}}$$

$$\frac{1}{K_{\text{AL}}} = \frac{1}{k_{\text{AL}}} + \frac{1}{H_{\text{A}} \cdot k_{\text{Ag}}}$$

where pressures and concentrations refer to equilibrium, that is:

$$p_{\text{A}}^* = H_{\text{A}} \cdot C_{\text{A}}$$

$$C_{\text{A}}^* = p_{\text{A}}/H_{\text{A}}$$

In this way:

$$\left( -r_{\text{A}}''' \right) = (-r_{\text{A}}) \cdot a = K_{\text{Ag}} \cdot a \cdot \left( p_{\text{A}} - p_{\text{A}}^* \right) = K_{\text{AL}} \cdot a \cdot \left( C_{\text{A}}^* - C_{\text{A}} \right)$$

So we have that:

$$\frac{1}{K_{\text{Ag}}} = \frac{1}{k_{\text{Ag}}} + \frac{H_{\text{A}}}{k_{\text{AL}}}$$

$$\frac{1}{K_{\text{AL}}} = \frac{1}{k_{\text{AL}}} + \frac{1}{H_{\text{A}} \cdot k_{\text{Ag}}}$$

Doing the corresponding calculations, we find that:

$$\frac{1}{K_{AL}} - \frac{1}{k_{AL}} = \frac{1}{H_A \cdot k_{Ag}}$$

$$k_{Ag} = H_A \left( \frac{1}{K_{AL}} - \frac{1}{k_{AL}} \right) = 0.002\,02 \cdot \left( \frac{1}{2.7 \cdot 10^{-5}} - \frac{1}{1.44 \cdot 10^{-4}} \right) = 0.0165\,\frac{m}{s}$$

The percentage of resistance in the gas film can be estimated by the proportion of the two summands in the former expressions, in such a way that:

$$\% \text{ gas resistance} = \frac{\frac{1}{H_A k_{Ag}}}{\frac{1}{k_{AL}} + \frac{1}{H_A k_{Ag}}} \cdot 100 = \frac{\frac{1}{H_A k_{Ag}}}{\frac{1}{K_{AL}}} \cdot 100 = \frac{K_{AL}}{H_A k_{Ag}} \cdot 100 = 81.5$$

**Problem 7.16**   At a temperature of $15\,°C$, the solubility of $NH_3$ in water is represented by $H_A = p_A/C_A = 0.461\,\text{atm·m}^3/\text{kmol}$. Additionally, the film coefficients of mass transfer for gas and liquid phases are given as $k_G = 1.10\,\text{kmol/m}^2\text{·h·atm}$ and $k_L = 0.34\,\text{m/h}$, respectively. We need to estimate the value of the overall mass transfer coefficient, $K_G$, in units of m/h and the percentage of the gas film resistance.

**Solution to Problem 7.16**

For details refer the Wiley website at http://www.wiley-vch.de/ISBN9783527354115

The mass transfer rate is usually calculated using individual mass transfer coefficients:

$$(-r''_A) = (-r_A) \cdot a = k_{Ag} \cdot a \cdot (p_A - p_{Ai}) = k_{AL} \cdot a \cdot (C_{Ai} - C_A)$$

but can also be calculated using global mass transfer coefficients, with driving force $(p_A - p_A^*)$ and $(C_A^* - C_A)$, using the equations:

$$\frac{1}{K_{Ag}} = \frac{1}{k_{Ag}} + \frac{H_A}{k_{AL}}$$

$$\frac{1}{K_{AL}} = \frac{1}{k_{AL}} + \frac{1}{H_A \cdot k_{Ag}}$$

where pressures and concentrations refer to equilibrium, that is:

$$p_A^* = H_A \cdot C_A$$

$$C_A^* = p_A/H_A$$

In this way:

$$(-r''_A) = (-r_A) \cdot a = K_{Ag} \cdot a \cdot (p_A - p_A^*) = K_{AL} \cdot a \cdot (C_A^* - C_A)$$

So we have that:

$$\frac{1}{K_{Ag}} = \frac{1}{k_{Ag}} + \frac{H_A}{k_{AL}}$$

$$\frac{1}{K_{AL}} = \frac{1}{k_{AL}} + \frac{1}{H_A \cdot k_{Ag}}$$

Doing the corresponding calculations, we find that:

---

**Data**

---

| | |
|---|---|
| Temperature $(T)$ $(^\circ\text{C})$ | 15.00 |
| Constant Henry $(H_{\text{NH}_3})$ $(\text{atm}\cdot\text{m}^3/\text{kmol})$ | 0.4610 |
| Mass transfer coefficient of gas film $(k_\text{G})$ $(\text{kmol/m}^2\cdot\text{h}\cdot\text{atm})$ | 1.100 |
| Mass transfer coefficient of liquid film $(k_\text{L})$ $(\text{m/h})$ | 0.3400 |
| | |
| Global mass transfer coefficient of gas film $(K_\text{G})$ $(\text{m/h})$ | 0.4415 |
| Global mass transfer coefficient of liquid film $(K_\text{L})$ $(\text{m/h})$ | 0.2035 |
| | |
| Percentage resistance of gas film (%) | 40.14 |

---

The percentage of resistance in the gas film can be estimated by the proportion of the two summands in the former expressions in such a way that:

$$\% \text{ gas resistance} = \frac{\dfrac{1}{k_{\text{Ag}}}}{\dfrac{1}{k_{\text{Ag}}} + \dfrac{H_\text{A}}{k_{\text{AL}}}} \cdot 100 = \frac{\dfrac{1}{k_{\text{Ag}}}}{\dfrac{1}{K_{\text{Ag}}}} \cdot 100 = \frac{K_{\text{Ag}}}{k_{\text{Ag}}} \cdot 100 = 40.14$$

**Problem 7.17**  A gas component A in air is being absorbed into water under conditions of 1 atm and 20 °C. The Henry's law constant, $H_\text{A}$, for this system is $1.67 \cdot 10^3$ Pa$\cdot$m$^3$/kmol. Additionally, the liquid film mass transfer coefficient, $k_\text{L}$, and the gas film coefficient, $k_\text{G}$, are given as $2.50 \cdot 10^{-6}$ and $3.00 \cdot 10^{-3}$ m/s, respectively. We need to determine the following:

(a) The overall coefficient of gas–liquid mass transfer, $K_\text{L}$, in units of m/s.
(b) When the bulk concentrations of A in the gas and liquid phases are $1.013 \cdot 10^4$ Pa and 2.00 kmol/m$^3$, respectively, calculate the molar flux of A.

**Solution to Problem 7.17**
The mass transfer rate is usually calculated using individual mass transfer coefficients:

$$\left(-r''_\text{A}\right) = (-r_\text{A}) \cdot a = k_{\text{Ag}} \cdot a \cdot (p_\text{A} - p_{\text{Ai}}) = k_{\text{AL}} \cdot a \cdot (C_{\text{Ai}} - C_\text{A})$$

but can also be calculated using global mass transfer coefficients, with driving force $(p_\text{A} - p_\text{A}^*)$ and $(C_\text{A}^* - C_\text{A})$, using the equations:

$$\frac{1}{K_{\text{Ag}}} = \frac{1}{k_{\text{Ag}}} + \frac{H_\text{A}}{k_{\text{AL}}}$$

$$\frac{1}{K_{\text{AL}}} = \frac{1}{k_{\text{AL}}} + \frac{1}{H_\text{A} \cdot k_{\text{Ag}}}$$

where pressures and concentrations refer to equilibrium, that is:

$$p_A^* = H_A \cdot C_A$$

$$C_A^* = p_A/H_A$$

In this way:

$$(-r_A'') = (-r_A) \cdot a = K_{Ag} \cdot a \cdot (p_A - p_A^*) = K_{AL} \cdot a \cdot (C_A^* - C_A)$$

So we have that:

$$\frac{1}{K_{Ag}} = \frac{1}{k_{Ag}} + \frac{H_A}{k_{AL}}$$

$$\frac{1}{K_{AL}} = \frac{1}{k_{AL}} + \frac{1}{H_A \cdot k_{Ag}}$$

(a) Doing the corresponding calculations, we find that:

$$\frac{1}{K_{AL}} = \frac{1}{2.5 \cdot 10^{-6}} + \frac{1}{3 \cdot 10^{-3}}$$

Note that, due to the units of the constants, in this case, the statement gives the value of the product $H_A \cdot k_{Ag}$. From the former expression: $K_{AL} = 2.49 \cdot 10^{-6}$ m/s

(b)

$$(-r_A) = k_{AL} \cdot (C_{Ai} - C_A) = k_{AL} \cdot (p_{Ai}/H_A - C_A) = K_{AL} \cdot (p_A/H_A - C_A)$$

$$(-r_A) = 2.49 \cdot 10^{-6}\,\frac{m}{s} \left(\frac{1.013 \cdot 10^4}{1.63 \cdot 10^3} - 2\right)\frac{kmol}{m^3} = 10.49 \cdot 10^{-6}\,\frac{kmol}{s \cdot m^2}$$

$$= 10.49 \cdot 10^{-3}\,\frac{mol}{s \cdot m^2}$$

**Problem 7.18** The second-order reaction $A \to R$ is studied in a recycle reactor using a very large recycle ratio. The following data are recorded:

> Empty volume of the reactor: 1 l
> Weight of catalyst used: 3 g
> Feed to the reactor: $C_{A0} = 2$ mol/l
> $Q_0 = 1$ l/h
> Output current conditions: $C_{Aout} = 0.5$ mol/l

(a) Calculate the rate constant for this reaction (give units).
(b) Calculate the amount of catalyst needed in a fixed-bed reactor for a conversion of 80% of a feed of 1000 l/h of concentration $C_{A0} = 1$ mol/l. There is no recirculation.
(c) Repeat part (b) if the reactor is filled with one part of catalyst and four parts of inert solid. This addition of inert helps maintain isothermal conditions and eliminates possible temperature rises.

*Note:* Assume isothermal conditions.

**Solution to Problem 7.18**

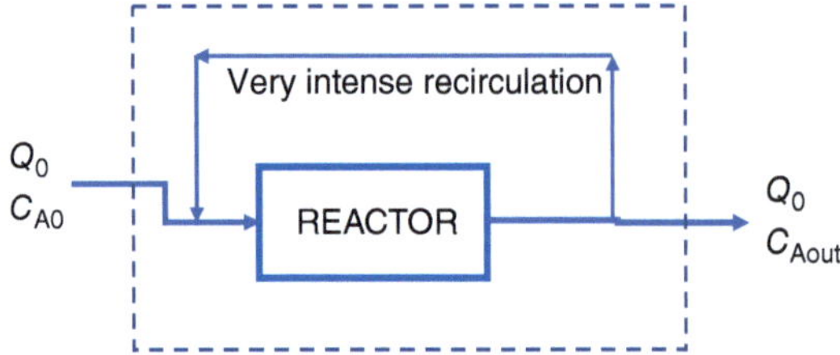



(a) This reactor can be considered as complete mixing, since the recirculation is very large, and we are going to have the entire reactor at the same concentration. We do a mass balance in the zone marked in the draw:

$$\text{Input} + \text{Generation} = \text{Output} + \text{Accumulation}$$

$$n_{A0} + r''_A V = n_{Aout}$$

$$Q_0 C_{A0} + \left(-k'' C^2_{Aout}\right) V = Q_0 C_{Aout}$$

$$1\,\frac{1}{h}2\,\frac{\text{mol}}{1} - k0.5^2\,\frac{\text{mol}^2}{1^2} \cdot 21 = 1\,\frac{1}{h}\,0.5\,\frac{\text{mol}}{1}$$

We obtain $k'' = 3\,1/\text{mol/h}$. If we consider the catalyst weight as the design variable:

$$Q_0 C_{A0} + \left(-k' C^2_{Aout}\right) W = Q_0 C_{Aout}$$

And $k' = 0.6\,1^2/\text{h}/\text{g}_{\text{cat}}/\text{mol}$.

(b) Now we have a fixed-bed reactor that probably will be similar to a PFR:



The differential mass balance is:

$$dn_A = r'_A\,dW$$

$$Q_0 dC_A = \left(-k' C^2_A\right) dW$$

$$\frac{dC_A}{-k' C^2_A} = \frac{dW}{Q_0}$$

$$W = \frac{Q_0}{k'}\left(\frac{1}{C_{Aout}} - \frac{1}{C_{A0}}\right) = \frac{1000}{0.6}\left(\frac{1}{0.2} - \frac{1}{1}\right) = 6666.6\,\text{g}$$

(c) No changes. The reactor will be larger, but the same amount of catalyst is needed.

**Problem 7.19**  A reactor with a volume of 50 ml is completely filled with spherical catalyst particles of 3 mm diameter. The bed porosity is 0.6, and the reactor can be considered an ideal plug flow system. The decomposition of pure A in the reactor follows first-order kinetics, given by the equation:

$$r_S = k_S \cdot C_A$$

where $r_S$ is the reaction rate per unit catalyst surface area (mol/s·m²$_{cat}$), $C_A$ is the concentration of A (mol/m³), and $k_S$ is the rate constant of the reaction ($2\cdot10^{-6}$ m³/s·m²$_{cat}$).

For the catalyst, the following data are provided:

$$\rho_{PARTICLES} \quad = \quad 1000\,\text{kg/m}^3$$
$$\text{Specific surface area "a"} \quad = \quad 400\,\text{m}^2/\text{g}$$
$$D_{e,A} \quad = \quad 2\cdot10^{-6}\,\text{m}^2/\text{s}$$

Now, 'let us address the different parts of the problem:

(a) Give the expression to calculate the efficiency factor and the design equation for an isothermal PFR with a given amount of this catalyst.
(b) In the above situation, the conversion of A in the reactor is 0.4. Calculate the efficiency factor of the catalyst and the volumetric flow rate through the reactor.
(c) The reaction tube is subjected to a vibration system, causing the density of the bed to increase, decreasing the porosity to 0.52. Calculate the conversion that will result.
(d) The 3 mm particles are replaced by the same mass of catalyst, but now with a diameter of 0.03 mm. The porosity of this bed is also 0.52. What is the new conversion?

**Solution to Problem 7.19**

(a) For the calculation of the efficiency factor, we have the expression:

$$\eta = \frac{\tanh(mL)}{mL}$$

With:

$$m = \sqrt{\frac{k'''}{D_e}}$$

and

$$L = \frac{\text{Volume of the particles}}{\text{External surface available for entry and difussion of the reagent}}$$

Usually, for spheres, $L = d_p/6$, as the external area coincides with $4\pi R^2$ and the volume is $4/3\pi R^3$. Nevertheless, for porous particles, we should account for the surface of the pores, so the best form to calculate the characteristic length in this case is:

$$L = \frac{1000\,\frac{\text{kg}_{cat}}{\text{m}^3_{cat}}}{400\,\frac{\text{m}^2_{cat}}{\text{g}_{cat}} \cdot 1000\,\frac{\text{g}}{\text{kg}}} = 0.0025\,\text{m}$$

We have:

$$m = \sqrt{\dfrac{2 \cdot 10^{-6}\,\frac{m^3}{m^2 s} \cdot 400\,\frac{m^2}{g} \cdot 1000\,\frac{kg}{m^3} \cdot 10^{-3}\,\frac{g}{kg}}{2 \cdot 10^{-6}\,\frac{m^2}{s}}} = 20\,\mathrm{m}^{-1}$$

And so, mL = 0.05 and $\eta = 0.9991$.

(b) In the PFR:

$$dn_A = r'_A dW$$

$$Q_0 dC_A = (-k'\eta C_A)dW$$

$$C_A = C_{A0} \cdot \exp\left(-k\eta\frac{W}{Q_0}\right)$$

As we have $C_A = (1-0.4)\cdot C_{A0}$, we obtain $W/Q_0 = 255\,625\,\mathrm{g\cdot s/m^3}$.
In the bed:

$$W = 50\,\mathrm{mL_{bed}}(1-0.6)\frac{mL_{cat}}{mL_{bed}}1000\,\frac{kg_{cat}}{m^3_{cat}} \cdot \frac{10^{-6}\,m^3_{cat}}{mL_{cat}} \cdot 1000\frac{g}{kg} = 20\,\mathrm{g}$$

And, finally, $Q_0 = 0.000\,078\,2\,\mathrm{m^3/s} = 78.2\,\mathrm{ml/s}$.

(c) In the bed, now:

$$W = 50\,\mathrm{mL_{bed}}(1-0.52)\frac{mL_{cat}}{mL_{bed}}1000\,\frac{kg_{cat}}{m^3_{cat}} \cdot \frac{10^{-6}\,m^3_{cat}}{mL_{cat}} \cdot 1000\,\frac{g}{kg} = 24\,\mathrm{g}$$

$$C_A = C_{A0} \cdot \exp\left(-2 \cdot 10^{-6} \cdot 0.9991 \cdot \frac{24}{0.000\,078\,2}\right) = 0.540 C_{A0}$$

$$X_A = 1 - \frac{C_A}{C_{A0}} = 0.458$$

(d) Now the effectiveness factor should be quite close to unity, and the conversion
is:

$$C_A = C_{A0} \cdot \exp\left(-2 \cdot 10^{-6} \cdot 1 \cdot \frac{24}{0.000\,078\,2}\right) = 0.541 C_{A0}$$

$$X_A = 1 - \frac{C_A}{C_{A0}} = 0.459$$

**Problem 7.20**  A recycle reactor containing 101 g of catalyst is used in an experimental study. The catalyst is packed into a segment of the reactor having a volume of 125 cm³. The recycle lines and pump have an additional volume of 150 cm³. The particle density of the catalyst is 1.12 g/cm³, its internal void fraction is 0.505, and its surface area is 400 m²/g. A gas mixture is fed to the system at 150 cm³/s. The inlet concentration of reactant A is 1.6 mol/m³. The outlet concentration of reactant A is 0.4 mol/m³.

(a) Determine the intrinsic pseudo-homogeneous reaction rate, the rate per unit mass of catalyst, and the rate per unit surface area of catalyst. The reaction is A → P.

(b) Suppose that the reaction is first-order. Determine the pseudo-homogeneous rate constant, the rate constant based on catalyst mass, and the rate constant based on catalyst surface area.

(c) The piping in the recycle reactor has been revised to lower the recycle line and pump volume to $100\,\text{cm}^3$. What effect will this have on the exit concentration of component A if all other conditions are held constant?

**Solution to Problem 7.20**

(a) The volume of the gas phase is the entire reactor except for the volume taken up by the mechanical parts and by the skeleton of the catalyst particles:

$$V_{\text{gas}} = 125 + 150 - \frac{101}{1.12}(1 - 0.505) = 230\,\text{cm}^3$$

For this:

$$\varepsilon_{\text{total}} = \frac{230}{275} = 0.836$$

and:

$$\bar{t} = \frac{230}{150}\left(\frac{\text{cm}^3}{\frac{\text{cm}^3}{\text{s}}}\right) = 1.53\,\text{seconds}$$

The intrinsic pseudo-homogeneous rate, as determined using the CSTR is:

$$r_A'' = \frac{(0.4 - 1.6)}{1.53} = -0.783\,\frac{\text{mol}}{\text{m}^3 \cdot \text{s}}$$

The average density of the catalyst is:

$$\rho_{\text{cat}} = \frac{101}{275} = 0.367\,\frac{\text{g}}{\text{cm}^3} = 367\,\frac{\text{kg}}{\text{m}^3}$$

and:

$$r_A' = \varepsilon_{\text{total}}\frac{r_A''}{\rho_{\text{cat}}} = -1.78 \cdot 10^{-3}\,\frac{\text{mol}}{\text{kg} \cdot \text{s}}$$

$$r_A = \varepsilon_{\text{total}}\frac{r_A''}{\rho_{\text{cat}} \cdot a_c} = -4.45 \cdot 10^{-9}\,\frac{\text{mol}}{\text{m}^2 \cdot \text{s}}$$

(b) Since the reaction rate is $r_A = -k''' C_{A,\text{out}}$, the rates in the previous problem are just divided by the appropriate exit concentration to obtain the values of $k$. The ordinary concentration is used for the $r_A$ ($\text{mol/m}^2 \cdot \text{s}$):

$$k''' = \frac{0.783}{0.4} = 1.96\,\text{s}^{-1}$$

The reactant concentration per unit mass is used for the rate based on catalyst mass:

$$(C_{A,\text{out}})_{\text{cat.mass}} = C_{A,\text{out}}\frac{\varepsilon_{\text{total}}}{\rho_{\text{cat}}} = 9.11 \cdot 10^{-4}\,\frac{\text{mol}}{\text{kg}}$$

and then:

$$k''' = \frac{r_A'}{(C_{A,\text{out}})_{\text{cat.mass}}} = 1.96\,\text{s}^{-1}$$

In a similar way:

$$(C_{A,out})_{surface\ area} = C_{A,out}\frac{\varepsilon_{total}}{\rho_{cat}a_c} = 2.28 \cdot 10^{-9}\ \frac{mol}{m^2}$$

$$k''' = \frac{r_A}{(C_{A,out})_{surface\ area}} = 1.96\ s^{-1}$$

(c) The catalyst charge is unchanged. If the reaction is truly heterogeneous and there are no mass transfer resistances, the rate of reaction of the component should be unchanged. More specifically, the pseudo-homogeneous rate for the CSTR will change since the gas phase volume and residence time change, but the heterogeneous rate should be the same:

$$V_{gas} = 125 + 100 - \frac{101}{1.12}(1 - 0.505) = 180\ cm^3$$

For this:

$$\varepsilon_{total} = \frac{180}{225} = 0.8$$

and:

$$\bar{t} = \frac{180}{150}\left(\frac{cm^3}{\frac{cm^3}{s}}\right) = 1.2\ seconds$$

The average density of the catalyst is:

$$\rho_{cat} = \frac{101}{225} = 0.449\ \frac{g}{cm^3} = 449\ \frac{kg}{m^3}$$

Assume $r'_A = -1.78 \cdot 10^{-3}\ \frac{mol}{kg \cdot s}$, the intrinsic pseudo-homogeneous rate is:

$$r''_A = \rho_{cat}r'_A/\varepsilon_{total} = -0.999\ \frac{mol}{m^3 \cdot s}$$

**Problem 7.21**  The reaction $(CH_3)_2C=CH_2 + H_2O \rightarrow (CH_3)_3C–OH$ with excess water is conducted in a fixed-bed reactor using a spherical catalyst. The catalyst has a radius of 0.213 cm and a density of 2 g/cm³. The reaction is carried out under isothermal conditions at 100 °C and 1 atm. It is a reversible reaction, and the final conversion achieved is 80% of the equilibrium conversion. At 100 °C, the equilibrium constant is measured to be 16.6. The reaction rate is $1.1 \cdot 10^{-5}$ mol/s $g_{cat}$, and the external concentration of the limiting reactant is $1.65 \cdot 10^{-5}$ mol/cm³. The mean effective diffusivity of the reactants in the catalyst is $2 \cdot 10^{-2}$ cm²/s. The task is to determine the effectiveness factor of the catalyst for this reaction.

**Solution to Problem 7.21**

We would assume a reversible first-order reaction forward and reverse since the solution is dilute, and so:

$$A \leftrightarrow R$$

The reaction rate will be:

$$(-r'_A) = (k'_1 \cdot C_A - k'_2 C_R)\,\eta = k'_1\eta\left(C_A - \frac{C_R}{K}\right)$$

being $k_1$ the kinetic constant of the forward reaction, $k_2$ the kinetic constant of the reverse reaction, and $K$ the equilibrium constant. At the equilibrium, the resulting rate is zero, so:

$$K = \frac{C_{Req}}{C_{Aeq}} = \frac{X_{Aeq}}{1 - X_{Aeq}} = 16.6$$

The equilibrium conversion would be $X_{Aeq} = 0.94$, and $X_A = 0.8 \cdot X_{Aeq} = 0.75$.

In the case of reversible reactions, the following expression is introduced for a first-order reaction:

$$m = \sqrt{\frac{k'''_{(forward\ reaction)}}{X_{Aeq}D_e}}$$

Using the definition of $X_A$, we can write:

$$(-r'_A) = k'_1\eta\left(C_{A0}(1 - X_A) - \frac{C_{A0}X_A}{K}\right) = k'_1\eta C_{A0}\left((1 - X_A) - \frac{X_A}{K}\right)$$

$$= k'_1\eta C_{A0}\left((1 - X_A) - \frac{X_A}{X_{Aeq}/(1 - X_{Aeq})}\right) = k'_1\eta C_{A0}\left(\frac{X_{Aeq} - X_A}{X_{Aeq}}\right)$$

Substituting the known parameters:

$$(-r'_A) = k'_1\eta C_{A0}\left(\frac{X_{Aeq} - X_A}{X_{Aeq}}\right) = 1.1 \cdot 10^{-5}\ \frac{mol}{s \cdot g_{cat}}$$

$$= \frac{k'_1\eta\left(1.65 \cdot 10^{-5}\ \frac{mol}{cm^3}\right)(0.94 - 0.75)}{0.94}$$

We obtain that:

$$k'_1\eta = 3.2982\ \frac{cm^3}{s \cdot g_{cat}}$$

Using the catalyst density:

$$k'''_1\eta = 1.6419\ s^{-1}$$

In the present case:

$$m = \sqrt{\frac{k'''_1}{X_{Aeq}D_e}} = \sqrt{\frac{k'''_1\eta/\eta}{X_{Aeq}D_e}} = \sqrt{\frac{(1.6419/\eta)\ s^{-1}}{0.94 \cdot 2 \cdot 10^{-2}\ \frac{cm^2}{s}}} = \frac{9.365}{\eta^{0.5}}$$

and:

$$\eta = \frac{\tanh(mL)}{mL}$$

being:

$$L = \frac{0.213}{3} = 0.071\ cm$$

Solving, by iteration, for the effectiveness factor, we find:

$$\eta = 0.855$$

**Problem 7.22**   In the previous problem, a new experiment was conducted using a different catalyst consisting of spherical particles with a diameter of 0.13 cm. The observed rate of the reaction was measured to be $0.8 \cdot 10^{-5}$ mol/s $g_{cat}$. All other values remain the same as in the previous problem. The objective is to determine the particle size required to eliminate the effects of diffusion in the reaction.

**Solution to Problem 7.22**

Using the Weisz modulus:

$$We = \frac{\left(-r_A'''\right)_{observed} \cdot L^2}{D_e \cdot C_{As}}$$

We can see that:

$$\frac{\left(-r_{A1}'''\right)_{observed} \cdot L_1^2}{We_1} = \frac{\left(-r_{A2}'''\right)_{observed} \cdot L_2^2}{We_2}$$

Substituting radius and reaction rate values:

$$\frac{1.1 \cdot 10^{-5} \cdot \left(\frac{0.213}{3}\right)^2}{We_1} = \frac{0.8 \cdot 10^{-5} \cdot \left(\frac{0.13}{6}\right)^2}{We_2}$$

And then:

$$\frac{We_1}{We_2} = 14.76$$

We can also calculate:

$$We_1 = \frac{\left(-r_A'''\right)_{observed} \cdot L^2}{D_e \cdot C_{As}} = \frac{1.1 \cdot 10^{-5} \frac{mol}{s \cdot g_{cat}} \cdot 2 \frac{g_{cat}}{cm^3} \cdot \left(\frac{0.213}{3}\right)^2 cm^2}{0.02 \frac{cm^2}{s} \cdot 1.65 \cdot 10^{-5} \frac{mol}{cm^3}} = 0.168$$

Thus:

$$We_2 = \frac{0.168}{14.76} = 0.0113$$

We see that there are no diffusion limitations when the particle diameter is smaller than that informed.

**Problem 7.23**   An irreversible reaction is carried out in a PFR using catalyst particles with a diameter of 3 mm. Pure reactant A is fed into the reactor at a flow rate of 1000 l/min, operating at a pressure of 1 atm and a constant temperature of 300 °C. The desired conversion at the outlet of the reactor is 80%. The task is to calculate the mass of catalyst required to fill the reactor. The following additional data are provided for the calculation:

$$k' = 7 \cdot 10^{-5} \ cm^3/cm^2/s \quad D_e = 0.01 \ cm^2/s \quad S_S = 3000 \ cm^2/g \ (\text{specific surface})$$
$$k_C \cdot a_C = 4.0 \ cm^3/s/g \quad \rho_s = 2 \ g/cm^3$$

**Solution to Problem 7.23**

Assuming that both diffusion and chemical steps are important, the rate of the process can be expressed by:

$$(-r_A) = \frac{(C_{Ag} - C_{As})}{\frac{1}{k_C}} = \frac{C_{As}}{\frac{1}{k\eta}} = \frac{C_{Ag}}{\frac{1}{k_C} + \frac{1}{k\eta}} = \frac{kC_{Ag}}{\frac{k}{k_C} + \frac{1}{\eta}} = k\Omega C_{Ag}$$

being $\Omega$ the efficiency including both effects. In this way:

$$\left(-r'_A\right) = kS_S C_{A0}(1 - X_A)\Omega$$

The molar balance in the PFR will be:

$$\frac{W}{n_{A0}} = \int_0^{X_A} \frac{dX_A}{(-r'_A)} = \int_0^{X_A} \frac{dX_A}{kS_S C_{A0}(1 - X_A)\Omega} = -\frac{1}{kS_S C_{A0}\Omega} \ln(1 - X_A)$$

For calculating the effectiveness factor, we previously determined the Thiele modulus:

$$mL = \sqrt{\frac{k'''}{D_e}}\left(\frac{R}{3}\right) = \sqrt{\frac{7\cdot 10^{-5}\frac{cm}{s} \cdot 3000\frac{cm^2}{g} \cdot 2\frac{g}{cm^3}}{0.01\frac{cm^2}{s}}}\left(\frac{0.3}{6}\right) cm = 0.324$$

And then:

$$\eta = \frac{\tanh(mL)}{mL} = 0.96$$

From that, we can determine the overall mass transfer limitation effect:

$$\Omega = \frac{1}{\frac{k}{k_C} + \frac{1}{\eta}} = \ldots = 0.1931$$

With:

$$C_{A0} = \frac{P}{RT} = 2.12 \cdot 10^{-2} \frac{mol}{l}$$

We can now calculate reaction rate constant:

$$\left(-r'_A\right) = kS_S C_{A0}(1 - X_A)\Omega = 7 \cdot 10^{-5}\frac{cm}{s} \cdot 3000\frac{cm^2}{g}$$

$$\cdot 2.12 \cdot 10^{-5}\frac{mol}{cm^3} \cdot (1 - X_A) \cdot 0.1931 = 1.352 \cdot 10^{-5} \cdot (1 - X_A)\frac{mol}{s \cdot g}$$

Then:

$$\frac{W}{n_{A0}} = -\frac{1}{kS_S C_{A0}\Omega} \ln(1 - X_A) = -\frac{1}{1.352 \cdot 10^{-5}} \ln(1 - 0.8) = 119\,018 \frac{s \cdot g}{mol}$$

with:

$$n_{A0} = C_{A0} \cdot Q_0 = 2.12 \cdot 10^{-2}\frac{mol}{l} \cdot \frac{1000}{60}\frac{l}{s} = 0.3533 \frac{mol}{s}$$

And then:

$$\frac{W}{0.3533} = 119\,018$$

So:

$$W = 420\,53\,g = 42.05\,kg$$

**Problem 7.24**  A gas-phase reaction $A \to R$ is being conducted in an adiabatic fixed-bed reactor. The reaction follows first-order kinetics at the surface, with a reaction rate given by $r'_A = k' \cdot C_{As}$. The reactor is fed with pure reactant A at a volumetric flow rate of $100\,l/s$ and a temperature of $300\,K$. The Reynolds and Sherwood numbers for the system are known. The initial concentration of A is $1\,M$. The objective is to calculate the mass of catalyst required to achieve a conversion of 60%. The given information is as follows:

$$\text{Sh} = 100 \cdot \text{Re}^{1/2} \qquad \text{Kinematic viscosity, } v = 0.02\,\text{cm}^2/\text{s}$$
$$d_p = 0.1\,\text{cm} \qquad u = 10\,\text{cm/s} \qquad D_e = 0.01\,\text{cm}^2/\text{s}$$
$$\rho = 2\,\text{g/cm}^3 \qquad k'_{300} = 0.01\,\text{cm}^3/\text{g}_{cat}\cdot\text{s}$$
$$S_s = 60\,\text{cm}^2/\text{g}_{cat} \qquad \Delta H = 0$$

**Solution to Problem 7.24**

For details refer the Wiley website at http://www.wiley-vch.de/ISBN9783527354115

We can first calculate Re and Sh in order to obtain the mass transfer coefficient, $k_C$:

$$\text{Re} = \frac{d_p \cdot \rho \cdot u}{\mu} = \frac{d_p \cdot u}{v} = 0.1 \cdot \frac{10}{0.02} = 50$$

and:

$$\text{Sh} = 100\,\text{Re}^{\frac{1}{2}} = 707.1$$

So:

$$k_C = \text{Sh} \cdot \frac{D_e}{d_p} = 70.7\,\frac{\text{cm}}{\text{s}}$$

changing units:

$$k'_C = k_C \cdot S_S = 4242\,\frac{\text{cm}^3}{\text{s} \cdot \text{g}_{cat}}$$

The global reaction rate is:

$$\left(-r''_A\right) = \frac{C_{Ag}}{\frac{1}{k_C} + \frac{1}{k\eta}} = \frac{k' C_{Ag}}{\frac{k'}{k'_C} + \frac{1}{\eta}} = k'\Omega C_{A0}(1 - X_A)$$

With:

$$W = n_{A0} \int_0^{X_A} \frac{dX_A}{-r''_A} = n_{A0} \int_0^{X_A} \frac{dX_A}{\Omega C_{A0}(1 - X_A)} = G \int_0^{X_A} \frac{dX_A}{\Omega(1 - X_A)}$$

For the effectiveness factor, we do:

$$\text{mL} = \sqrt{\frac{k''}{D_e}}\left(\frac{d_p}{6}\right) = 1.41$$

and $\eta = 0.9988$.

With the shown data, we can finally calculate $\Omega$:

$$\frac{k'}{\dfrac{k'}{k'_c} + \dfrac{1}{\eta}} = k'\Omega = 0.009\,988\,2$$

Integrating:

$$W = G \int_0^{X_A} \frac{\mathrm{d}X_A}{k'\Omega(1 - X_A)} = 100 \int_0^{0.6} \frac{\mathrm{d}X_A}{0.009\,988\,2(1 - X_A)} = 9164\,\mathrm{g}$$

**Problem 7.25**   In a liquid-phase catalytic reactor (PFR) for the hydrogenation of oil, spherical catalytic particles with a diameter of 1 cm are used. The external concentration of the reactant is 1 kmol/l, while the concentration on the particle surface is 0.1 kmol/l. The superficial velocity of the liquid is 0.1 m/s. The objective is to verify if there are any mass effects in the system. Additionally, it needs to be determined whether there would be any change in the presence of a particle diameter of 0.5 cm. The effects of diffusive pores can be neglected. The provided additional data are as follows:

> Viscosity kinematic $= 0.5 \cdot 10^{-6}\ \mathrm{m^2/s}$
> Surface area $= 100\ \mathrm{m^2/g}$
> Diffusivity $= 10^{-10}\ \mathrm{m^2/s}$
> Consider: $\mathrm{Sh} = 2 + 0.6\mathrm{Re}^{0.5}\mathrm{Sc}^{0.33}$
> where Re = Reynolds number, Sc = Schmidt number

**Solution to Problem 7.25**

For details refer the Wiley website at http://www.wiley-vch.de/ISBN9783527354115

The rate of mass transfer is:

$$(-r''_A)\left(\frac{\mathrm{mol}}{1 \cdot \mathrm{s}}\right) = k_C a S_S(C_{Ag} - C_{As})$$

To calculate the mass transfer coefficient, we will use the given correlation. For the Reynolds number:

$$\mathrm{Re} = \frac{d_p u}{v} = 2000$$

$$\mathrm{Sc} = \frac{v}{D_e} = 5000$$

$$\mathrm{Sh} = 2 + 0.6 \cdot (2000)^{0.5} \cdot 5000^{0.33} = 460$$

And then:

$$(k_C a) \cdot \frac{d_p}{D_e} = 400$$

Obtaining: $k_C a = 4.6 \cdot 10^{-6}\ (1/1 \cdot \mathrm{s})$.

The rate of reaction is then:

$$(-r''_A)\left(\frac{\mathrm{mol}}{1 \cdot \mathrm{s}}\right) = k_C a S_S(C_{Ag} - C_{As}) = 4.6 \cdot 10^{-6} \cdot 100 \cdot 0.9 \cdot 1000 = 0.414\,\frac{\mathrm{mol}}{1 \cdot \mathrm{s}}$$

With the particle diameter equal to 0.5 cm, Re $= 1000$, Sh $= 32.5$, $k_C a = 6.5 \cdot 10^{-6}$ (1/1·s), and $(-r_A) = 0.585$ mol/1·s.

**Problem 7.26** Partial oxidation of *n*-butane (B) to maleic anhydride (MA) is carried out at 1 atm in a commercial reactor, the packed tubes of which contain a VPO catalyst ($1.2\,\mathrm{g_{cat}/cm^3_{bed}}$) and are cooled by alkali metal nitrate. The inlet and outlet temperatures are 390 and 410 °C; desired B conversion and selectivity to MA are 80% and 60%, respectively.

An applicable rate equation for conversion of B to MA is:

$$r'_B = \frac{k \cdot K_B \cdot C_B \cdot C_O^{0.23}}{1 + K_B C_B}$$

where *r* has units of mol/$\mathrm{g_{cat}}$·s and $C_B$ units of mol/l; $k = 3.36 \cdot 10^{-7}$ mol$^{0.77}$·l$^{0.23}$/$\mathrm{g_{cat}}$·s at 300 °C; $K_B = 261$ l/mol; and $E = 45.1$ kJ/mol.

Given a feed of 1.8% butane in air and a gas-feed rate of 1500 m$^3$ (STP)/h, determine:

(a) How much kg per hour of MA is produced?
(b) How many reactor tubes of 3 cm diameter and 12 m length will be needed?

**Solution to Problem 7.26**

For details refer the Wiley website at http://www.wiley-vch.de/ISBN9783527354115

(a) The concentration of the gas at the inlet, assuming an isothermal bed at 400 °C, is:

$$C_{T0} = \frac{P}{RT} = \frac{1}{0.082 \cdot 673} = 0.018\,12\,\frac{\text{mol}}{\text{l}}$$

Only 1.8% of this gas is B, so:

$$C_{B0} = 1.8 \cdot \frac{C_{T0}}{100} = 3.26 \cdot 10^{-4}\,\frac{\text{mol}}{\text{l}}$$

The concentration of oxygen (21% of the air) will be:

$$C_O = C_{T0}\left(1 - \frac{1.8}{100}\right) \cdot 0.21 = 3.74 \cdot 10^{-3}\,\frac{\text{mol}}{\text{l}}$$

At the exit:

$$C_B = C_{B0}(1 - X_B) = 3.26 \cdot 10^{-4}(1 - 0.8) = 6.52 \cdot 10^{-5}\,\frac{\text{mol}}{\text{l}}$$

and:

$$C_{MA} = (C_{B0}X_B)S_{MA} = 3.26 \cdot 10^{-4} \cdot 0.8 \cdot 0.6 = 1.57 \cdot 10^{-4}\,\frac{\text{mol}}{\text{l}}$$

So the production of MA is:

$$\text{MA exit} = C_{MA} \cdot G = 1.57 \cdot 10^{-4} \cdot 1500 \cdot 1000 = 235\,\frac{\text{mol}}{\text{h}}$$

On a weight basis ($\mathrm{MW_{MA}} = 98.1$ g/mol):

$$\text{MA exit} = 23\text{ kg/h}$$

(b) Let us calculate reaction rate. First, the kinetic constant can be calculated using:

$$k_{400°C} = k_{300°C} \cdot \exp\left(\frac{E}{R}\left(\frac{1}{573} - \frac{1}{673}\right)\right) = \ldots = 1.37 \cdot 10^{-6}\ \frac{\text{mol}^{0.77} \cdot \text{l}^{0.23}}{\text{g}_{\text{cat}} \cdot \text{s}}$$

With the values indicated in part (a), we can calculate $r_B$:

$$r'_B = \frac{k \cdot K_B \cdot C_B \cdot C_O^{0.23}}{1 + K_B C_B} = \ldots = 6.35 \cdot 10^{-9}\ \frac{\text{mol}}{\text{g}_{\text{cat}} \cdot \text{s}}$$

For calculating the weight of catalyst needed:

$$W = \frac{(C_{B0} - C_B)\left(\frac{\text{mol}}{\text{l}}\right) \cdot G\left(\frac{\text{l}}{\text{s}}\right)}{r_B \left(\frac{\text{mol}}{\text{g}_{\text{cat}} \cdot \text{s}}\right)} = 4.22 \cdot 10^7\ \text{g}_{\text{cat}} = 4.22 \cdot 10^4\ \text{kg}_{\text{cat}}$$

Each tube has a volume of $8.48 \cdot 10^{-3}$ m$^3$, i.e., 8.48 l, which using the bed density gives a total of $10\,178$ g$_{\text{cat}}$/tube, so a total of 4147 tubes are necessary. This calculation is made assuming that the conditions of the bed are those with butane totally converted, i.e., a CSTR assumption. If, on the contrary, we use the conditions of the gases at the input condition:

$$r'_{B0} = \frac{k \cdot K_B \cdot C_{B0} \cdot C_O^{0.23}}{1 + K_B C_{B0}} = \ldots = 2.98 \cdot 10^{-8}\ \frac{\text{mol}}{\text{g}_{\text{cat}} \cdot \text{s}}$$

And the weight of catalyst needed:

$$W = 9.01 \cdot 10^6\ \text{g}_{\text{cat}} = 9.01 \cdot 10^3\ \text{kg}_{\text{cat}}$$

This is a total of 885 tubes. A good approximation to the real conditions is to use a log mean of the concentration:

$$C_{\text{Bmean}} = \frac{C_{B0} - C_{\text{Bexit}}}{\ln\left(\frac{C_{B0}}{C_{\text{Bexit}}}\right)} = 1.62 \cdot 10^{-4}\ \frac{\text{mol}}{\text{l}}$$

This gives a mean reaction rate of $1.54 \cdot 10^{-8}\ \frac{\text{mol}}{\text{g}_{\text{cat}} \cdot \text{s}}$ and the total number of tubes will be 1710.

**Problem 7.27** Your supervisor has assigned you to design and construct a continuous-flow reactor system for hydrogenation of fatty esters to fatty alcohols, which he hopes will significantly increase plant capacity and productivity. He insists that you try to use "existing surplus facilities," which consist of (i) a 5 m$^3$ tubular reactor and (ii) a 50 m$^3$ CSTR. Both reactors are rated for pressure service to 350 atm of $H_2$. A finely powdered $CuO/Cr_2O_3$ catalyst having a fixed-bed density of 2.3 kg/l (when coated on a monolith support) and a slurry-bed density of 0.4 kg/l is available. Assume isothermal operation at 300 °C and 300 atm of $H_2$ with a liquid feed consisting of 15 kmol/h of fatty ester. A rate equation determined from a batch slurry reactor under these conditions in terms of disappearance of ester as a function of its fractional conversion is:

$$\left(-r'_E\right) = 2 \cdot 10^{-3}\left(1 - X_E\right)/(1 + X_E)$$

expressed in $\text{kmol/kg}_{\text{cat}} \cdot \text{h}$. The desired conversion of ester is 85%. Assume the stoichiometry is $1:1$.

Can either of these reactors be used by themselves with the available catalyst to meet the conversion requirements? Or will it be necessary to use both reactors connected to each other in a series? If so, which reactor should be positioned first? Explain and document your answers where possible with calculations.

**Solution to Problem 7.27**

For details refer the Wiley website at http://www.wiley-vch.de/ISBN9783527354115

Let us assume a PFR for the tubular one. In this situation:

$$\left(-r'_{\text{E}}\right) dW = n_{\text{E0}} dX_{\text{E}}$$

$$\frac{k(1 - X_{\text{E}})}{1 + X_{\text{E}}} dW = n_{\text{E0}} dX_{\text{E}}$$

$$\frac{(1 - X_{\text{E}})}{1 + X_{\text{E}}} \frac{k dW}{n_{\text{E0}}} = dX_{\text{E}}$$

This is an ODE (ordinary differential equation) that can be solved by using the initial condition $X_{\text{E0}} = 0$. The bed has a total volume of $5\,\text{m}^3$ that will be used for incrementing the weight of the catalyst using the bed density. We should prepare a table such as:

| $V_{\text{bed}}$ (m³) | $W_{\text{bed}}$ (kg$_{\text{cat}}$) | $X_{\text{E}}$ | $(-r_{\text{E}})$ (kmol/kg$_{\text{cat}}$·h) | $W_{\text{bed}}/n_{\text{E0}}$ |
|---|---|---|---|---|
| 0 | 0 | 0 | $2.00\cdot10^{-3}$ | 0 |
| 0.05 | 115 | $4.05\cdot10^{-2}$ | $1.84\cdot10^{-3}$ | $2.03\cdot10$ |
| 0.1 | 230 | $7.79\cdot10^{-2}$ | $1.71\cdot10^{-3}$ | $4.05\cdot10$ |
| ... | | | | |
| 4.95 | 11 385 | $9.18\cdot10^{-1}$ | $8.58\cdot10^{-5}$ | $2.01\cdot10^3$ |
| 5 | 11 500 | $9.19\cdot10^{-1}$ | $8.39\cdot10^{-5}$ | $2.03\cdot10^3$ |

In the form of a graph, we have:

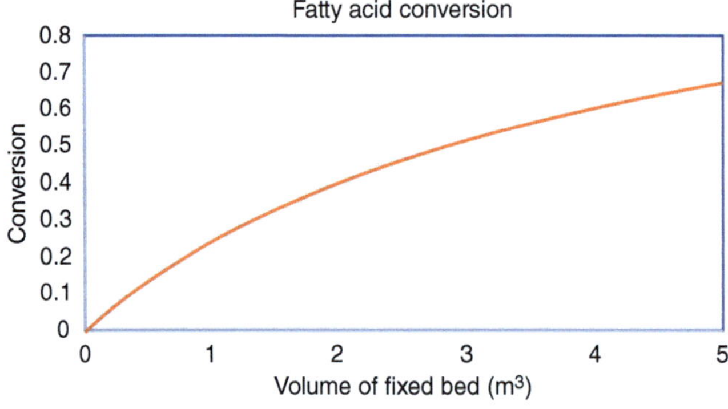

As we can check, the final conversion in this bed is 0.67, so it does not reach the desired conversion.

In the slurry reactor, we will assume a CSTR behavior, so:

$$\text{Exit} - \text{Input} = \text{Generation}$$

$$n_E - n_{E0} = \left(-r'_E\right) W$$

$$n_{E0}(1 - X_E) - n_{E0} = \frac{k'(1 - X_E)}{1 + X_E} W$$

$$\frac{X_E(1 + X_E)}{1 - X_E} = \frac{k'W}{n_{E0}}$$

$$X_E = \frac{-\left(1 + \frac{k'W}{n_{E0}}\right) + \left(\left(1 + \frac{k'W}{n_{E0}}\right)^2 + 4\frac{k'W}{n_{E0}}\right)^{0.5}}{2}$$

Using the data in the problem, we have that $\frac{k'W}{n_{E0}} = 2.67$ and $X_E = 0.622$, so this reactor neither reaches the 0.85 conversion.

Let us try to combine both reactors. As a first thought, surely, we should place bed reactor first and later the slurry reactor. In this situation, the conversion in the first reactor is exactly the same as in the previous case, i.e., 0.67. The exit of this reactor is:

$$n_E \text{ (exit of the PFR)} = n_{E0}(1 - X_E) = 4.95 \frac{\text{kmol}}{\text{h}}$$

that will enter the slurry reactor, so:

$$\frac{k'W}{n_{E0}} = 8.07$$

and the new conversion in the second reactor is 0.816. The global conversion can be calculated:

$$n_E \text{ (exit of the CSTR)} = n_{E0}(1 - X_E) = 4.95\,(1 - 0.816) = 0.91$$

$$X_{E\text{ global}} = 1 - \frac{0.91}{15} = 0.939$$

In the second combination, CSTR is sited before the PFR. In this case, the conversion in the first reactor is 0.622 (nothing changes with respect to the isolated reactor). Now, the input to the second (tubular) reactor is:

$$n_E \text{ (exit of the CSTR)} = n_{E0}(1 - X_E) = 5.67 \frac{\text{kmol}}{\text{h}}$$

bearing in mind this new value of flowrate entering the PFR, we obtain a conversion of 0.919 in the second reactor, so:

$$n_E \text{ (exit of the PFR)} = n_{E0}(1 - X_E) = 5.67 \cdot (1 - 0.919) = 2.15 \frac{\text{kmol}}{\text{h}}$$

and:

$$X_{E\text{ global}} = 1 - \frac{2.15}{15} = 0.857$$

**Problem 7.28**   In a fixed-bed catalytic reactor, methanol will be produced, using excess hydrogen from an electrolytic process in reaction with carbon dioxide from the air. The reaction takes place in the gas phase in the presence of a commercial TMC-31 catalyst:

$$3H_2 + CO_2 \leftrightarrow CH_3OH + H_2O$$

This is an equilibrium reaction, whose constant $K_p$ is given by:

$$\log(K_p) = \frac{3066}{T} - 10\,592$$

The kinetics of the reaction are well represented by the Van den Busche (VDB) model, with the following equation:

$$r_M = \frac{(k_d * p_{CO_2} * p_{H_2} * \left(1 - \frac{1}{K_p} * \frac{p_{H_2O} \cdot p_{CH_3OH}}{p_{H_2}^3 \cdot p_{CO_2}}\right)}{\left(1 + k_c * \frac{p_{H_2O}}{p_{H_2}} + k_a * p_{H_2}^{0.5} + k_b * p_{H_2O}\right)^3}$$

In which rate is given in (mol of methanol/h·$g_{cat}$) when the following values are used:

| Parameter | A | E (kJ/mol) |
|---|---|---|
| $k_a$ | $0.499\,\text{bar}^{-0.5}$ | $-17\,197$ |
| $k_b$ | $6.62 \cdot 10^{-11}\,\text{bar}^{-1}$ | $-124\,119$ |
| $k_c$ | $3453.8$ | – |
| $k_d$ | $3.862\,\text{mol/g·h·bar}^2$ | $-36\,696$ |

since each constant follows the Arrhenius-type behavior ($k = A \cdot \exp[-E/RT]$). Be careful that the activation energies in this correlation are negative, which is quite rare.

100 kmol/h of hydrogen is available, and it will be mixed in a stoichiometric proportion with $CO_2$, and we aim to reach a conversion of hydrogen near 0.2. Determine the amount of catalyst needed, as well as the best inlet pressure and temperature conditions, in the range where the catalyst is active (140–280 °C).

**Solution to Problem 7.28**

Let us call "A" to the hydrogen and "B" to the $CO_2$. In that situation, we have:

$$3A + B \leftrightarrow \text{Methanol (C)} + \text{water (D)}$$

$n_{B0} = n_{A0}/3$

Catalyst

$X_A = 0.2$ (desired)

$n_{A0} = 140$ kmol/h

$T = 140$ to $280$ °C

Design eq.: $r_M \cdot dM = dn_C$

This is a reaction among gases, and we should account for the expansion of the gas. The expansion factor can be calculated using:

$$\varepsilon = \frac{n_{A0}}{\sum n_{j0}} \sum \frac{\alpha_{ij}}{-\alpha_{ik}} = \frac{n_{A0}}{n_{A0} + n_{B0}} \cdot \left( \frac{-3 + 1 + 1 + 1}{-(-3)} \right) = -\frac{2n_{A0}}{3(n_{A0} + n_{B0})} = -0.5$$

where species "$j$" refers to all components and "$k$" refers to the key component, in this case, the hydrogen. $\alpha_i$ are the stoichiometric numbers in the reaction (negative for the reactants and positive for the products).

In that way, we can do:

$$Q = Q_0(1 + \varepsilon X_A)$$

$$n_A = n_{A0}(1 - \varepsilon X_A)$$

$$n_B = n_{B0} + n_{A0}\left(+\frac{1}{3}\varepsilon X_A\right)$$

$$n_C = n_D = n_{A0}\left(+\frac{1}{3}\varepsilon X_A\right)$$

$$n_{total} = n_A + n_B + n_C + n_D$$

In terms of the molar fractions:

$$\chi_i = \frac{n_i}{n_{total}}$$

And partial pressures are:

$$p_i = \chi_i \cdot P$$

In the PFR, we have:

$$r_M \cdot dM = dn_C = d\left(n_{A0}\left(+\frac{1}{3}\varepsilon X_A\right)\right) = 0.5\frac{n_{A0}}{3}dX_A$$

$$\frac{(k_d * p_{CO_2} * p_{H_2} * \left(1 - \frac{1}{K_p} * \frac{p_{H_2O} \cdot p_{CH_3OH}}{p_{H_2}^3 \cdot p_{CO_2}}\right)}{\left(1 + k_c * \frac{p_{H_2O}}{p_{H_2}} + k_a * p_{H_2}^{0.5} + k_b * p_{H_2O}\right)^3} \cdot dM \cdot \frac{6}{n_{A0}} = dX_A$$

With initial condition $X_A = 0$ for $M = 0$, we can simulate the system for different values of pressure and temperature and choose the best one for achieving the objective of converting 20% of the hydrogen. Let us assume a total pressure of 70 bar and a temperature of 280 °C. With these values, we can calculate the following at the input:

$$n_{A0} = 100 \frac{kmol}{h}$$

$$n_{B0} = 100/3 \frac{kmol}{h}$$

$$n_{total0} = 400/3 \frac{kmol}{h}$$

$$X_{A0} = \frac{n_{A0}}{n_{total0}} = 3/4$$

$$X_{B0} = \frac{n_{A0}}{n_{total0}} = 1/4$$

$$p_{A0} = \frac{3}{4} \cdot P = \frac{210}{4} \text{ bar}$$

$$p_{B0} = \frac{1}{4} \cdot P = \frac{70}{4} \text{ bar}$$

With these variables, we can calculate $r_M$, in this case, giving 1.71 mol/g$_{cat}$·h, and taking a value of $\Delta W$, we will calculate $\Delta X_A$. For example, if $\Delta W = 4000$ g, the corresponding $\Delta X_A$ in this differential is 0.0488. With this, product formed and reactants consumed can also be calculated.

We can do this calculation repetitively until the conversion gets its maximum value, or equals the desired 0.2 value. In the spreadsheet corresponding to this problem, we find the following:

| Catalyst (gr) | $P(H_2)$ (bar) | $P(CO_2)$ (bar) | $p_C = p_D$ (bar) | $P_{total}$ (bar) | $r_M$ (mol/g$_{cat}$·h) | Moles H$_2$ (kmol/h) | Moles CO$_2$ (kmol/h) | Moles methanol (kmol/h) | kmol/h total | $X_A$ | Flow rate (l/h) |
|---|---|---|---|---|---|---|---|---|---|---|---|
| 0 | 52.50 | 17.50 | 0.00 | 70.00 | 1.7111 | 140 | 46.7 | 0 | 186.7 | 0 | 120 922.7 |
| 4000 | 49.93 | 16.64 | 0.86 | 68.29 | 0.3936 | 133.15 | 44.39 | 2.281 | 182.1 | 0.0488 | 120 922.66 |
| 8000 | 50.58 | 16.86 | 1.08 | 69.60 | 0.2991 | 131.58 | 43.86 | 2.806 | 181.1 | 0.0601 | 117 966.69 |
| … | … | | | | | | | | | | |
| 392 000 | 47.41 | 15.80 | 3.39 | 69.99 | 0.008952227 | 115.27 | 38.42 | 8.241 | 170.2 | 0.1766 | 110 260.6 |
| 396 000 | 47.40 | 15.80 | 3.39 | 69.99 | 0.008807336 | 115.23 | 38.41 | 8.253 | 170.2 | 0.1768 | 110 244.9 |

Graphically, we found:

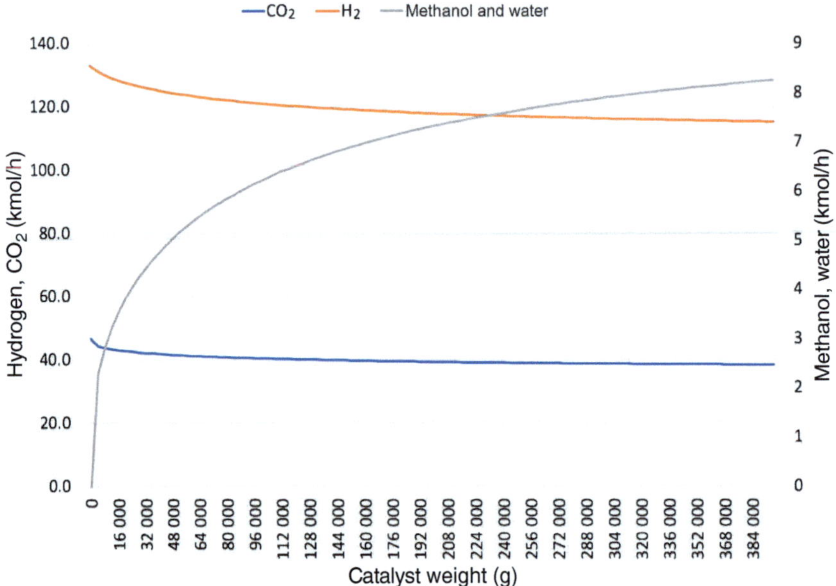

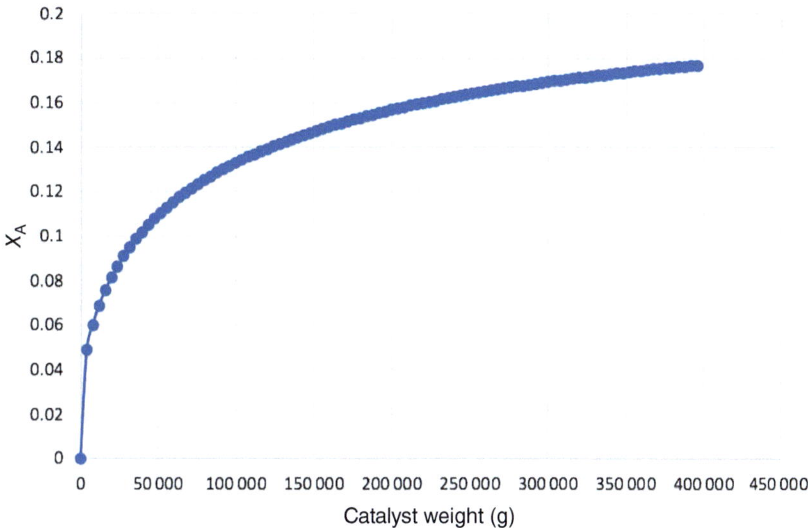


As we can see, the conversion does not reach 0.2 at any point, as the existing equilibrium does not permit more reactants to be converted.

We can try to change total pressure and temperature of the operation, but it is almost impossible to get a higher conversion. A total amount of nearly 400 kg of catalyst will be needed in these conditions.

# 8

# Multiphase Reactor Design

## Summary of Rate Expressions

### Process of Absorption (No Reaction)

$$(-r_A) = \frac{p_A}{\frac{1}{k_{Ag}} + \frac{H_A}{k_{Al}}}$$

$$(-r_A'') = (-r_A) \cdot a = k_{Ag} \cdot a \cdot (p_A - p_{Ai}) = k_{AL} \cdot a \cdot (C_{Ai} - C_A)$$

In the equation, valid units for $k_{Ag}$ are, for example, (kmol/s·m$^2_{\text{interface}}$·atm), whereas $k_{AL}$ can be used in m$^3_L$/m$^2_{\text{interface}}$·s = m/s.

Rate of the absorption process can also be calculated using global mass transfer coefficients, with driving force $(p_A - p_A^*)$ and $(C_A^* - C_A)$, using the equations:

$$\frac{1}{K_{Ag}} = \frac{1}{k_{Ag}} + \frac{H_A}{k_{AL}}$$

$$\frac{1}{K_{AL}} = \frac{1}{k_{AL}} + \frac{1}{H_A \cdot k_{Ag}}$$

where pressures and concentrations refer to equilibrium, that is:

$$p_A^* = H_A \cdot C_A$$

$$C_A^* = p_A/H_A$$

In this way:

$$(-r_A'') = (-r_A) \cdot a = K_{Ag} \cdot a \cdot (p_A - p_A^*) = K_{AL} \cdot a \cdot (C_A^* - C_A)$$

In the equation, valid units for $K_{Ag}$ are, for example, (kmol/s·m$^2_{\text{interface}}$·atm), whereas $K_{AL}$ can be used in m$^3_L$/m$^2_{\text{interface}}$·s = m/s.

### Fluid–Fluid Reaction

$$A(g) + bB\,(l) \rightarrow \text{Products}$$

$$(-r_A'') = \frac{p_A}{\frac{1}{k_{Ag} \cdot a} + \frac{H_A}{k_{AL} \cdot a \cdot E} + \frac{H_A}{k_{ap}''}}$$

In the previous equation, $k''_{ap}$ is expressed in units of $m^3_F/m^3_{reactor} \cdot s$ and can be related with other constants by:

$$k''_{ap} = k'''_{ap}\varepsilon_L$$

with $\varepsilon_L$ = liquid load $(m^3_L/m^3_{reactor})$ and $k_{ap}$ is expressed in units of inverse time.

$$E = \left[\frac{\text{actual adsorption rate of component A}}{\text{adsorption rate of A without chemical reaction}}\right]_{\text{at the same } C_{Ai},C_A,C_{Bi},C_B}$$

$$E_i = \frac{\text{reaction rate with instantaneous reaction}}{\text{mass transfer rate}} = \left(1 + \frac{D_{BL} \cdot C_B}{b \cdot D_{AL} \cdot C_{Ai}}\right)$$

$$M_H = \frac{\sqrt{D_{AL} \cdot k'''}}{k_{AL}}$$

Note that this equation is nonlinear and must be solved by iterative methods. Next figure shows the equation in graphic form. Note that valid units of the diffusion coefficient, by definition, are $m^3_F/m_c \cdot s$, which are usually indicated as $m^2/s$. Correspondingly, valid units for $k_{AL}$ are $m^3_F/m^2_c \cdot s$, usually m/s. In this way, units of $k'''$ should be $m^3_F/m^3_c \cdot s$.

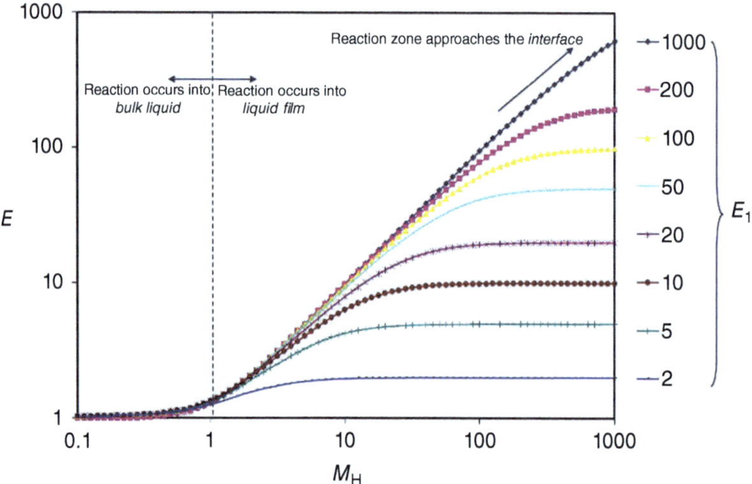


A good correlation for that graph is:

$$E = \frac{\sqrt{\dfrac{M_H^2(E_i-E)}{E_i-1}}}{\tanh\left(\sqrt{\dfrac{M_H^2(E_i-E)}{E_i-1}}\right)}$$

$k''_{ap}$ is the first-order apparent rate constant. In many systems, where chemical reaction is second order, we will do:

$$\left(-r''_A\right) = k''C_A C_B = (k''C_B)C_A = k''_{ap}C_A$$

In general, for any kinetic law, we will group in $k''_{ap}$ all the variables except $C_A$:

$$\left(-r''_A\right) = k'' C_A^n C_B^m = \left(k'' C_A^{n-1} C_B^m\right) C_A = k''_{ap} C_A$$

## Fluid–Fluid (Gas–Liquid) Reaction in Catalysts

$$\text{A (g} \xrightarrow{\text{disolution}} \text{l)} + \text{bB (l)} \xrightarrow{\text{in the catalyst}} \text{products}$$

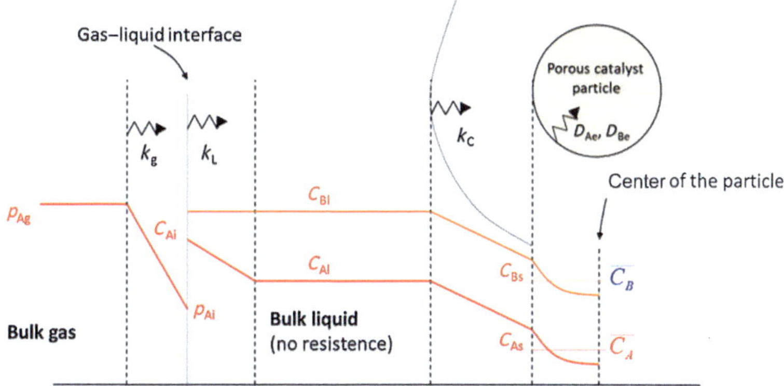


$k_g$ = mass transfer coefficient in the gas film (mol/Pa·m²·s)
$k_L$ = mass transfer coefficient in the liquid film (m³_L/m²·s)
$k_c$ = mass transfer coefficient within the solid film (m³_L/m²_cat·s)
$D_{Ae}, D_{Be}$ = effective diffusion coefficients (m³_L/m_cat·s)

- Transfer through the gaseous film: $k_{Ag} a_i (p_A - H_A C_{Ai})$
- Transfer through the liquid film: $k_{Al} a_i (C_{Ai} - C_{AL})$
- Transfer through the solid layer: $k_{Ac} a_c (C_{AL} - C_{As})$
- Surface chemical reaction: $k_{ap} \eta_A f_s C_{As}$

$$\left(-r''_A\right) = \frac{p_{Ag}}{\dfrac{1}{k_{Ag} \cdot a_i} + \dfrac{H_A}{k_{Al} \cdot a_i} + \dfrac{H_A}{k_{Ac} \cdot a_c} + \dfrac{H_A}{k_{ap_A} \cdot \eta_A \cdot \varepsilon_s}}$$

$$\left(-r''_B\right) = \frac{C_{BL}}{\dfrac{1}{k_{Bc} \cdot a_c} + \dfrac{1}{k_{ap_B} \cdot \eta_B \cdot \varepsilon_s}}$$

$\varepsilon_s$ = solid load (m³_solid/m³_reactor)
$k_{ap_A}$ is the first-order apparent rate constant for A, so the rate law would be:

$$\left(-r''_A\right) = k''_{ap_A} C_A$$

and $k_{ap_B}$ is the first-order apparent rate constant for B:

$$\left(-r''_B\right) = k''_{ap_B} C_B$$

$\eta_A$ and $\eta_B$ are efficiency factors.
Remember that, in any case:

$$\left(-r''_A\right) = \left(-r''_B\right)/b$$

Schematically:

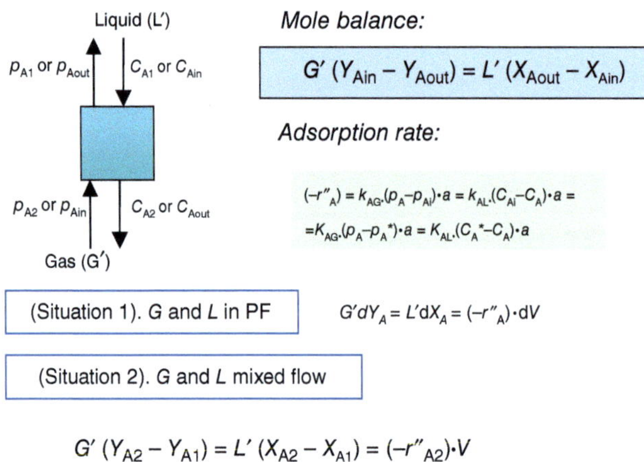



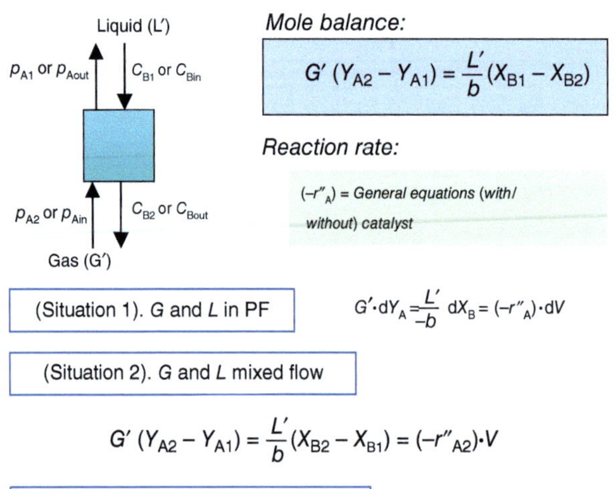



For more details, please consult Conesa (2019).

**Problem 8.1**  At a temperature of 25 °C, a rotating drum absorber is utilized to facilitate the absorption of $CO_2$ into a 2.5 M mono-ethanolamine solution. The absorber features a contact surface area of 1885 $cm^2$ and a contact time of 0.2 seconds. In the gas phase, the partial pressure of $CO_2$ is 0.1 atm. The absorption reaction involved follows the equation: $CO_2 + 2R_2NH \rightarrow R_2NCOO^- + R_2NH^+$. Experimental measurements reveal that the absorption rate under these conditions amounts to $3.26 \times 10^{-4}$ mol/s. Determine the rate coefficient when the gas-phase resistance is neglected, and the reaction is assumed to follow a pseudo-first-order behavior.

Additional data:

$D_A = 1.4\ 10^{-5}$ $cm^2$/s; $D_B = 0.77 \cdot 10^{-5}$ $cm^2$/s; $K_{AL} = 0.98$ cm/s; Henry constant, $H_A = 29.8 \cdot 10^3$ atm·$cm^3$/mol.

**Solution to Problem 8.1**

For details refer the Wiley website at http://www.wiley-vch.de/ISBN9783527354115

In the complete system of mass transfer and reaction, we have:

$$\left(-r_A''\right) = \frac{p_A}{\dfrac{1}{k_{Ag} \cdot a} + \dfrac{H_A}{k_{AL} \cdot a \cdot E} + \dfrac{H_A}{k_{ap}''}}$$

where "A" refers to the species in the gas phase, i.e., $CO_2$. As we have no gas phase transfer problems, the first sum in the denominator is zero:

$$\left(-r_A''\right) = \frac{p_A/H_A}{\dfrac{1}{k_{AL} \cdot a \cdot E} + \dfrac{1}{k_{ap}''}}$$

In terms of surface:

$$\left(-r_A\right) = \frac{p_A/H_A}{\dfrac{1}{k_{AL} \cdot E} + \dfrac{1}{k_{ap}''}} = \frac{3.26 \cdot 10^{-4}}{1885}\ \frac{\text{mol}}{\text{s} \cdot \text{cm}^2}$$

We can try to calculate the increase factor, $E$. For that purpose, we first calculate $M_H$ and $E_i$.

$$E_i = \left(1 + \frac{D_{BL} \cdot C_B}{b \cdot D_{AL} \cdot C_{Ai}}\right) = \left(1 + \frac{D_{BL} \cdot C_B}{b \cdot D_{AL} \cdot p_{Ai}/H_A}\right)$$

$$= \left(1 + \frac{7.7 \cdot 10^{-6} \cdot 2.5}{2 \cdot 1.4 \cdot 10^{-5} \cdot 0.1/29\,800}\right) = 205.87$$

$$M_H = \frac{\sqrt{D_{AL} \cdot k''}}{k_{AL}} = \frac{\sqrt{4 \cdot 10^{-5}\ \frac{\text{cm}^2}{\text{s}} \cdot k''\ (\text{s}^{-1})}}{0.98\ \text{cm/s}} = 0.006\,454\sqrt{k''}$$

$$E = \frac{\sqrt{\dfrac{M_H^2(E_i - E)}{E_i - 1}}}{\tanh\left(\sqrt{\dfrac{M_H^2(E_i - E)}{E_i - 1}}\right)}$$

Assuming a value of $k_{ap}''$, we next calculate $M_H$ and then $E$. With this, a value of the reaction rate is calculated and is compared to the actual one. Finally, a good value of

$k$ should be obtained:

$$k''_{ap} = 0.05\ \text{s}^{-1}$$

$$E = 1$$

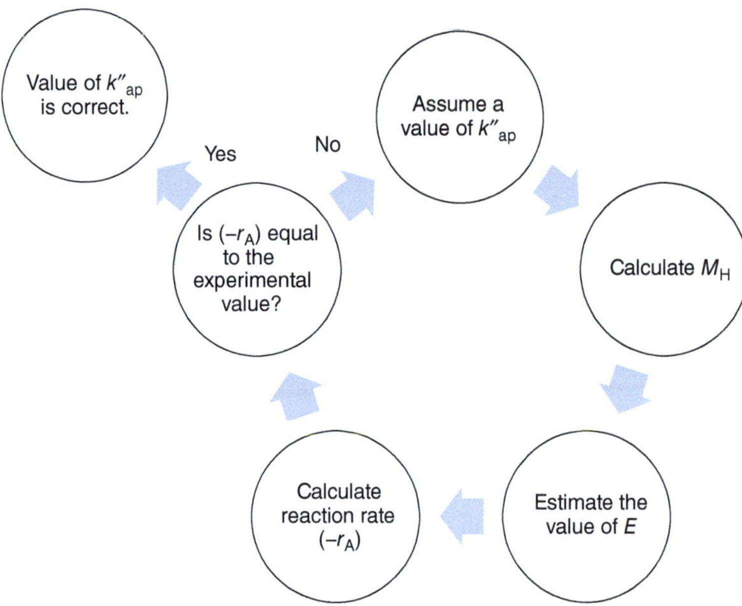



**Problem 8.2**  The heterogeneous reaction A (liq) + 2B (gas) → Products is carried out in a suspension reactor with spherical particles of 0.004 cm in diameter and a particle density (dry) of 1 g/cm$^3$. The reactor operates at 2 atm and 30 °C, considering negligible effect of hydrostatic pressure. The intrinsic kinetics in the liquid phase embedded within the catalyst is given by $(-r''_B) = k'C_B^2$ where $k = 1.2\ \text{l}^2/\text{mol·g·s}$ and is zero-order with respect to liquid reactant A. The catalyst loading, $\varepsilon_s$, is 0.08 cm$^3$ catalyst/cm$^3$ free liquid, and the corresponding external mass transfer coefficient, $k_C$, is 0.03 cm/s. The effective diffusivity of B inside the catalyst is $2\cdot10^{-4}$ cm$^2$/s. The internal resistance is significant, and to calculate the effectiveness factor, we can use the following polynomial for mL < 5:

$$\eta = 1 - 0.046(\text{mL}) - 0.1(\text{mL})^2 + 0.029(\text{mL})^3 - 0.0025(\text{mL})^4$$

200 l/s of gas is fed with $y_{B0} = 0.3$ and 1 l/s of liquid with $C_{A0} = 0.55$ M. The gas bubbles correspond to a retention of the gas of 0.10. Henry's constant for B is 0.06 M liquid/M gas, and suppose that the liquid enters saturated with B at gas feed conditions.

(a) If a conversion of A to 65% is required. Calculate the volume of liquid required and the residence time of the liquid.

(b) What is the outlet concentration of B in the gas? What implications does this difference have for the concentration of B in the feed gas?

(c) Calculate the internal volume of the vessel if 30% of this volume is occupied by the stirrer, detectors, free space above the liquid level, and bubble distributor (assume that the liquid below this distributor can be neglected).

**Solution to Problem 8.2**

For details refer the Wiley website at http://www.wiley-vch.de/ISBN9783527354115

(a) In the liquid phase, we have:

$$(-r_B) = kC_B^2 \eta$$

with:

$$C_{B0} = 0.3 \cdot \frac{P_{total}}{RT} = 0.0242 \, \frac{mol}{l}$$

being the effectiveness factor as a function of the Thiele modulus (mL). For a second-order reaction, we can write:

$$mL = \left[\frac{3k \cdot C_{Bs}}{2 \cdot D_{eB}}\right]^{\frac{1}{2}} \left(\frac{R}{3}\right)$$

$$mL = \sqrt{\frac{3 \cdot \left(1.2 \, \frac{l_{liq}^2}{mol \cdot g \cdot s} \cdot 0.08 \, \frac{l_{cat}}{l_{liq}} \cdot 1 \, \frac{g_{cat}}{cm_{cat}^3} \cdot 10 \, \frac{cm^3}{l}\right) \cdot 0.0242 \, \frac{mol}{l_{liq}}}{2 \cdot 2 \cdot 10^{-4} \, \frac{cm^2}{s}}} \left(\frac{0.004}{6}\right) cm = 0.087$$

giving a value of $\eta = 0.995$ in the conditions at the input of the gas. Considering a PFR for the liquid, the mass balance of species A is:

$$r_A'' dV = dn_A = -n_{A0} dX_A$$

$$r_A = r_B/2$$

$$\frac{r_B''}{2} dV = -n_{A0} dX_A = -C_{A0} L dX_A$$

being "$L$" the flow rate of the liquid. From the expression:

$$\frac{-k'' C_B^2 \eta}{2} dV = -C_{A0} L dX_A$$

$$V = \frac{2 C_{A0} L}{k''} \int_0^{X_A} \frac{dX_A}{C_B \cdot \eta}$$

The relationship between $X_A$ and $C_B$ can be obtained as follows:

$$p_B \cdot G = n_B RT$$

$$C_B = \frac{n_B}{G} = \frac{p_B}{RT} = \frac{(y_b \cdot P_{total})}{RT}$$

By stoichiometry:

$$2G' dY_B \left(\frac{\text{moles of B reacted}}{s}\right) = (C_{A0} L) dX_A \left(\frac{\text{moles of A reacted}}{s}\right)$$

$$2G'(Y_B - Y_{B0}) = C_{A0}L(X_A - 0) = C_{A0}LX_A$$

$$Y_B = Y_{B0} + \frac{C_{A0}LX_A}{2G'}$$

$$C_B = \frac{P_{total}}{RT}Y_B$$

We do the numerical integration of the equations by using a spreadsheet. For example:

| $X_A$ | $y_B$ | $C_B$ (mol/l) | mL | $\eta$ | $2 \cdot C_{A0} \cdot L/(kC_B^2\eta)$ | Integration |
|---|---|---|---|---|---|---|
| 0.0000 | 0.3000 | 0.0241 | 0.0879 | 0.9952 | 19.7433 | |
| 0.0325 | 0.3000 | 0.0242 | 0.0879 | 0.9952 | 19.7374 | 0.6416 |
| 0.0650 | 0.3001 | 0.0242 | 0.0879 | 0.9952 | 19.7316 | 1.2829 |
| ... | | | | | | |
| 0.6175 | 0.3008 | 0.0242 | 0.0880 | 0.9952 | 19.6322 | 12.1571 |
| 0.6500 | 0.3009 | 0.0242 | 0.0880 | 0.9952 | 19.6264 | 12.7951 |

This gives a total volume of approximately 12.8 l and a residence time for the liquid of 12.8 seconds (as the flow rate of the liquid is 1 l/s).

(b) In the previous table, we can see that $C_B$ at the exit is 0.0242 mol/l, a value very similar to that of the input. This means that actually it can be considered constant, and the reaction is apparently zero-order.

(c)

$$V_{recip} = \frac{V_L(1 + f_G + f_S)}{f_L} = \frac{V_L(1 + 0.1 + 0.08)}{0.7} = 21.56\,l$$

**Problem 8.3**  The reaction between hydrogen and sulfur hydrocarbons is being studied for the elimination of sulfur in the form of $SH_2$. This reaction takes place in cylindrical $SCo.SMo_3/Al_2O_3$ catalysts, with a diameter of 0.08 cm and a particle density of $0.9\,g/cm^3$. The reaction conditions are 25 atm and 300 °C. The gaseous reactant reacts with the hydrocarbon in the liquid phase. The intrinsic catalytic rate is pseudo-first-order with respect to dissolved hydrogen and is given by:

$$\tfrac{1}{2}H_2\ (g) + R\text{-}S\ (l) \rightarrow R\text{-}H\ (l) + SH_2\ (g) \quad r = kC_{H_2}$$

With $k = 0.02\,l/g{\cdot}s$. The diffusivity of $H_2$ in the solvent is $5{\cdot}10^{-5}\ cm^2/s$. 1.1 l/s of liquid are fed with 0.2 mol of reagent/l and free of hydrogen. The fed gas contains 2% $H_2$ by volume, and its volumetric flow can be considered constant and equal to 20 l/s. The transport parameters for hydrogen are:

$$k_L a_L = 5.1 \cdot 10^{-5}\,l/g\,s$$

$$k_c a_c = 8 \cdot 10^{-4}\,l/g \cdot s$$

We will assume that the pressure drop and the transfer in the gas phase are negligible, that isothermal conditions exist throughout the system, and that the gas behaves ideally.

(a) Calculate $C_{H_2-G}$, $C_{H_2-L}$, and $C_{H_2-S}$ (in the gas, liquid, and catalyst surface, respectively) under the mentioned conditions.
(b) Calculate the rate of reaction.
(c) If the reaction is carried out in a reactor that could be considered completely mixed, both for the gas and for the liquid, with 5000 kg of catalyst, calculate the outlet concentration of hydrogen (or its partial pressure) in the reactor.

**Solution to Problem 8.3**

(a) We should calculate the Thiele modulus and the effectiveness:

$$L = \frac{R}{2} = 0.02 \text{ cm}$$

$$m = \sqrt{\frac{k'''}{D_e}} = \sqrt{\frac{0.02\frac{1}{g\cdot s} \cdot 1000\frac{cm^3}{l} \cdot 0.9\frac{g}{cm^3}}{5 \cdot 10^{-5}\frac{cm^2}{s}}} = 600 \text{ cm}^{-1}$$

So mL = 12 and:

$$\eta = \frac{\tanh(mL)}{mL} = 0.083\,33$$

In the system, we have:

$$(-r'_A) = (k_{AL}a_i)(C_{Ag} - C_{AL}) = (k_{Ac}a_c)(C_{AL} - C_{As}) = k'C_{As}\eta_A$$

$$\begin{cases} 2 \cdot 10^{-5}(1.06 \cdot 10^{-2} - C_{AL}) = 8 \cdot 10^{-4} \cdot (C_{AL} - C_{As}) \\ \quad 8 \cdot 10^{-4} \cdot (C_{AL} - C_{As}) = 0.02 C_{As} \cdot 0.083\,33 \end{cases}$$

where $C_{Ag}$ has been calculated by:

$$C_{Ag} = \frac{P_{Ain}}{RT} = 1.06 \cdot 10^{-2}\,\frac{mol}{l}$$

We have a system of two equations with two unknowns, and finally: $C_{AL} = 6.23 \cdot 10^{-4}$ mol/l, $C_{As} = 2.0 \cdot 10^{-4}$ mol/l

(b)

$$(-r''_A) = \frac{P_{Ag}}{\dfrac{1}{k_{Ag}\cdot a_i} + \dfrac{H_A}{k_{a_L}\cdot a_i} + \dfrac{H_A}{k_{Ac}\cdot a_c} + \dfrac{H_A}{k''_{ap}\cdot \eta_A\cdot \varepsilon_s}}$$

With:

$$\frac{1}{k_{Ag}\cdot a_i} = 0$$

$$\frac{1}{k_{a_L}\cdot a_i} = 2000\,\frac{g\cdot s}{l}$$

$$\frac{1}{k_{Ac} \cdot a_C} = 1250 \, \frac{g \cdot s}{l}$$

$$\frac{1}{k_{ap}'' \cdot \eta_A \cdot \varepsilon_s} = \frac{1}{0.02 \cdot 0.083\,33} = 600 \, \frac{g \cdot s}{l}$$

The sum of terms is 21 850 g·s/l and:

$$\left(-r_A'\right) = \frac{p_{Ag}/H_A}{21\,850} = 4.85 \cdot 10^{-7} \, \frac{mol}{s \cdot g_{cat}}$$

(c) With 50 kg of catalyst, the "A" converted is $4.85 \cdot 10^{-7} \, \frac{mol}{s \cdot g_{cat}} \cdot 50\,000\,g_{cat} = 0.0241 \frac{mol}{s}$

$$0.0241 \, \frac{mol}{s} = 20 \, \frac{l}{s} \cdot (C_{A0} - C_A) \, \frac{mol}{l} = 20 \cdot (1.06 \cdot 10^{-2} - C_A)$$

and then: $C_A = 0.0093 \, mol/l$.

**Problem 8.4**   A fermenter is used to carry out an enzymatic reaction. The enzyme, E, is supported on spherules with a density similar to that of water. The suspension reactor is perfectly stirred. The internal resistance to mass transfer can be neglected since the spherules are very small. However, the external resistance is significant. The mass transfer coefficients are:

$$(k_C \cdot a)_A = 0.54 \, l/min \cdot l_{liquid}$$

$$(k_C \cdot a)_B = 0.36 \, l/min \cdot l_{liquid}$$

The overall reaction and the expression of the intrinsic kinetics are:

$$2A + B \rightarrow Products \quad \left(-r_B'''\right) = \frac{k''' C_A C_B}{1 + K \cdot C_A}$$

where $k''' = 8.7 \, l^2/min \cdot mol \cdot l_{catalyst}$ and $K = 9.1 \, M^{-1}$. The internal volume of the tank is $10 \, m^3$. The accessories, including the stirrer and nozzles to retain the spherules, uniformly occupy 10% of the internal volume. The upper 20% of the entire internal volume of the reactor must be free of suspension. The spherules occupy 15% of the volume of the suspension.

(a) Calculate the volume of liquid inside the reactor, in $m^3$.
    Two independent streams with equal volumetric flows ($G$) are fed. The first contains a 3.0 M aqueous solution of A, and the second contains another 1.2 M aqueous solution of B.
(b) If 80% of the B fed reacts, calculate the concentrations at the outlet.
(c) Estimate the rate of reaction for B, referring to the volume of liquid.
(d) Calculate the volumetric flow rate, in l/min, which must feed on each solution.

**Solution to Problem 8.4**
(a) The volume of the liquid can be easily obtained.
$$V_L = V_{reactor} \cdot (1 - 0.2) \cdot (1 - 0.08 - 0.15) = 6.16 \, m^3.$$

(b) If 80% of the B fed is converted:

$$n_B = n_{B0}(1 - X_B) = 0.2 n_{B0} = 0.2 \cdot (C_{B0}G) = 0.24G \, \frac{\text{mol B}}{\text{h}}$$

$$n_A = n_{A0} - n_{B0} \cdot X_B \cdot 2 = 1.08G \, \frac{\text{mol A}}{\text{h}}$$

The concentrations will be:

$$C_B = \frac{n_B}{\text{total flow rate}} = 0.24 \frac{G}{2G} = 0.12 \, \frac{\text{mol}}{\text{l}}$$

$$C_A = 1.08 \frac{G}{2G} = 0.54 \, \frac{\text{mol}}{\text{l}}$$

(c) The external mass transfer resistance is significant, so:

$$(k_C \cdot a)_A (C_A - C_{As}) = \left(-r_A'''\right) = 2 \frac{k''' C_{As} C_{Bs}}{1 + K \cdot C_{As}}$$

$$(k_C \cdot a)_B (C_B - C_{Bs}) = \left(-r_B'''\right) = \frac{k''' C_{As} C_{Bs}}{1 + K \cdot C_{As}}$$

Note that the reaction rates are calculated at the concentration present on the surface of the catalyst, not at the global concentrations. In the previous equations, the only unknowns were $C_{As}$ and $C_{Bs}$. We can check that $C_{As} = 0.43$ M and $C_{Bs} = 0.04$ M. With these values $(-r_B''') = 0.03$ mol/min·l

(d) For example, for B we have:

$$\text{moles of B reacted} = n_{B0} \cdot X_B = (C_{B0} \cdot G) \cdot X_B = 0.96G$$

$$\text{moles of B reacted} = (-r_B) \cdot V = 0.03 \cdot 6.16 = 0.1848$$

So finally, $G = 0.24 \, \text{m}^3/\text{min}$.

**Problem 8.5**  40 l/s of a gas containing 0.05 M of A is continuously fed to a semi-continuous reactor with three phases. This flow rate can be assumed to be constant within the reactor. The reactor is charged with 2400 l of a 1 M liquid solution of B. For the liquid phase, the reactor operates in batches. The solubility of A is 0.02 M liquid/M gas. The relevant reaction is:

$$A \,(g) + B \,(l) \rightarrow \text{products}$$

with:

$$(-r_A) = k_{ap} C_A C_B = 2 \, \frac{1}{\text{mol} \cdot \text{s}} C_A C_B$$

where the resulting velocity is based on the free volume of liquid, and the $k_{ap}$ already includes the effects of internal resistances to mass transfer. During discharge, the solid catalyst remains soaked inside the reactor. Under the operating conditions, the parameters that describe the external mass transfer are:

$$(k_{AL} \cdot a_L) = 0.3 \, \text{s}^{-1}, (k_{As} \cdot a_s) = 0.1 \, \text{s}^{-1}, (k_{Bs} \cdot a_s) = 0.01 \, \text{s}^{-1}$$

The residence time of the bubbles inside the reactor is very short compared to the retention time of the liquid. Since the rate of reaction is relatively slow, it is a reasonable assumption to consider $C_{AL}$ to be "constant" but only in relation to the time a bubble travels between its inlet and the upper level of the liquid.

(a) Determine the time needed to achieve a conversion fraction of 0.99 for B, and
(b) Check which resistances had a significant effect.

**Solution to Problem 8.5**

For details refer the Wiley website at http://www.wiley-vch.de/ISBN9783527354115

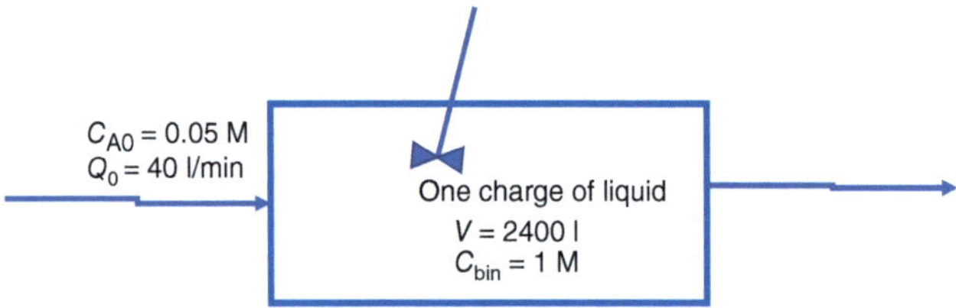


(a) For the component in the gas, we have:

$$(-r''_A) = \frac{(C_{Ai} - C_{AL})}{\dfrac{1}{k_{Al} \cdot a_i}} = \frac{(C_{AL} - C_{As})}{\dfrac{1}{k_{Ac} \cdot a_c}} = \frac{C_{As}}{\dfrac{1}{k''_{ap} \cdot \eta_A \cdot f_s}} = \frac{C_{As}}{\dfrac{1}{k_{Al} \cdot a_i} + \dfrac{1}{k_{Ac} \cdot a_c} + \dfrac{1}{k''_{ap} \cdot \eta_A \cdot \varepsilon_s}}$$

$\left(k''_{ap}\right) \cdot \eta_A \cdot f_s$ is the $k_{ap}$ mentioned in the statement. In the previous equation, $\overline{C}_B$ corresponds to $C_{BL}$. In the gas–liquid interface:

$$C_{Ai} = C_{Ag} \cdot \text{Solubility} = 0.05 \cdot 0.02 = 1 \cdot 10^{-3} \, \frac{\text{mol}}{\text{m}^3}$$

Once we know the reaction rate, we can do:

$$C_{AL} = C_{Ai} - \frac{\left(-r''_A\right)}{k_{Al} \cdot a_i}$$

To confirm if "A" is limiting the reaction, we can also calculate the reaction rate for the other species:

$$(-r''_B)\left(\frac{\text{mol}}{\text{s} \cdot \text{m}^3_R}\right) = \frac{C_{BL}}{\dfrac{1}{k_{Bc} \cdot a_c} + \dfrac{1}{k''_{ap} \cdot \eta_B \cdot \varepsilon_s}}$$

$k''_{ap} \cdot \eta_B \cdot f_s$ is $k_{ap}$ and $\overline{C}_A = C_{AL}$. We can also do a mass balance (batch reactor for the liquid), obtaining:

$$r_B = \frac{dC_{BL}}{dt}$$

$$C^{t+1}_{BL} = C^t_{BL} + r_B \cdot \Delta t$$

We should do the integration until we reach the desired conversion. With the use of a spreadsheet, we find that the total time needed is 258 min. The scheme of the calculation is:

| | Gas | Interface | Liquid | | | | |
| | $C_A$ | $C_{Ai}$ | $C_A$ | $C_B$ | | | |
| $t$ (s) | (mol/m$^3$) | (mol/m$^3$) | (mol/m$^3$) | (mol/m$^3$) | $(-r_B)$ | $(-r_A)$ | Conversion |
|---|---|---|---|---|---|---|---|
| 0 | 0.05 | 0.001 | $1.000\,10^{-3}$ | 1 | 0 | 0 | 0.000 000 |
| 2.5 | 0.05 | 0.001 | $2.771\,10^{-4}$ | 1.000 | 0.000 | $7.229\,10^{-5}$ | 0.000 181 |
| 5 | 0.05 | 0.001 | $2.771\,10^{-4}$ | 1.000 | 0.001 | $7.229\,10^{-5}$ | 0.000 361 |
| 7.5 | 0.05 | 0.001 | $2.771\,10^{-4}$ | 0.999 | 0.001 | $7.229\,10^{-5}$ | 0.000 542 |
| 10 | 0.05 | 0.001 | $2.771\,10^{-4}$ | 0.999 | 0.001 | $7.229\,10^{-5}$ | 0.000 723 |
| ... | | | | | | | |
| 15 495 | 0.05 | 0.001 | $8.405\,10^{-4}$ | 0.010 | 0.000 | $1.595\,10^{-5}$ | 0.989 91 |
| 15 497.5 | 0.05 | 0.001 | $8.410\,10^{-4}$ | 0.010 | 0.000 | $1.590\,10^{-5}$ | 0.989 95 |
| 15 500 | 0.05 | 0.001 | $8.415\,10^{-4}$ | 0.010 | 0.000 | $1.585\,10^{-5}$ | 0.989 99 |
| 15 502.5 | 0.05 | 0.001 | $8.420\,10^{-4}$ | 0.010 | 0.000 | $1.580\,10^{-5}$ | 0.990 03 |

(b) The main resistance is the one contributing the more to the total value of the denominator of the rate equation, that in this case is the mass transfer to the surface.

**Problem 8.6**   The process for the elimination of chlorinated compounds (R-Cl) (chlorobenzene, chloroform, and carbon tetrachloride) from polluting gases is being studied at 300 K on a catalyst formed by microporous silica ($SiO_2$ and $Ag/SiO_2$) in the presence of aqueous solutions of MEA (methyl ethyl amine). In a packed tower, the liquid is fed from the top and the gas is fed in countercurrent. Under these conditions, $(k_g\,a)$ for these R-Cl is estimated at 0.3 mol/min·atm·l, $(k_L\,a)$ is 10 min$^{-1}$, and there is an elimination of 60% (molar) for a gas residence time of 3 min. The kinetics of the equimolar reaction between R-Cl and MEA is first order with respect to R-Cl.

(a) Determine the value of the rate constant of the reaction, clearly indicating its units. Is it a real or an apparent constant?
(b) If the gas flow is 1000 l/min, what volume should the reactor have? Clearly specify if the calculated volume is for gas, liquid, solid, or full reactor.

   *Data:* $H_A = 0.7$ atm·l/mol, $R = 0.082$ atm·l/mol·K. State any necessary assumptions.

**Solution to Problem 8.6**
(a) This is a triphasic catalytic process, and we will assume plug flow for both the liquid and the gas. We can do a scheme of the process:

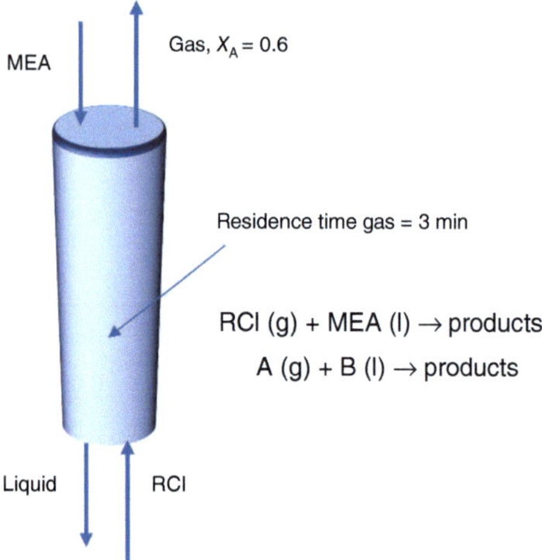


A balance of R-Cl in a differential volume is:

$$dn_A = r_A dV$$

In isothermal conditions and diluted gases:

$$G dC_A = r_A dV$$

$$\int_{C_{A0}}^{C_A} \frac{dC_A}{r_A} = \frac{\int_0^V dV}{G} = \frac{V}{G} = \overline{t_{gas}}$$

with:

$$(-r_A'') = \frac{p_{Ag}}{\dfrac{1}{k_{Ag} \cdot a_i} + \dfrac{H_A}{k_{Al} \cdot a_i} + \dfrac{H_A}{k_{Ac} \cdot a_c} + \dfrac{H_A}{k_{ap}'' \cdot \eta_A \cdot \varepsilon_s}}$$

$$\int_{C_{A0}}^{C_A} \frac{1}{C_{Ag} \cdot RT} \left( \frac{1}{k_{Ag} \cdot a_i} + \frac{H_A}{k_{Al} \cdot a_i} + \frac{H_A}{k_{Ac} \cdot a_c} + \frac{H_A}{k_{ap}'' \cdot \eta_A \cdot \varepsilon_s} \right) dC_A = \overline{t_{gas}}$$

All the parameters that are in parentheses are constant:

$$\left( \frac{1}{k_{Ag} \cdot a_i} + \frac{H_A}{k_{Al} \cdot a_i} + \frac{H_A}{k_{Ac} \cdot a_c} + \frac{H_A}{k_{ap}'' \cdot \eta_A \cdot \varepsilon_s} \right) \int_{C_{A0}}^{C_A} \frac{1}{C_{Ag} \cdot RT} dC_A = \overline{t_{gas}}$$

$$\left( \frac{1}{k_{Ag} \cdot a_i} + \frac{H_A}{k_{Al} \cdot a_i} + \frac{H_A}{k_{Ac} \cdot a_c} + \frac{H_A}{k_{ap}'' \cdot \eta_A \cdot \varepsilon_s} \right) \ln \left( \frac{C_{A0}}{(1 - X_A)C_{A0}} \right) = \overline{t_{gas}} \cdot RT$$

Substituting in the known values:

$$\left( \frac{1}{0.3} + \frac{0.7}{10} + \frac{0.7}{k_{Ac} \cdot a_c} + \frac{0.7}{k_{ap}'' \cdot \eta_A \cdot \varepsilon_s} \right) \ln \left( \frac{C_{A0}}{(1 - X_A)C_{A0}} \right) = 3 \cdot 0.082 \cdot 300$$

Assuming that $k_{Ac} \cdot a_c = \infty$ we get $k_{ap}'' \cdot \eta_A \cdot \varepsilon_s = 0.0165 \text{ min}^{-1}$. This is an apparent constant that bears in mind the process of diffusion through the particle of catalyst. Furthermore, we have assumed that the external mass transfer is instantaneous.

We can also talk about a constant that encompasses all three phenomena (external, internal diffusion, and chemical reaction):

$$\left( \frac{1}{0.3} + \frac{0.7}{10} + \frac{0.7}{k''_{ap}} \right) \ln \left( \frac{C_{A0}}{(1 - X_A)C_{A0}} \right) = 3 \cdot 0.082 \cdot 300$$

And $k''_{ap} = 0.165\ \text{min}^{-1}$

(b) As $G = 1000\,\text{l/min}$ and $\overline{t}_{gas} = 3\ \text{min}$, the resulting volume of gas is $3000\,\text{l}$.

**Problem 8.7** The concentration of $SO_2$ in the air (at $1\,\text{bar} = 10^5$ Pa) must be reduced from 0.1% (i.e. 100 Pa) to 0.02% (i.e. 20 Pa). Calculate the required height of a tower that works in countercurrent flow if it is carried out:

(a) With pure water.
(b) With water to which KOH is added at a rate of $800\ \text{mol/m}^3$. This reacts with $SO_2$ in such a way that:

$$2KOH + SO_2 \rightarrow K_2SO_3 + H_2O$$

(c) In both situations, indicate which resistance is the one that controls the global process.

*Data:* ("A" refers to $SO_2$ and "B" to KOH)
For the filling: $k_{Ag}\, a = 0.32\ \text{mol/h m}^3$ Pa, $k_{Al}\, a = 0.1\ \text{h}^{-1}$
The solubility of $SO_2$ in water is given by the Henry's law constant:

$$H_A = p_{Ai}/C_{Ai} = 12.5\ \text{Pa m}^3/\text{mol}$$

Flows per square meter of section of the tower are:
Gas $= 1\ 10^5\ \text{mol/h m}^2$
Liquid $= 7\ 10^5\ \text{mol/h m}^2$
The molar density of the liquid under all conditions is $56.000\ \text{mol/m}^3$.
KOH reacts with $SO_2$ extremely quickly. We will assume that the diffusivities of the two components in water are equal.

**Solution to Problem 8.7**
For details refer the Wiley website at http://www.wiley-vch.de/ISBN9783527354115

(a) Our strategy is to first solve the mass balance and then determine the height of the tower. Since we are dealing with dilute solutions, we can use the simplified form of the material balance. Thus, for any point on the tower $p_A$ and $C_A$ are related by:

$$p_A - p_{A1} = \frac{\left( \frac{L}{S} \right)}{\left( \frac{G}{S} \right)} \cdot \frac{P_T}{C_T} \cdot (C_A - C_{A1})$$

where $p_{A1}$ is the partial pressure of A at the outlet and $C_{A1}$ is the inlet concentration (countercurrent). In this system:

$$p_A - 20 = \frac{(7 \cdot 10^5)}{(1 \cdot 10^5)} \cdot \frac{10^5}{56\,000} \cdot (C_A - 0)$$

This provides:

$$C_A = 0.08 \cdot p_A - 1.6$$

From this equation, the concentration of A in the liquid leaving the tower is:

$$C_{A2} = 0.08 \cdot 100 - 1.6 = 6.4 \, \frac{\text{mol}}{\text{m}^3}$$

The expression for the height of the tower is:

$$h = \frac{\frac{G}{S}}{P_T \cdot (K_{Ag} \cdot a)} \cdot \int_{20}^{100} \frac{dp_A}{p_A - p_A^*}$$

We have to:

$$\frac{1}{K_{Ag} \cdot a} = \frac{1}{k_{Ag} \cdot a} + \frac{H_A}{k_{AL} \cdot a} = \frac{1}{0.32} + \frac{12.5}{0.1} = 3.12 + 125 = 128.12$$

In this expression, it is seen that the controlling resistance is the liquid film (part c)).

$$p_A - p_A^* = p_A - H_A \cdot C_A = p_A - 1.25 \cdot (0.08 p_A - 1.6) = 20 \, \text{Pa}$$

Integrating, we obtain: $h = 512.5 \, \text{m}$.

(b) The strategy to solve the problem is the following:

Step 1. Write the material balance and calculate $C_{B2}$ in the output stream.
Step 2. Determine which form of the rate equation should be used.
Step 3. Calculate the height of the tower.
Let us begin:
Step 1. Material balance. For rapidly reacting dilute solutions, the following equation provides for any point of the tower, $p_A$, $C_B$:

$$p_A - p_{A1} = \frac{\left(\frac{L}{S}\right)}{\left(\frac{G}{S}\right)} \cdot \frac{P_T}{bC_T} \cdot (C_{B1} - C_B)$$

Substituting: $C_B = 800 - \frac{p_A - 20}{6.25}$ ($b = 2$)

At the bottom of the tower (exit, $p_A = 100$): $C_{B2} = 787.2 \, \frac{\text{mol}}{\text{m}^3}$

Step 2. Form of the rate equation to be used. We can use the general equation:

$$h = \frac{\frac{G}{S}}{P_T} \cdot \int_{20}^{100} \frac{dp_A}{(-r_A''')} = \frac{\frac{G}{S}}{P_T} \cdot \int_{20}^{100} \frac{dp_A}{\frac{1}{k_{Ag} \cdot a} + \frac{H_A}{k_{AL} \cdot a \cdot E} + \frac{H_A}{k_{ap}'' \cdot \varepsilon_L}}$$

In the third term, $k = \text{infinity}$, so the third addend disappears. In case of instantaneous reaction, $E = E_i = \left(1 + \frac{D_{BL} \cdot C_B}{b \cdot D_{AL} \cdot C_{Ai}}\right)$, and so:

$$h = \frac{\frac{G}{S}}{P_T} \cdot \int_{20}^{100} \frac{dp_A}{\frac{1}{k_{Ag} \cdot a} + \frac{H_A}{k_{AL} \cdot a \cdot \left(1 + \frac{D_{BL} \cdot C_B}{b \cdot D_{AL} \cdot C_{Ai}}\right)}}$$

Substituting $C_B = 800 - \frac{p_A - 20}{6.25}$, and $C_{Ai} = p_A/H_A$ gives:

$$h = \frac{\frac{G}{S}}{P_T} \cdot \int_{20}^{100} \frac{\mathrm{d}p_A}{\dfrac{1}{k_{Ag} \cdot a} + \dfrac{\frac{p_A}{H_A}}{k_{AL} \cdot a \cdot \left(1 + \dfrac{D_{BL} \cdot \left(800 - \frac{p_A - 20}{6.25}\right)}{b \cdot D_{AL} \cdot p_A/H_A}\right)}}$$

The integration of this expression is easier to be done numerically. Using a spreadsheet, we can easily calculate the height of the tower:

| $p_A$ (bar) | $p_A$ (Pa) | $C_B$ (mol/m³) | $E' = "E_i"$ | $C_{Ai}$ (mol/m³) | Integral |
|---|---|---|---|---|---|
| 20 | 2 000 000 | 800.00 | 1.003 | 160 000 | 0.156 |
| 21 | 2 100 000 | 799.84 | 1.002 | 168 000 | 0.164 |
| 22 | 2 200 000 | 799.68 | 1.002 | 176 000 | 0.172 |
| 23 | 2 300 000 | 799.52 | 1.002 | 184 000 | 0.180 |
| 24 | 2 400 000 | 799.36 | 1.002 | 192 000 | 0.188 |
| … | | | | | |
| 98 | 9 800 000 | 787.52 | 1.001 | 784 000 | 0.765 |
| 99 | 9 900 000 | 787.36 | 1.000 | 792 000 | 0.773 |
| 100 | 10 000 000 | 787.20 | | | |
| | | | | $\Sigma =$ | 37.18 |

The final value for the height of the tower is 37.18 m.

(c) When only water is used for removing the $SO_2$, the controlling step is resistance in the liquid film, as we have shown. When using KOH to help, we must compare both addends of the denominator of the rate expression. If we do this, we will see that the gas resistance $(1/k_{Ag} \cdot a)$ is approximately 2.4% of the total in every section of the tower.

**Problem 8.8**  A reaction is being carried out for the elimination of chlorinated compounds in water by reaction with a nucleophile compound ($NaNH_2$) on a catalyst in the form of small spheres. The suspension reactor is perfectly agitated. The internal resistance to mass transfer can be neglected since the spherules are very small. However, external resistance is significant. The mass transfer coefficients are:

$$(k_C \, a_c)_A = 0.5 \text{ min}^{-1}$$
$$(k_C \, a_c)_B = 0.3 \text{ min}^{-1}$$

where "A" is the compound to be eliminated and "B" is the nucleophile. The overall reaction and expression of the kinetics of the chemical step are:

$$2A + B \rightarrow \text{Products} \quad (-r_B) = kC_A C_B$$

where $k = 18.7 \, \text{l}^2_{\text{liquid}}/\text{min} \cdot \text{mol} \cdot \text{l}_{\text{catalyst}}$. The volume of liquid in the tank is 6 l. The spherules occupy 15% of the volume of the suspension. Two independent currents with equal volumetric flows are fed. The first contains a 3.0 M aqueous solution of A, and the second contains another 1.2 M aqueous solution of B.

(a) If the conversion of B is 0.8, calculate the concentrations at the outlet.
(b) Estimate the reaction rate for B, referring to the volume of liquid.
(c) Calculate the volumetric flow, in l/min, that must be fed with each solution.

**Solution to Problem 8.8**

(a)

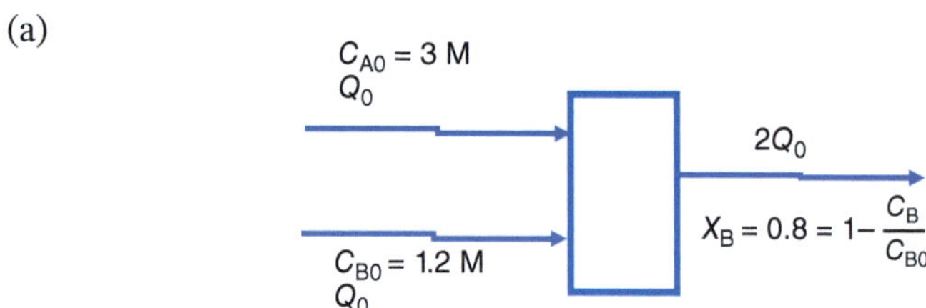


We can calculate $C_{B0} = 1.2/2 = 0.6\,\text{mol/l}$ and $C_B = C_{B0}(1 - X_B) = 0.12\,\text{mol/l}$. Respect to "A," $C_{A0} = 3/2 = 1.5\,\text{mol/l}$ and $C_A = C_{A0} - C_{B0}\cdot 2X_B = 1.5 - 2\cdot 0.6\cdot 0.8 = 0.54\,\text{mol/l}$.

(b) For estimating the reaction rate for B, referring to the volume of liquid, we see in the statement that the important steps in the reaction are the mass transfer in the catalyst film and the chemical reaction itself. In this way:

$$(-r_B) = (k_{Bc}a_c)(C_B - C_{Bs}) = kC_{As}C_{Bs}$$

$$(-r_A) = (k_{Ac}a_c)(C_A - C_{As}) = 2kC_{As}C_{Bs}$$

Values of $C_B$ and $C_A$ refer to a point far from the surface of the catalyst and coincide with the "global" concentrations calculated before. Note that the chemical rate is evaluated at the concentration of each reactant on the surface (as internal diffusion is fast). Substituting the known variables:

$$(-r_B) = 0.36 \, \frac{1}{\text{min} \cdot 1\text{L}_{\text{liquid}}} (0.12 - C_{Bs}) \frac{\text{mol}}{\text{l}_{\text{liquid}}}$$

$$= 18.7 \, \frac{\text{l}^2_{\text{liquid}}}{\text{min} \cdot \text{mol} \cdot \text{l}_{\text{catalyst}}} C_{As} \cdot C_{Bs} \frac{\text{mol}^2}{\text{l}^2_{\text{liquid}}}$$

$$(-r_A) = 0.54 \, \frac{1}{\text{min} \cdot \text{l}_{\text{liquid}}} (0.54 - C_{As}) \frac{\text{mol}}{\text{l}_{\text{liquid}}}$$

$$= 2 \cdot 18.7 \, \frac{\text{l}^2_{\text{liquid}}}{\text{min} \cdot \text{mol} \cdot \text{l}_{\text{catalyst}}} C_{As} \cdot C_{Bs} \frac{\text{mol}^2}{\text{l}^2_{\text{liquid}}}$$

And we get $C_{As} = 0.256\,\text{mol/l}$ and $C_{Bs} = 0.013\,\text{mol/l}$. Finally:

$$(-r_B) = kC_{As}C_{Bs} = 18.7 \, \frac{\text{l}^2_{\text{liquid}}}{\text{min} \cdot \text{mol} \cdot \text{l}_{\text{catalyst}}} 0.256 \cdot 0.013 \, \frac{\text{mol}^2}{\text{l}^2}$$

$$= 0.0635 \, \frac{\text{mol}}{\text{l}_{\text{catalyst}} \cdot \text{min}}$$

As we know that in the reactor $V_{catalyst}/(V_{liquid} + V_{catalyst}) = 0.15$, we can obtain that $V_{catalyst}/V_{liquid} = 0.15/(1 - 0.15) = 0.176$, so:

$$(-r_B'') = 0.0635 \frac{mol}{l_{catalyst} \cdot min} \cdot 0.176 \frac{l_{catalyst}}{l_{liquid}} = 0.0112 \frac{mol}{l_{liquid} \cdot min}$$

(c) We can relate the reaction rate with the moles reacted:

$$(-r_B'') = (C_{B0} - C_B) \frac{mol}{l} \cdot (2Q_0) \frac{1}{min} = (0.6 - 0.12) \cdot 2Q_0 = 0.0112$$

Obtaining $Q_0 = 0.011\ 685\ l/min$

The residence time is then $\bar{t} = \frac{V_{liquid}}{Q_0} = \frac{6}{0.011\ 685} = 513.44\ min = 8.55\ h.$

**Problem 8.9**  You are requested to provide a recommendation, supported by justification, for the most appropriate equipment type to carry out a gas–liquid reaction involving gas A and a solution of reactant B. Here are the relevant system details:

- Concentration of reactant B in the solution: $5\ kmol/m^3$
- Diffusivity of gas A in the solution: $1.5 \cdot 10^{-9}\ m^2/s$
- Reaction rate constant (A + B → Products; second-order overall): $0.03\ m^3/kmol\ s$
- Considering different equipment options, we have the following ranges for the liquid-side mass transfer coefficient ($k_{AL}$):
  - Bubble dispersions (plate columns, bubble columns, agitated vessels): $k_{AL}$ ranging from $2 \cdot 10^{-4}$ to $4 \cdot 10^{-4}\ m/s$.
  - Packed column: $k_{AL}$ ranging from $0.5 \cdot 10^{-4}$ to $1.0 \cdot 10^{-4}\ m/s$.

Based on these options, it is necessary to assess the suitability of each equipment type for facilitating the desired gas–liquid reaction. Please provide your recommendation along with the reasons supporting your choice.

**Solution to Problem 8.9**
We can calculate the Hatta modulus for each of the possible situations. This modulus will indicate whether the reaction is fast or slow:

$$M_H = \frac{\sqrt{D_{AL} \cdot k'''}}{k_{AL}} = \frac{\sqrt{1.5 \cdot 10^{-9} \cdot 0.03}}{k_{AL}}$$

For the bubble dispersions, $M_H$ will lie in the range of 0.075 to 0.038, and for the packed column, $M_H$ is between 0.03 and 0.15. These values indicate that the reaction is only moderately fast, and a relatively high holdup of liquid will be advisable. From there, an agitated tank or a simple bubble column will be desirable for performing the operation.

**Problem 8.10**  The objective is to oxidize *o*-Xylene to *o*-methylbenzoic acid in the liquid phase using air dispersion in an agitated vessel. The reaction can be represented as:

$$o\text{-Xylene} + 1.5O_2 \rightarrow \text{Benzoic acid} + H_2O$$

To maintain a constant low conversion, the reaction mixture in the vessel undergoes continuous withdrawal of the product stream while fresh *o*-Xylene is continuously added. Under these conditions, the reaction rate is approximately independent of the o-Xylene concentration and follows a pseudo-first-order kinetics with respect to the dissolved oxygen concentration. The rate equation can be expressed as:

$$(-r_O) = k_1 C_O$$

Here, $(-r_O)$ represents the rate of oxygen consumption in kmol $O_2$/s·m³ liquid, and $C_O$ is the concentration of dissolved oxygen in kmol $O_2$/m³.

The rate constant $k_1$ at the operating temperature of 160 °C is 1.05 s$^{-1}$. The reactor operates at a pressure that ensures the sum of the partial pressures of oxygen and nitrogen inside the vessel is 14 bar. The liquid volume in the reactor is maintained at 5 m³, and air is dispersed into the reactor at a flow rate of 450 m³/h, measured at reactor conditions.

The quantities to be determined are as follows:

(a) The fraction of the supplied oxygen that reacts.
(b) The corresponding rate of production of benzoic acid is in kilomoles per hour.
(c) Additionally, it is necessary to consider whether an agitated tank is the most suitable reactor for this process.

Further Data:

Physical properties of system:
Henry's law coefficient for $O_2$ dissolved in o-xylene $H_A = 127$ m³ bar/kmol (1.27·10⁷ Nm/kmol)
Diffusivity of $O_2$ in liquid xylene $D_A = 1.4 \cdot 10^{-9}$ m²/s
Equipment performance characteristics:
Gas volume fraction in the dispersion $(1 - \varepsilon_G) = 0.34$
Mean diameter of the bubbles present in the dispersion = 1.0 mm
Liquid phase mass transfer coefficient $k_{AL} = 4.1 \cdot 10^{-4}$ m/s

**Solution to Problem 8.10**
(a) First, we will calculate the oxygen fed:

$$G = 450 \, \frac{\text{m}^3}{\text{h}} = 0.125 \, \frac{\text{m}^3}{\text{s}}$$

$$n_0 = \frac{PV}{RT} = \frac{14 \cdot 10^5 \cdot 0.125}{8314 \cdot 433} = 0.0486 \, \frac{\text{kmol}}{\text{s}}$$

$$n_{A0} = 0.21 n_0 = 0.0102 \, \frac{\text{kmol A}}{\text{s}}$$

Using the conversion degree:

$$n_A = n_{A0}(1 - X_A)$$

We can calculate the reaction rate:

$$(-r_A'') = \frac{p_A}{\dfrac{1}{k_{Ag} \cdot a} + \dfrac{H_A}{k_{AL} \cdot a \cdot E} + \dfrac{H_A}{k_{ap}''}}$$

When we have a value of $M_H$ smaller than unity, a good approximation for the $E$ factor is:

$$E = 1 + \frac{M_H^2}{3} + \frac{M_H^3}{4} + \ldots$$

In the present case, $E = 1.0029$

For calculating the interface present, we can assume that bubbles are spheres, in such a way that:

$$a = \frac{S_{\text{interphase}}}{V_{\text{total}}} = \frac{S_{\text{spheres}}}{V_{\text{spheres}} + V_{\text{liquid}}} = \frac{S_{\text{spheres}}}{V_{\text{spheres}} + \left(\frac{1-\varepsilon_G}{\varepsilon_G}\right) V_{\text{spheres}}}$$

with:

$$\frac{S_{\text{spheres}}}{V_{\text{spheres}}} = \frac{\left(\pi \cdot d_b^2\right)}{\pi \cdot d_b^2/6} = \frac{6}{d_b}$$

From these equations:

$$a = \frac{6\varepsilon_G}{d_b} = 2040 \ \frac{\text{m}^2}{\text{m}^3}$$

Partial pressure at the exit will be calculated bearing in mind that, for each mole of air entering at the exit, we will have 0.79 mol of nitrogen and $0.21(1 - X_A)$ moles of oxygen, so:

$$y_{\text{Aexit}} = \frac{0.21(1 - X_A)}{0.79 + 0.21\,(1 - X_A)}$$

Also:

$$p_{\text{Aexit}} = 14 \cdot y_{\text{Aexit}}$$

(b) In the system, the following is fulfilled:

$$0.0102(1 - X_A) = 0.004\,521 \cdot 14 \cdot \frac{0.21(1 - X_A)}{0.79 + 0.21(1 - X_A)}$$

And finally, we obtain: $X_A = 0.9996$. The hourly rate of production of the acid will then be:

$$0.0102(1 - X_A) = 0.091 \ \frac{\text{kmol oxygen reacted}}{\text{s}}$$

$$\frac{0.091 \ \text{kmol oxygen reacted/s}}{1.5} = 0.090 \ \frac{\text{kmol acid produced}}{\text{s}} = 21.6 \ \frac{\text{kmol}}{\text{h}}$$

(c) A stirred tank, where there is a large volume of liquid, is suitable when the speed of the chemical step is slow, since the conversion process takes place in the entire volume of liquid. This is the case that corresponds to this problem, since $M_H$ is less than unity.

**Problem 8.11** Carbon dioxide ($CO_2$) needs to be removed from an air stream with an inlet concentration of 0.1 mol% and a total pressure of 1 bar. This will be achieved by bubbling the air through a NaOH solution with a concentration of

0.05 M ($0.05\ \text{kmol/m}^3$) at $20\,°C$ in a bubble column. The flow rate of the caustic soda solution is designed to ensure that its composition remains relatively unaffected by the absorption process. The gas phase will have a superficial velocity of 0.06 m/s. In each case, the height of the gas–liquid dispersion will be 1.5 m.

The objective is to calculate the ratio of the outlet concentration of $CO_2$ in the air to the inlet concentration in the following cases:

(a) In the first case, the diameter of the column is 0.20 m, and the gas flows through the column in plug flow.
(b) In the second case, the diameter of the column is 0.20 m, and the dispersed plug flow model is applicable to the gas.
(c) In the third case, the diameter of the column is 1.5 m, and the dispersed plug flow model is applicable to the gas.

For each case, the ratio of outlet concentration of $CO_2$ to the inlet concentration will be determined based on the given parameters and the specific flow characteristics of the gas and liquid phases in the bubble column:

▸ Second-order rate constant for the reaction is $k = 9.5 \cdot 10^3\ \text{m}^3\ \text{/kmol·s}$

$$CO_2 + OH^- \rightarrow HCO_3^{2-}$$

▸ Diffusivity of $CO_2$ in the liquid: $1.8 \cdot 10^{-9}\ \text{m}^2/\text{s}$
▸ Solubility of $CO_2$: Henry's law constant $H_A = 25\ \text{bar m}^3/\text{kmol}\ (2.5 \cdot 10^6\ \text{Nm/kmol})$
▸ Bubble column mass transfer characteristics: The following estimates for a superficial gas velocity of 0.06 m/s will be used for both columns:
  • Gas holdup: $\varepsilon_G = 0.23$
  • Gas–liquid interfacial area per unit volume of dispersion: $a = 195\ \text{m}^2/\text{m}^3$
  • Liquid-side mass transfer coefficient: $k_{AL} = 2.5 \cdot 10^{-4}\ \text{m/s}$

**Solution to Problem 8.11**

First, we can calculate Hatta modulus:

$$M_H = \frac{\sqrt{D_{AL} \cdot k'''_{ap}}}{k_{AL}} = \frac{\sqrt{1.8 \cdot 10^{-9} \cdot (9.5 \cdot 10^3 \cdot 0.05)}}{2.5 \cdot 10^{-4}} = 3.7$$

Note that, in the second-order reaction, $k'''_{ap} = k''' \cdot C_B$. As the Hatta modulus is quite high, surely all the reactions are taking place in the film because chemical step is quite fast. The concentration of dissolved $CO_2$ at the interface in contact with the gas entering with a partial pressure of 0.001 bar will be:

$$C_{Ai} = \frac{0.001}{H_A} = 4 \cdot 10^{-5}\ \frac{\text{kmol}}{\text{m}^3}$$

In general, the rate of transfer per unit volume of dispersion is then:

$$r_A = k_{AL} \cdot a \cdot (C_{Ai} - 0) = k_{AL} \cdot a \cdot \frac{p_{Ai}}{H_A} = k_{AL} \cdot a \cdot \frac{RTC_{Ag}}{H_A}$$

$$= k_{AL} \cdot a \cdot \frac{RT}{H_A} C_{Ag} = k'_{AL} C_{Ag} = \ldots = 0.176 C_{Ag}$$

(a) For a pseudo-first-order reaction such as that being produced in the reactor, we can use the simple mass balance for the PFR, and so:

$$\frac{C_{Aout}}{C_{Ain}} = \exp\left(-k'_{AL}\,\overline{t}_{gas}\right)$$

where $\overline{t}$ represents the mean residence time of the gas phase: $\overline{t}_{gas} = \dfrac{Length}{u_G} = \dfrac{1.5}{0.06} = 25\,s$

$$\frac{C_{Aout}}{C_{Ain}} = \exp\left(-k'_{AL}\,\overline{t}_{gas}\right) = \exp(-0.176 \cdot 25)$$

We can obtain: $C_{Aout} = 0.0123 C_{Ain}$

(b) In a reactor where dispersion exists, we can do a mass balance in a differential volume (consult Chapter 1), finding that for a pseudo-first-order reaction with first-order constant $k_{ap}$:

$$\varepsilon_G D_A \frac{d^2 C_{Ag}}{dz^2} - u_G \frac{dC_{Ag}}{dz} - k_{ap} C_{Ag} = 0$$

Solution to that equation gives the ratio of the concentration of the reactant A at the exit and the inlet:

$$\frac{C_{Aout}}{C_{Ain}} = \frac{4A \exp\left(\frac{u_G L}{2\varepsilon_G D_A}\right)}{(1+A)^2 \exp\left(\frac{A u_G L}{2\varepsilon_G D_A}\right) - (1-A)^2 \exp\left(\frac{A u_G L}{2\varepsilon_G D_A}\right)}$$

where

$$A = \left(1 + \frac{4 k_{ap} \varepsilon_G D_A}{u_G^2}\right)^{\frac{1}{2}}$$

In this case, $k_{ap} = k'_{AL}$ and the value of $D_A$ can be estimated by using an empirical equation for the gas-phase dispersion coefficient:

$$D_A = 50\left(\frac{u_G}{\varepsilon_G}\right)^3 d_C^{1.5}$$

where the units are $D_A$ (m$^2$/s), $u_G$ (m/s), and $d_C$ (m) (diameter column). Using the data, we can check that $D_A = 0.0794\,\mathrm{m^2/s}$, and $\left(\frac{u_G L}{2\varepsilon_G D_A}\right) = 2.464$, so $A = 2.138$ and $C_{Aout} = 0.0526 C_{Ain}$

(c) For a 1.5 m diameter column, using the same equations, we find that:

$$D_A = 50\left(\frac{u_G}{\varepsilon_G}\right)^3 d_C^{1.5} = 1.63\,\frac{\mathrm{m^2}}{\mathrm{s}}$$

$$\left(\frac{u_G L}{2\varepsilon_G D_A}\right) = 0.120$$

$$A = \left(1 + \frac{4 k'_{AL} \varepsilon_G D_A}{u_G^2}\right)^{\frac{1}{2}} = 8.63$$

$$\frac{C_{Aout}}{C_{Ain}} = 0.162$$

**Problem 8.12**  During the hydrogenation of α-methyl styrene using palladium deposited on porous alumina as the catalyst, experimental findings suggest that the reaction follows a pseudo-first-order kinetics with respect to hydrogen, characterized by a rate constant $k_1$ of $16.8\,\text{s}^{-1}$ at $50\,°\text{C}$. The effective diffusivity of hydrogen within the catalyst bed is estimated to be $0.11\cdot10^{-8}\ \text{m}^2/\text{s}$.

For the design of a stirred tank reactor with a "slurry" configuration (catalyst in suspension), spherical particles with a size of $100\,\mu\text{m}$ are considered. The Thiele modulus can be calculated using the standard procedure, and the effectiveness factor is estimated using the equation:

$$\eta = \frac{1}{\text{mL}}\left\{\coth(3\text{mL}) - \frac{1}{3\text{mL}}\right\}$$

where $L$ represents the corresponding characteristic length ($R/3$) of the particles.

Additional parameters for the design are provided: bubble size ($d_\text{b}$) is $0.8\,\text{mm}$, gas fraction ($\varepsilon_\text{G}$) is $0.20$, liquid film mass transfer coefficient ($k_\text{AL}$) is $1.23\cdot10^{-3}$ m/s, catalyst film mass transfer coefficient ($k_\text{Ac}$) is $0.54\cdot10^{-3}$ m/s, and the solid/liquid ratio is $0.10$ (by volume) to ensure an excess of catalyst. The operation is carried out at 30 bars and $50\,°\text{C}$, with the Henry's law constant for dissolved hydrogen at this temperature being $285\,\text{m}^3$ bar/kmol. Considering these values, the overall reaction rate can be calculated, assuming instantaneous mass transfer in the gas phase.

**Solution to Problem 8.12**
In the agitated tank slurry reactor, we have:

$$\text{mL} = \sqrt{\frac{k}{D_\text{e}}}\,L = \sqrt{\frac{k}{D_\text{e}}}\,\frac{R}{3} = 2.067$$

And so:

$$\eta = \frac{1}{\text{mL}}\left\{\coth(3\text{mL}) - \frac{1}{3\text{mL}}\right\} = 0.407$$

For one sphere:

$$a = \frac{S_\text{spheres}}{V_\text{spheres}} = \frac{\left(\pi\cdot d_\text{b}^2\right)}{\pi\cdot d_\text{b}^2/6} = \frac{6}{d_\text{b}}$$

and the interfacial area between gas and liquid is:

$$a_i = \frac{6\varepsilon_\text{G}}{d_\text{b}} = 1500\ \text{m}^{-1}$$

$d_\text{b}$ being the diameter of the bubbles. In the solid–liquid interface:

$$a_\text{C} = \frac{6\varepsilon_\text{s}}{d_\text{p}} = 4363\ \text{m}^{-1}$$

$d_\text{p}$ being the diameter of the particles. For the distribution of phases in the reactor, we have:

$$\varepsilon_G + \varepsilon_L + \varepsilon_S = 1$$

$$\varepsilon_G = 0.2$$

$$\varepsilon_S/\varepsilon_L = 0.1$$

So, $\varepsilon_L = 0.727$ and $\varepsilon_S = 0.0727$
The global reaction rate is:

$$\left(-r_A''\right) = \cfrac{p_{Ag}}{\cfrac{1}{k_{Ag} \cdot a_i} + \cfrac{H_A}{k_{Al} \cdot a_i} + \cfrac{H_A}{k_{Ac} \cdot a_c} + \cfrac{H_A}{(k_{ap}) \cdot \eta_A \cdot \varepsilon_S}}$$

$$\left(-r_A''\right) = \cfrac{p_{Ag}}{\cfrac{1}{\infty} + \cfrac{285}{1.23 \cdot 10^{-3} \cdot 1500} + \cfrac{285}{5.4 \cdot 10^{-4} \cdot 4362} + \cfrac{285}{(18.8) \cdot 0.407}}$$

$$= \frac{p_{Ag}}{1.2} = \frac{30}{1.2} = 0.036 \; \frac{\text{kmol}}{\text{m}_R^3 \cdot \text{s}}$$

**Problem 8.13**  The aim is to evaluate the performance of a three-phase fluidized bed for the hydrogenation of α-methyl styrene at 50 °C, using the same catalyst as described in Problem 8.12, but with spherical particles of 2.0 mm diameter. The liquid and gas flow rates are adjusted to achieve a volume fraction of particles in the bed of 0.50 and a volume fraction of gas of 0.20. The estimated bubble size in the bed is 5.0 mm.

The task at hand is to calculate the overall rate of reaction for this system.

**Solution to Problem 8.13**
Because the conditions for the reaction are the same as in former problem, the pseudo-first-order rate constant $k$ will again be $16.8 \, \text{s}^{-1}$, and the effective diffusivity of hydrogen in the liquid-filled pores of the catalyst $D_e$ will be $0.11 \cdot 10^{-8} \, \text{m}^2/\text{s}$. Also, because the transfer coefficients $k_{AL}$ and $k_S$ depend mainly on the physical properties of the system, the same values, namely, $k_{AL} = 1.23 \cdot 10^{-3} \, \text{m/s}$ and $k_S = 0.54 \cdot 10^{-3} \, \text{m/s}$, will be used.

In this problem, the aim is again to evaluate the individual resistance terms in the general equation rate and to examine them in turn. In addition, the overall rate of hydrogenation at the same hydrogen pressure of 30 bar will be calculated.

(a) Gas–liquid mass transfer resistance $1/k_{AL}a$:

Because $\varepsilon_G = 0.20$ and $d_b = 0.005$ m, $a = 6\varepsilon_G/d_b = 240 \, \text{m}^{-1}$ and:

$$\frac{1}{k_{AL}a} = \frac{1}{1.23 \cdot 10^{-3} \cdot 240} = 3.39 \, \text{s}$$

(b) Liquid–solid mass transfer resistance $1/k_C a\varepsilon_L$:

$$a\_c = 6/d_p = 6/0.002 = 3000 \, \text{m}^{-1}$$

$$\frac{1}{k_C a_c \varepsilon_S} = \frac{1}{0.54 \cdot 10^{-3} \cdot 3000 \cdot 0.50} = 1.23 \, \text{s}$$

(c) Reaction within the particle $(k_{ap}\eta\varepsilon_S)^{-1}$:

As in previous problem, the effectiveness factor for spherical particles is given by:

$$\eta = \frac{1}{mL}\left\{\coth(3mL) - \frac{1}{3mL}\right\}$$

In this case, $mL = 41.2$ and $\eta = 0.0241$ and $(k_{ap}\eta\varepsilon_S)^{-1} = 4.94\,\text{s}$

The overall rate of reaction is therefore obtained:

$$r_A = \frac{C_{Ai}}{3.39 + 1.23 + 4.94} = \frac{0.105}{3.39 + 1.23 + 4.94} = 0.0110\,\frac{\text{kmol}}{\text{m}^3 \cdot \text{s}}$$

**Problem 8.14**  In the slurry bed reactor, ethanol (B) diluted in water (2–3% w/w) is oxidized to acetic acid in the presence of pure oxygen (A) bubbled at 10 atm over catalyst pellets $(\text{Pd}/\text{Al}_2\text{O}_3)$. The reaction takes place at 30 °C and is described by the equation:

$$\text{O}_2\,(\text{g}) + \text{CH}_3\text{CH}_2\text{OH} \rightarrow \text{CH}_3\text{COOH} + \text{H}_2\text{O}$$

The rate of the reaction is given by the expression:

$$-r'_A = k'C_A$$

We need to determine the conversion of ethanol in this process.
Here are the relevant data:

- $k' = 1.77 \cdot 10^{-5}\,\text{m}^3/\text{kg·s}$
- $d_p = 10^{-4}\,\text{m}$ (diameter of catalyst pellets)
- Solid density $= 1800\,\text{kg/m}^3$
- $D_e = 4.16 \cdot 10^{-10}\,\text{m}^2/\text{s}$ (effective diffusivity of ethanol)
- Mass transfer coefficient for oxygen: $k_{Ac} = 4 \cdot 10^{-4}\,\text{m/s}\,\text{O}_2$
- Reactor dimensions: Volume $= 5.0\,\text{m}$ (height)$\cdot 0.1\,\text{m}^2$ (cross-section)
- Volume fraction: $\varepsilon_G = 0.05$, $\varepsilon_L = 0.75$, $\varepsilon_S = 0.2$
- Feed rate: $G = 0.01\,\text{m}^3/\text{s}$
- Henry's constant: $H_A = 86\,000\,\text{Pa} \cdot \text{m}^3/\text{mol}$
- Liquid flow rate: $L = 2 \cdot 10^{-4}\,\text{m}^3/\text{s}$
- Initial concentration of ethanol: $C_{B0} = 400\,\text{mol/m}^3$

**Solution to Problem 8.14**

In this problem, we have:

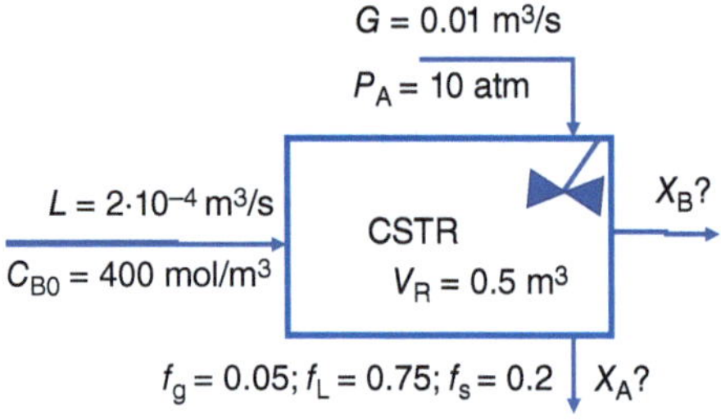

The concentration of oxygen in the liquid is:

$$C_{\mathrm{AL}} = \frac{p_{\mathrm{A}}}{H_{\mathrm{A}}} = 11.77 \; \frac{\mathrm{mol}}{\mathrm{m}^3}$$

So in the present case, $C_{\mathrm{B0}} \gg C_{\mathrm{AL}}$, and we are working in excess of B. The reaction rate will be given by:

$$\left(-r_{\mathrm{A}}''\right) = \frac{p_{\mathrm{Ag}}}{\dfrac{1}{k_{\mathrm{Ag}} \cdot a_i} + \dfrac{H_{\mathrm{A}}}{k_{\mathrm{Al}} \cdot a_i} + \dfrac{H_{\mathrm{A}}}{k_{\mathrm{Ac}} \cdot a_c} + \dfrac{H_{\mathrm{A}}}{k_{a p_{\mathrm{A}}} \cdot \eta_{\mathrm{A}} \cdot \varepsilon_s}}$$

For the spheres of catalyst:

$$a_{\mathrm{C}} = \frac{6\varepsilon_{\mathrm{S}}}{d_{\mathrm{p}}} = 12\,000 \; \frac{\mathrm{m}^2_{\mathrm{cat}}}{\mathrm{m}^3}$$

and then:

$$(k_{\mathrm{Ac}} \cdot a_c) = 4.8 \; \mathrm{s}^{-1}$$

Let us calculate Thiele modulus and effectiveness:

$$\mathrm{mL} = \sqrt{\frac{k'''}{D_e}} \left(\frac{d_p}{6}\right) = \sqrt{\frac{1.77 \cdot 10^{-5} \, \frac{\mathrm{m}^3}{\mathrm{kg} \cdot \mathrm{s}} \cdot 1800 \, \frac{\mathrm{kg}}{\mathrm{m}^3}}{4.16 \cdot 10^{-10} \, \frac{\mathrm{m}^2}{\mathrm{s}}}} \left(\frac{10^{-4}}{6}\right) = 0.1458$$

$$\eta = \frac{\tanh(\mathrm{mL})}{\mathrm{mL}} = 0.9933$$

So, the reaction rate is (assumed constant in the reactor, as both phases of flow are completely stirred):

$$\left(-r_{\mathrm{A}}''\right) = \frac{p_{\mathrm{Ag}}}{\dfrac{1}{k_{\mathrm{Ag}} \cdot a_i} + \dfrac{H_{\mathrm{A}}}{k_{\mathrm{Al}} \cdot a_i} + \dfrac{H_{\mathrm{A}}}{k_{\mathrm{Ac}} \cdot a_c} + \dfrac{H_{\mathrm{A}}}{k_{a p_{\mathrm{A}}} \cdot \eta_{\mathrm{A}} \cdot \varepsilon_s}}$$

$$= \frac{10 \cdot 10 \, 1300}{0 + 0 + \dfrac{86\,000}{4.8} + \dfrac{86\,000}{1.77 \cdot 10^{-5} \cdot 1800 \cdot 0.9933 \cdot 0.2}} = 0.0745 \; \frac{\mathrm{mol}}{\mathrm{m}^3 \cdot \mathrm{s}}$$

In the reactor of $0.5 \, \mathrm{m}^3$, the total mole converted will be $0.0745 \, \frac{\mathrm{mol}}{\mathrm{m}^3 \cdot \mathrm{s}} \cdot 0.5 \, \mathrm{m}^3 = 0.037\,72 \, \frac{\mathrm{mol}}{\mathrm{s}}$

With the feed gas concentration:

$$p_{\mathrm{A}} \cdot G = n_{\mathrm{A}} \cdot RT$$

$$C_{\mathrm{Ag}} = \frac{n_{\mathrm{A}}}{G} = \frac{p_{\mathrm{A}}}{RT} = \frac{10}{0.082 \cdot 303} = 0.4024 \; \frac{\mathrm{mol}}{\mathrm{l}} = 402.4 \; \frac{\mathrm{mol}}{\mathrm{m}^3}$$

And then:

$$n_{\mathrm{A0}} = p_{\mathrm{A}} \cdot \frac{G}{RT} = \frac{10 \, \mathrm{atm} \cdot 10 \, \frac{\mathrm{l}}{\mathrm{s}}}{0.082 \, \mathrm{atm} \cdot \frac{\mathrm{l}}{\mathrm{mol} \cdot \mathrm{K}} \cdot 303 \, \mathrm{K}} = 4.02 \; \frac{\mathrm{mol}}{\mathrm{s}}$$

So the conversion is:

$$X_{\mathrm{A}} = \frac{0.037\,72}{4.02} = 0.0092$$

With respect to species in the liquid phase, as the stoichiometry is $1:1$, we can do:

$$\left(-r_B''\right) = \left(-r_A''\right) = 0.0745 \,\frac{\text{mol}}{\text{m}^3 \cdot \text{s}}$$

Then:

$$X_B = \frac{n_{B0} - n_B}{n_{B0}} = \frac{0.0745\frac{\text{mol}}{\text{m}^3 \cdot \text{s}} \cdot 0.5\,\text{m}^3}{2 \cdot 10^{-4}\,\frac{\text{m}^3}{\text{s}} \cdot 400\,\frac{\text{mol}}{\text{m}^3}} = \frac{0.037\,72\,\frac{\text{mol}}{\text{s}}}{0.08\,\frac{\text{mol}}{\text{s}}} = 0.4653$$

**Problem 8.15**  The following data were obtained in a slurry bed reactor during the hydrogenation of compound A. The table provides information on the solubility of $H_2$ (S) in the liquid mixture (in mol/l) based on the mass of catalyst (in g/l) and the reaction rate (in mol/l·min):

| Particle | $S/(-r_A)$ (min) | $1/\varepsilon_s$ (l/g) |
|---|---|---|
| A | 4.2 | 0.01 |
| A | 7.5 | 0.02 |
| B | 1.5 | 0.01 |
| B | 2.5 | 0.03 |
| B | 3.0 | 0.04 |

We need to determine which particle size has the lowest effectiveness factor.

**Solution to Problem 8.15**
Solubility of hydrogen is given by the concentration of $H_2$ ("A") in the liquid, in such a way that:

$$S = C_{AL} = p_{Ag}/H_A$$

The reaction rate is given by:

$$\left(-r_A''\right) = \frac{p_{Ag}}{\dfrac{1}{k_{Ag} \cdot a_i} + \dfrac{H_A}{k_{Al} \cdot a_i} + \dfrac{H_A}{k_{Ac} \cdot a_c} + \dfrac{H_A}{k_{ap_A} \cdot \eta_A \cdot \varepsilon_s}}$$

So, the data given are then:

$$\frac{p_{Ag}}{-r_A''} = \frac{1}{k_{Ag} \cdot a_i} + \frac{H_A}{k_{Al} \cdot a_i} + \frac{H_A}{k_{Ac} \cdot a_c} + \frac{H_A}{k_{ap_A} \cdot \eta_A \cdot \varepsilon_s}$$

$$\frac{C_{AL}}{-r_A''} = \frac{1}{H_A \cdot k_{Ag} \cdot a_i} + \frac{1}{k_{Al} \cdot a_i} + \frac{1}{k_{Ac} \cdot a_c} + \frac{1}{k_{ap_A} \cdot \eta_A \cdot \varepsilon_s}$$

The area of spheres is proportional to the amount of catalyst present in the following form:

$$a_c = \frac{6\varepsilon_s}{d_p}$$

So:

$$\frac{C_{AL}}{-r_A''} = \frac{1}{H_A \cdot k_{Ag} \cdot a_i} + \frac{1}{k_{Al} \cdot a_i} + \frac{1}{k_{Ac} \cdot \frac{6\varepsilon_s}{d_p}} + \frac{1}{k_{ap_A} \cdot \eta_A \cdot \varepsilon_s} = R_{G+L} + \frac{R_{S+R}}{f_s}$$

where $R_{G+L}$ is the resistance to mass transfer from the gas to the liquid, and $R_{S+R}$ is the resistance to the transfer into the solid catalyst and the chemical reaction. By doing the corresponding plots, we can calculate the slopes and intercepts, obtaining, for catalyst A:

$$(R_{G+L})_{\text{catalyst A}} = 0.9 \text{ min}$$

$$(R_{S+R})_{\text{catalyst A}} = 300 \frac{g}{l \cdot \text{min}}$$

And for catalyst B:

$$(R_{G+L})_{\text{catalyst B}} = 1.0 \text{ min}$$

$$(R_{S+R})_{\text{catalyst B}} = 50 \frac{g}{l \cdot \text{min}}$$

The values shown for $(R_{G+L})$ should be equal for both particles, as they depend on the bubbles but not on the catalyst. The differences should be attributed to experimental errors. From the values of $(R_{S+R})$ one can conclude that the particle diameter A is larger than that of B (as A has a lower effectiveness). In fact, if $R_S$ is considered negligible, we can do:

$$(R_R)_{\text{catalyst A}} = \frac{1}{k_{ap} \cdot \eta_A} = 300$$

$$(R_R)_{\text{catalyst B}} = \frac{1}{k_{ap} \cdot \eta_B} = 50$$

And by comparison: $\eta_B = 6\eta_A$

**Problem 8.16**   The hydrogenation of a diol is carried out in a slurry bed reactor by introducing 0.5 g of catalyst powder into the reactor. The initial concentration of the diol is 2.5 mol/cm$^3$. The reaction follows first-order kinetics with respect to both $H_2$ and the diol. Hydrogen is supplied to the reactor through a distributor at a pressure of 1 atm and a temperature of 35 °C. The equilibrium concentration of $H_2$ is 0.01 mol/cm$^3$, and the specific reaction rate constant is $4.8 \cdot 10^{-5}$ cm$^3$/cm$^2 \cdot$s. The catalyst has a solid loading $(\varepsilon_s)$ of 0.1 kg/m$^3$, a particle diameter of 0.01 cm, and a density of 1.5 g/cm$^3$. We need to calculate the overall reaction rate and determine whether pore diffusion can be neglected. Additionally, we need to verify if mass transfer in the pores and external transfer can be considered negligible.

Data:

$$(k_{Ag}a_i) = 0.3 \text{ cm}^3/\text{s}$$
$$k_{AC} = 0.005 \text{ cm/s}$$
$$D_e = 10^{-5} \text{ cm}^2/\text{s}$$
$$S_s = 25 \text{ m}^2/\text{g}$$

**Solution to Problem 8.16**

In the reactor, the reaction rate is as follows:

$$\left(-r''_A\right) = \frac{p_{Ag}}{\dfrac{1}{k_{Ag}\cdot a_i} + \dfrac{H_A}{k_{Al}\cdot a_i} + \dfrac{H_A}{k_{Ac}\cdot a_c} + \dfrac{H_A}{k_{ap_A}\cdot \eta_A\cdot \varepsilon_s}}$$

where the second term can be neglected, i.e.,

$$\left(-r''_A\right) = \frac{p_{Ag}}{\dfrac{1}{k_{Ag}\cdot a_i} + \dfrac{H_A}{k_{Ac}\cdot a_c} + \dfrac{H_A}{k_{ap_A}\cdot \eta_A\cdot \varepsilon_s}}$$

We can calculate $a_C$:

$$a_C = \frac{\text{area}}{\text{mass}} = \frac{6}{d_p \cdot \rho_S} = 400 \ \frac{\text{cm}^2}{\text{g}}$$

So:

$$k'_{Ac} \cdot a_C = 2.0 \ \frac{\text{cm}^3}{\text{g}\cdot\text{s}}$$

Bearing in mind the total amount of catalyst:

$$k_{Ac} \cdot a_C = 2.0 \ \frac{\text{cm}^3}{\text{g}\cdot\text{s}} \cdot 0.5\,\text{g} = 1.0 \ \frac{\text{cm}^3}{\text{s}}$$

Let us calculate Thiele modulus and effectiveness:

$$mL = \sqrt{\frac{k'''}{D_e}}\left(\frac{d_p}{6}\right) = \sqrt{\frac{4.8\cdot 10^{-5}\frac{\text{cm}^3}{\text{cm}^2\cdot\text{s}} \cdot 250\,000\,\frac{\text{cm}^2}{\text{g}} \cdot 1.5\,\frac{\text{g}}{\text{cm}^3}}{1\cdot 10^{-5}\,\frac{\text{cm}^2}{\text{s}}}}\left(\frac{0.01}{6}\right) = 2.23$$

$$\eta = \frac{\tanh(mL)}{mL} = 0.447$$

For a system with $p_A = 1\,\text{atm}$ and equilibrium concentration of $0.01\,\text{mol/cm}^3$, we have:

$$H_A = \frac{p_A}{C_A} = \frac{1}{0.01} = 100\,\text{atm}\cdot\frac{\text{cm}^3}{\text{mol}}$$

And so:

$$\left(-r''_A\right) = \frac{1\,\text{atm}/100\,\text{atm}\cdot\frac{\text{cm}^3}{\text{mol}}}{\dfrac{1}{0.3\frac{\text{cm}^3}{\text{s}}} + \dfrac{1}{1.0\frac{\text{cm}^3}{\text{s}}} + \dfrac{1}{4.8\cdot10^{-5}\frac{\text{cm}^3}{\text{cm}^2\cdot\text{s}}\cdot 250\,000\,\frac{\text{cm}^2}{\text{g}}\cdot 0.5\,\text{g}\cdot 0.447}} = 0.002\,23\ \frac{\text{mol}}{\text{s}\cdot\text{cm}^3}$$

**Problem 8.17**  It is proposed to reduce the concentration of NO in the effluent of a plant by passing it through a packed bed of spherical, porous, solid carbonaceous granules. A mixture of 2% NO and 98% air flows with a velocity of $1\cdot10^{-6}\,\text{m}^3/\text{s}$ through a tube with an internal diameter of 5 cm packed with the porous solid at a temperature of 1173 K and a pressure of 101.3 kPa. The reaction:

$$NO + C \rightarrow CO + \tfrac{1}{2}N_2$$

It is first order with respect to NO and occurs mainly in the pores, within the granule where the internal surface area is $530\,\text{m}^2/\text{g}$ and the rate constant is $4.42\cdot10^{-10}\,\text{m}^3/\text{m}^2\cdot\text{s}$.

(a) Estimate the value of the efficiency factor in this system.
(b) Calculate the value of the external matter transport coefficient.
(c) Calculate the weight of the porous solid needed to reduce the NO concentration to a level of 0.004%.

Additional data. At 1173 K, the properties of the fluid are:
Kinematic viscosity $= v = 1.53 \; 10^{-8} \; \text{m}^2/\text{s}$
Gas phase diffusivity $= 2 \; 10^{-8} \; \text{m}^2/\text{s}$
Effective diffusivity $= 1.89 \; 10^{-8} \; \text{m}^2/\text{s}$
The properties of the catalyst and the bed are:
Catalyst particle density $= 2.8 \; \text{g/cm}^3$
Bed porosity $= 0.5$
Volumetric density of the bed $= 1.4 \; 10^6 \; \text{g/m}^3$
Granule radius $= 3 \; 10^{-3} \; \text{m}$
In the bed, the Thoenes–Kramers relation is fulfilled: $\text{Sh} = (\text{Re})^{(1/2)}(\text{Sc})^{(1/3)}$
If necessary, consider that there is counterdiffusion in the reaction.

Remember that:

$$\text{Sh} = (k_C \, d_p)/D_{AB}$$

$$\text{Re} = (\rho u d_p)/\mu = (u d_p)/v$$

$$\text{Sc} = \mu/(D_{AB}\rho)$$

**Solution to Problem 8.17**
(a) The desired conversion is:

$$X_A = \frac{C_{A0} - C_A}{C_{A0}} = 0.998$$

In a PFR, the mass balance of NO would be:

$$n_A - (n_A + dn_A) + (-k' C_A \eta dW) = 0$$

$$\frac{dn_A}{dW} = \frac{Q_0 dC_A}{dW} = -k' C_A \eta$$

with $k'$ in units of $\text{m}^3/\text{kg·s}$ and $Q_0$ in $\text{m}^3/\text{s}$. In terms of conversion, the mass balance is then:

$$\frac{Q_0 dX_A}{dW} = k'(1 - X_A)\eta$$

Integrating:

$$X_A = 1 - \exp\left(-\frac{k' \eta W}{Q_0}\right)$$

or:

$$W = \frac{Q_0}{k' \eta} \ln\left(\frac{1}{1 - X_A}\right)$$

Thiele modulus can be calculated:

$$mL = \sqrt{\frac{k''}{D_e}}\left(\frac{d_p}{6}\right) = \sqrt{\frac{4.42\cdot 10^{-10}\,\frac{m^3}{m^2\cdot s}\ 530\,\frac{m^2}{g}\ 2.8\cdot 10^6\,\frac{g}{m^3}}{1.98\cdot 10^{-8}\,\frac{m^2}{s}}}\left(\frac{0.03}{3}\right)\ m = 5.75$$

The effectiveness is then, $\eta = \frac{1}{mL} = 0.17$

(b) and (c)

The mass of catalyst can then be calculated:

$$W = \frac{Q_0}{k'\eta}\ln\left(\frac{1}{1-X_A}\right) = \frac{10^{-6}\ m^3/s}{4.42\cdot 10^{-10}\frac{m^3}{m^2\cdot s}\cdot 530\frac{m^2}{g}\cdot 0.17}\ln\left(\frac{1}{1-0.998}\right) = 156\ kg$$

For the external mass transfer coefficient, we do:

$$Sh = (Re)^{\frac{1}{2}}(Sc)^{\frac{1}{3}} = \left(\frac{ud_p}{v}\right)^{0.5}\cdot\left(\frac{\mu}{D_{AB}\rho}\right)^{0.333} = \frac{k_C d_p}{D_{AB}}$$

From the equation, we obtain:

$$k_C = 6\cdot 10^{-5}\ \frac{m}{s}$$

**Problem 8.18**  We want to determine the necessary bed height in a 1 m diameter cylindrical bubbling tank in which the total hydrogenation of acetylene to ethane is carried out, bubbling both gases ($H_2$ and $C_2H_2$), in the necessary stoichiometric proportion, through a suspension in octane of solid particles of a certain metallic catalyst.

The gas distributor plate, which represents a total flow rate of 15.5 l/s, measured under the reactor's working conditions, has a porosity of 50%, with pores of 250 μm in average diameter.

The reactor works at a temperature of 77 °C and a pressure of 5 atm, conditions in which the overall rate is controlled by the gas–liquid mass transfer. In the reactor, the suspension is perfectly mixed, and the introduction of the gases causes an increase of 10% in its volume, creating bubbles that ascend in plug-type flow.

It is intended to have a sufficient bed height to achieve a fractional conversion of acetylene of 65%.

Data for acetylene in octane at 77 °C: Diffusivity $= 7.5\cdot 10^{-5}$ cm$^2$/s, Henry's constant $= 2.87\cdot 10^5$ atm/(mol/cm$^3$)

The density of octane under the conditions of the problem is 0.6844 g/cm$^3$ and its viscosity is $3.2\cdot 10^{-3}$ g/cm·s. The average between the two gases can be used as the density of the gas, $1.7\cdot 10^{-3}$ g/cm$^3$. The bubbles produced are estimated to be spherical with a mean diameter of 0.167 cm.

For the calculation of the gas–liquid mass transfer coefficient, the Calderbank correlation can be used:

$$k_{AL} = 0.31 Sc^{-\frac{2}{3}}\left[(\rho_L - \rho_G)\frac{\mu_L g}{\rho_L^2}\right]^{\frac{1}{3}}$$

which gives the coefficient in cm/s if the centimeter–gram–second system of units is used.

**Solution to Problem 8.18**

The global reaction rate is:

$$\left(-r''_A\right) = \cfrac{p_{Ag}}{\cfrac{1}{k_{Ag}\cdot a_i} + \cfrac{H_A}{k_{Al}\cdot a_i} + \cfrac{H_A}{k_{Ac}\cdot a_c} + \cfrac{H_A}{k_{ap_A}\cdot \eta_A\cdot \varepsilon_s}}$$

But many of the steps in the reaction are quite fast, and:

$$\left(-r''_A\right) = \cfrac{p_{Ag}}{0 + \cfrac{H_A}{k_{Al}\cdot a_i} + 0 + 0} = \cfrac{p_{Ag}\cdot k_{Al}\cdot a_i}{H_A}$$

In the feed, the proportion $H_2:C_2H_2 = 2:1$, being "B" the hydrogen and "A" acetylene, so:

$$p_{A0} = \frac{p_{B0}}{2}$$

$$P_{total} = 3p_{A0}$$

and the mole fraction of A at each point is:

$$y_A = \frac{p_{Ag}}{P_{total}} = \frac{p_{A0}(1-X_A)}{3p_{A0} - p_{A0}(1-X_A) - p_{B0}\cdot 2X_A}$$
$$= \frac{p_{A0}(1-X_A)}{3p_{A0} - p_{A0}(1-X_A) - p_{A0}\cdot 2\cdot 2X_A} = \frac{(1-X_A)}{3 - (1-X_A) - 4X_A} = \frac{(1-X_A)}{2 - 3X_A}$$

With this:

$$\left(-r''_A\right) = \frac{p_{Ag}\cdot k_{Al}\cdot a_i}{H_A} = \frac{k_{Al}\cdot a_i}{H_A}P_{total}\frac{(1-X_A)}{2-3X_A}$$

The mole balance in the reactor is:

$$G'dY_A = \left(-r''_A\right)\cdot dV$$

$$n_{A0}dX_A = \left(-r''_A\right)\cdot dV$$

That, for the whole system is:

$$V = \int_0^{X_A}\frac{n_{A0}dX_A}{-r''_A} = \int_0^{X_A}\frac{n_{A0}dX_A}{\frac{k_{Al}\cdot a_i}{H_A}P_{total}\frac{(1-X_A)}{2-3X_A}}$$

Using the Calderbank correlation, we can easily calculate:

$$k_{AL} = 0.033\,\frac{cm}{s}$$

and, for spheres:

$$a_i = \ldots = 3.6\,cm^{-1}$$

The mole flow rate at the input is:

$$n_{A0} = G'\cdot y_{A0}\cdot \frac{P_{total}}{RT} = 0.9\,\frac{mol}{s}$$

And finally:

$$V = \int_0^{0.65} \frac{0.9 \, dX_A}{\frac{0.033 \cdot 3.6}{2.87 \cdot 10^5} \cdot 5 \cdot \frac{(1-X_A)}{2-3X_A}} = 1.022 \text{ m}^3$$

Using the section given:

$$\text{height} = \frac{1.022 \text{ m}^3}{\pi \cdot \left(\frac{1}{2}\right)^2} = 130 \text{ m}$$

**Problem 8.19** A solute A, initially in the gas phase, reacts mole for mole with a compound B in the liquid phase by means of a second-order reaction. Previous results indicate that in this reaction, the increase factor $E$ conforms to the equation:

$$E = 0.96 + 0.80 \sqrt{C_B/C_A}$$

The equilibrium relationship is $C_{Ai} = 0.9 \cdot p_{Ai}$ ($C_A$ in mol/m$^3$, $p_A$ in atm). The operating conditions are given in the figure. The resistance of the gas film is negligible. The reaction is completed in the liquid film. Calculate the volume of the tower to carry out the required separation.

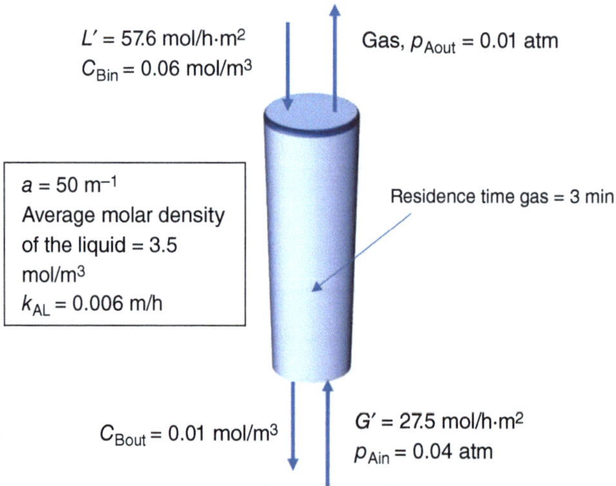


**Solution to Problem 8.19**

For details refer the Wiley website at http://www.wiley-vch.de/ISBN9783527354115

Usually, $E = f(M_H, E_i)$ but in the present case:

$$E = 0.96 + 0.80 \sqrt{C_B/C_A}$$

and we have that:

$$(-r_A'') = \frac{p_A}{\dfrac{1}{k_{Ag} \cdot a} + \dfrac{H_A}{k_{AL} \cdot a \cdot E} + \dfrac{H_A}{k_{ap}''}}$$

where the chemical reaction is supposed to be of pseudo-first order. In this case, we do:

$$(-r_A) = kC_A C_B = (k \cdot C_B)C_A = k_{ap}C_A$$

Note that the value of $H_A$ is:

$$H_A = \frac{p_{Ai}}{C_{Ai}} = 0.9 \frac{\text{atm} \cdot \text{mol}}{\text{m}^3}$$

Assuming plug flow both for gas and liquid phases, mole balance is:

$$\text{(A lost by the gas)} = (1/b)\,\text{(B lost by the liquid)} = \text{(A disappeared by reaction)}$$

$$G'dY_A = -\frac{L'dX_B}{b} = (-r_A'') \cdot dV$$

If we do a global balance:

$$G'(Y_{A2} - Y_{A1}) = \frac{L'}{b}(X_{B1} - X_{B2})$$

assuming a dilute system:

$$\frac{G'}{P_{total}}(p_{A2} - p_{A1}) = \frac{L'}{bC_{total}}(C_{B1} - C_{B2})$$

We can estimate the total pressure as follows:

$$\frac{27.5\frac{\text{mol}}{\text{h}\cdot\text{m}^2}}{P_{total}}(0.04 - 0.01)\,\text{atm} = \frac{57.6\frac{\text{mol}}{\text{h}\cdot\text{m}^2}}{1 \cdot 3.5\,\text{mol/m}^3}(0.06 - 0.01)\frac{\text{mol}}{\text{m}^3}$$

So: $P_{total} = 1.022$ atm.

In the same way, a balance between the input and a definite section is:

$$\frac{27.5\frac{\text{mol}}{\text{h}\cdot\text{m}^2}}{1.022\,\text{atm}}(0.04 - p_A)\,\text{atm} = \frac{57.6\frac{\text{mol}}{\text{h}\cdot\text{m}^2}}{1 \cdot 3.5\,\text{mol/m}^3}(0.06 - C_B)\,\text{mol/m}^3$$

obtaining:

$$C_B = 0.0768 - 1.67 p_A$$

As the resistance in the film is negligible, $\frac{1}{k_{Ag}\cdot a} = 0$ and assuming $k_{ap} = $ infinity, as the reaction is completed in the liquid film, we have:

$$(-r_A'') = \frac{p_A}{\dfrac{1}{k_{Ag}\cdot a} + \dfrac{H_A}{k_{AL}\cdot a \cdot E} + \dfrac{H_A}{k_{ap}''}} = \frac{p_A}{0 + \dfrac{H_A}{k_{AL}\cdot a \cdot E} + 0}$$

$$= \frac{p_A \cdot k_{AL} \cdot a \cdot E}{H_A} = \frac{p_A \cdot k_{AL} \cdot a}{H_A}\left(0.96 + 0.80\sqrt{\frac{C_B}{C_A}}\right)$$

Using the relationships:

$$(-r_A'') = \frac{p_A \cdot k_{AL} \cdot a}{H_A}\left(0.96 + 0.80\sqrt{\frac{C_B}{C_A}}\right)$$

$$= \frac{p_A \cdot k_{AL} \cdot a}{H_A}\left(0.96 + 0.80\sqrt{\frac{0.0768 - 1.67p_A}{p_A \cdot H_A}}\right)$$

We can use the mole balance:

$$G'\mathrm{d}Y_A = (-r_A'') \cdot \mathrm{d}V$$

$$V = G' \int_{Y_{A1}}^{Y_{A2}} \frac{\mathrm{d}Y_A}{(-r_A'')}$$

That, doing a numerical integration, gives:

| $Y_A$ | $C_B$ (mol/m³) | $p_A$ (atm) | $C_A$ (mol/m³) | $E$ | $(-r_A'')$ mol/h·m³ | $\Delta p_A/(-r_A'')$ (h·m³)/(mol·atm) | $\Delta p_A/(-r_A'')\cdot$ $G'/P_{total}$ (m³) |
|---|---|---|---|---|---|---|---|
| 0.040 | 0.010 | 0.040 | 0.036 | 1.378 | 0.018 | 0.218 | 5.97 |
| 0.036 | 0.017 | 0.036 | 0.032 | 1.531 | 0.018 | 0.218 | 5.97 |
| 0.032 | 0.023 | 0.032 | 0.029 | 1.678 | 0.018 | 0.223 | 6.13 |
| 0.028 | 0.030 | 0.028 | 0.025 | 1.831 | 0.017 | 0.234 | 6.42 |
| 0.024 | 0.037 | 0.024 | 0.022 | 2.001 | 0.016 | 0.250 | 6.86 |
| 0.020 | 0.043 | 0.020 | 0.018 | 2.200 | 0.015 | 0.273 | 7.48 |
| 0.016 | 0.050 | 0.016 | 0.014 | 2.449 | 0.013 | 0.306 | 8.40 |
| 0.012 | 0.057 | 0.012 | 0.011 | 2.791 | 0.011 | 0.179 | 4.92 |
| 0.010 | 0.060 | 0.010 | 0.009 | 3.024 | 0.010 | 0.992 | 27.22 |
| | | | | | | $\Sigma =$ | 79.38 |

$$V = 79.38 \text{ m}^3$$

# Part IV

# Biochemical Reactor Design

# 9

# Biochemical Reactor Design: Enzymatic Processes

## Summary of Kinetic Expressions

### Enzymatic Reactions

#### Michaelis–Menten Kinetics

$$r = r_m \frac{S}{K_M + S}$$

With:

$$r_m = K_{cat} \cdot E$$

This expression can be linearized:

$$\frac{1}{r} = \frac{1}{r_m} \frac{K_M + S}{S} = \frac{1}{r_m} + \frac{K_M}{r_m} \frac{1}{S}$$

In the presence of inhibition:

$$r = \frac{r_m \cdot S}{K_M \left(1 + \frac{1}{k_{i1}}\right) + S \left(1 + \frac{1}{k_{i2}}\right)}$$

For more details, please consult Conesa (2019).

**Problem 9.1**  Consider the enzyme kinetics:

$$E + S \underset{k_{-1}}{\overset{k_1}{\rightleftharpoons}} ES$$

$$ES \overset{k_2}{\rightarrow} P + E$$

in which the free enzyme E binds with substrate S to form bound substrate ES in the first reaction, and the bound substrate is converted to product P and releases free enzyme in the second reaction. If the rates of these two reactions are such that either the free or bound enzyme is present in small concentrations, the stationary concentration of ES is nil.

Assume $k_1 \gg k_{-1}$, $k_2$ so E is present in a small concentration. Apply the quasi-steady state assumption (QSSA) and show that the slow-time scale model reduces to a first-order, irreversible decomposition of S to P

$$S \to P$$

(a) For a well-stirred batch reactor, show that the total enzyme concentration satisfies, at any time:

$$E(t) + ES(t) = E(0) + ES(0)$$

(b) Find an expression for the concentration of E. What is the corresponding concentration of ES?
(c) Show that the rate expression for the reduced model's single reaction is:

$$r = \frac{k(S)}{1 + K(S)}, \quad k = k_2 K E(0), \quad K = \frac{k_{-1}}{k_{-1} + k_2}$$

which depends solely on the substrate concentration. The inverse of the constant $K$ is known as the Michaelis constant.
(d) Plot the concentrations versus time for the full model and QSSA model for the following values of the rate constants and initial conditions:

| | | | |
|---|---|---|---|
| $k_1 = 5$ | $k_{-1} = 1$ | $k_2 = 10$ | |
| $E(0) = 1$ | $ES(0) = 0$ | $S(0) = 50$ | $P(0) = 0$ |

**Solution to Problem 9.1**

For details refer the Wiley website at http://www.wiley-vch.de/ISBN9783527354115

(a) For the enzyme (E), we can write:

$$\frac{d(E)}{dt} = k_{-1}(ES) + k_2(ES) - k_1(E)(S)$$

For the active complex (ES), we do:

$$\frac{d(ES)}{dt} = -k_{-1}(ES) - k_2(ES) + k_1(E)(S)$$

In that way:

$$\frac{d(E + ES)}{dt} = k_{-1}(ES) + k_2(ES) - k_1(E)(S) - k_{-1}(ES) - k_2(ES) + k_1(E)(S) = 0$$

This is, the sum of the concentrations of (E) and (ES) at any time is constant and equals the initial values:

$$ES(0) + E(0) = (ES) + (E)$$

(b) If QSSA is assumed:

$$\frac{d(ES)}{dt} = -k_{-1}(ES) - k_2(ES) + k_1(E)(S) = 0$$

So:

$$k_{-1}(\text{ES}) + k_2(\text{ES}) = k_1(\text{E})(\text{S})$$

And:

$$(\text{ES}) = \frac{k_1(\text{E})(\text{S})}{k_{-1} + k_2}$$

And then:

$$\frac{(\text{ES})}{(\text{E})(\text{S})} = \frac{k_1}{k_{-1} + k_2} = \frac{1}{K_\text{M}}$$

Assuming that $\text{ES}(0) = 0$, we can write:

$$\text{E}(0) = (\text{ES}) + (\text{E})$$

So:

$$\text{E}(0) - (\text{ES}) = (\text{E})$$

And then:

$$K_\text{M} = \frac{(\text{E})(\text{S})}{(\text{ES})} = \frac{(\text{E}(0) - (\text{ES}))(\text{S})}{(\text{ES})}$$

So:

$$(\text{ES}) = \frac{\text{E}(0)}{1 + \frac{K_\text{M}}{(\text{S})}}$$

And $(\text{E}) = \text{E}(0) - (\text{ES}) = \text{E}(0) - \dfrac{\text{E}(0)}{1 + \frac{K_\text{M}}{(\text{S})}}$

(c)  Here, the production rate of this product is:

$$r = \frac{\text{d}P}{\text{d}t} = k_2(\text{ES}) = k_2\text{E}(0) \cdot \frac{(\text{S})}{(K_\text{M} + (\text{S}))} = k_2\text{E}(0)K\frac{(\text{S})}{(1 + K(\text{S}))}$$

(d)  With the QSSA:

$$\frac{\text{d}(\text{ES})}{\text{d}t} = 0$$

whereas with the full model:

$$\frac{\text{d}(\text{ES})}{\text{d}t} = -k_{-1}(\text{ES}) - k_2(\text{ES}) + k_1(\text{E})(\text{S})$$

$$\frac{\text{d}(\text{E})}{\text{d}t} = k_{-1}(\text{ES}) + k_2(\text{ES}) - k_1(\text{E})(\text{S})$$

$$\frac{\text{d}(\text{S})}{\text{d}t} = k_{-1}(\text{ES}) - k_1(\text{E})(\text{S})$$

$$\frac{\text{d}(\text{P})}{\text{d}t} = k_2(\text{ES})$$

We can use the finite differences method for integrating the above expressions, using the initial values shown in the statement:

$$\text{ES}^{t+1} = [-k_{-1}(\text{ES}^t) - k_2(\text{ES}^t) + k_1(\text{E}^t)(\text{S}^t)] \cdot \Delta t + \text{ES}^t$$

$$E^{t+1} = [k_{-1}(ES) + k_2(ES) - k_1(E)(S)] \cdot \Delta t + E^t$$

$$S^{t+1} = [k_{-1}(ES) - k_1(E)(S)] \cdot \Delta t + S^t$$

$$P^{t+1} = k_2(ES) \cdot \Delta t + P^t$$

We obtain, for the QSSA:

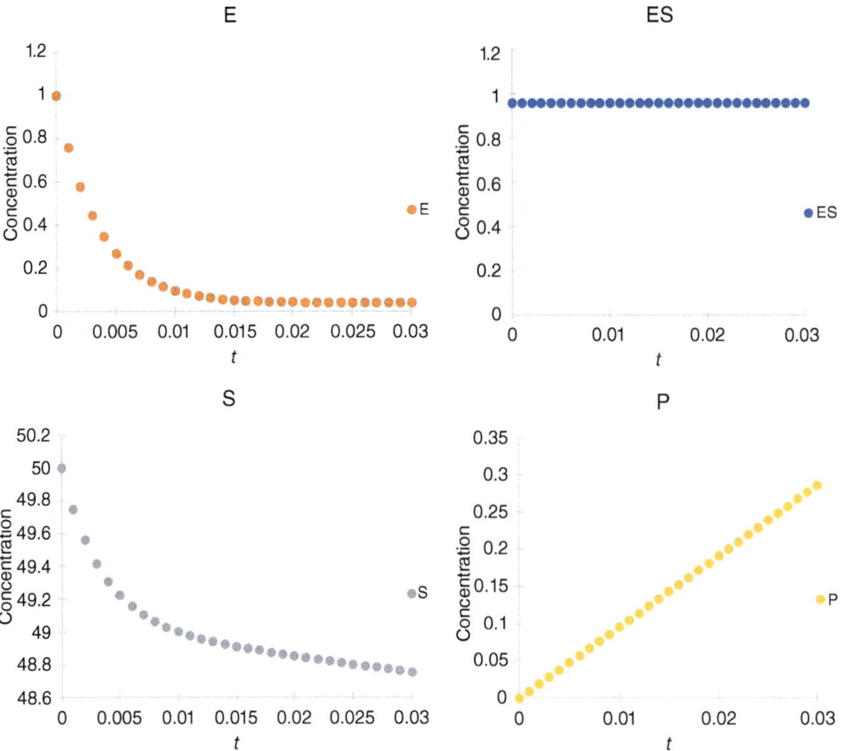

And for the full model:

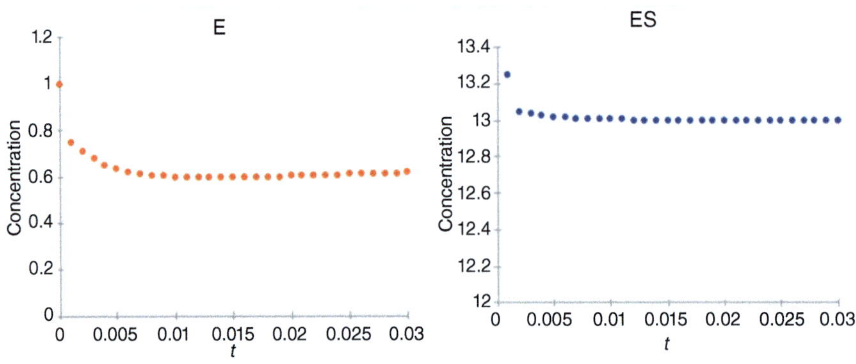

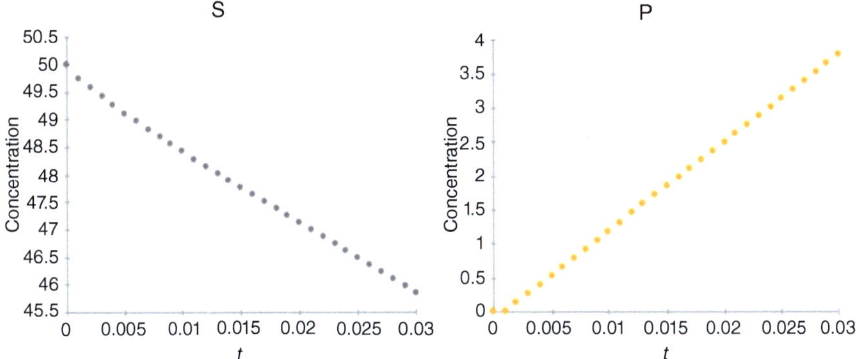


**Problem 9.2**  Consider the following data for the enzymatic hydrolysis of AEE (arginine ethyl ester) by trypsin bound to particles of porous glass in a fixed bed reactor (considered as a plug flow system):

| Mean residence time (min) | $C_{AEE}$ (mM) |
| --- | --- |
| 0.3 | 0.47 |
| 1 | 0.4 |
| 8 | 0.2 |
| 22 | 0.1 |
| 36 | 0.05 |
| 5 | 0.3 |

In the runs, the input concentration of AEE is 0.5 mM. For Michaelis–Menten kinetics, compute values for the constants.

**Solution to Problem 9.2**

For details refer the Wiley website at http://www.wiley-vch.de/ISBN9783527354115

In Michaelis–Menten kinetics:

$$r_S = r_m \frac{S}{K_M + S}$$

In this system, AEE is the substrate, and the trypsin bound to particles is the enzyme. In the plug flow reactor, we have:

$$r_S \cdot dV = -n_{S0} \cdot dX_S$$

where $X_S$ refers to the conversion of substrate, in such a way that $S = S_0(1 - X_S)$, and $n_{S0}$ is the flow of substrate (mol/min) at the input of the reactor.

$$r_m \frac{S}{K_M + S} \cdot dV = -n_{S0} \cdot dX_S = r_m \frac{S_0(1 - X_S)}{K_M + S_0(1 - X_S)} \cdot dV$$

Rearranging:

$$\int_0^{X_S} -\frac{K_M + S_0(1 - X_S)}{r_m \cdot S_0(1 - X_S)} dX_S = \int_0^V \frac{dV}{n_{S0}}$$

Finally:

$$\frac{V}{n_{S0}} = -\frac{X_S}{r_m} + \frac{K_M \ln(S_0 - S_0 X_S)}{r_M S_0}$$

Bearing in mind that:

$$\frac{V}{n_{S0}} = \frac{V}{Q_0 \cdot S_0} = \frac{\bar{t}}{S_0}$$

So:

$$\frac{\bar{t}}{S_0} = -\frac{X_S}{r_m} + \frac{K_M \ln(S_0 - S_0 X_S)}{r_M S_0}$$

$$\frac{\bar{t}}{X_S \cdot S_0} = -\frac{1}{r_m} + \frac{K_M \ln(S_0 - S_0 X_S)}{r_M S_0 \cdot X_S}$$

We can plot $\frac{\bar{t}}{X_S \cdot S_0}$ versus $\frac{\ln(S_0 - S_0 X_S)}{X_S}$ and we will get a straight line with slope $\frac{K_M}{r_M S_0}$ and intercept $-\frac{1}{r_m}$. In the spreadsheet:

| S (mM) | Time (min) | $X_S$ | $X_S \cdot S_0$ (mM) | $-\ln(1 - X_S)/(X_S)$ | $t/(S_0 \cdot X_S)$ (min·L/mmol) |
|---|---|---|---|---|---|
| 0.47 | 0.3 | 0.060 | 0.030 | 1.03 | 10.00 |
| 0.4 | 1 | 0.200 | 0.100 | 1.12 | 10.00 |
| 0.2 | 8 | 0.600 | 0.300 | 1.53 | 26.67 |
| 0.1 | 22 | 0.800 | 0.400 | 2.01 | 55.00 |
| 0.05 | 36 | 0.900 | 0.450 | 2.56 | 80.00 |
| 0.3 | 5 | 0.400 | 0.200 | 1.28 | 25.00 |
| | | | Slope | 46.437 | |
| | | | Intercept | −39.246 | |
| | | | Correlation ($r^2$) | 0.985 | |

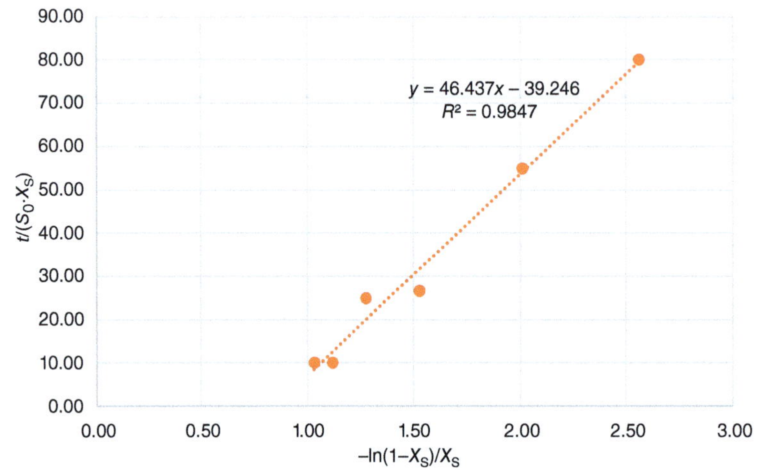

From the values of slope and intercept:

$$K_M = 0.592 \, \text{mM}$$
$$r_m = 0.025 \, \text{mM/min}$$

**Problem 9.3**  The enzyme triose phosphate isomerase catalyzes the interconversion of D-glyceraldehyde 3-phosphate and dihydroxyacetone phosphate:

$$CHO \cdot CH(OH) \cdot CH_2OPO_3{}^{2-} \Leftrightarrow CH_2OH \cdot CO \cdot CH_2OPO_3{}^{2-}$$

The following results refer to the initial reaction velocity, with glyceraldehyde 3-phosphate (S) as substrate at a total enzyme concentration of $E = 2.22 \cdot 10^{-4}$ mM, pH = 7.42, and 30 °C:

| $10^3$ [S] (M) | 0.071 | 0.147 | 0.223 | 0.310 | 0.602 | 1.47 | 2.60 |
|---|---|---|---|---|---|---|---|
| $10^7$ rate (M/s) | 1.31 | 2.45 | 3.37 | 3.90 | 5.63 | 7.47 | 8.17 |

Determine the Michaelis constant $K_M$ and the catalytic constant $k_{cat}$ for the enzyme under these conditions.

**Solution to Problem 9.3**

For details refer the Wiley website at http://www.wiley-vch.de/ISBN9783527354115

In Michaelis–Menten kinetics:

$$r = r_m \frac{S}{K_M + S}$$

Doing the inverse:

$$\frac{1}{r} = \frac{1}{r_m} + \frac{K_M}{r_m} \cdot \frac{1}{S}$$

Plotting $1/r$ versus $1/S$, we can do a linear fitting and obtain slope $(= K_M/r_m)$ and intercept $(= 1/r_m)$. Using the data, the plot is:

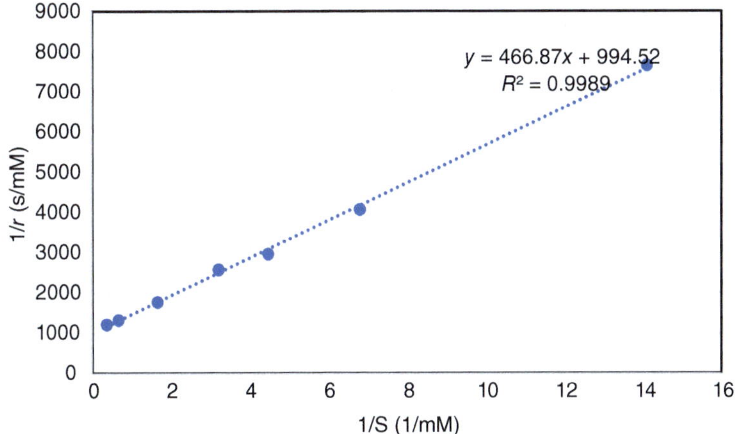

This indicates that:

$$466.87 = K_M/r_m \quad \text{and} \quad 994.52 = 1/r_m$$

So:

$$r_m = 0.00101\,\text{mM/s}$$

$$K_M = 0.4694\,\text{mM}$$

Finally, bearing in mind that: $r_m = k_{cat}E$

$$k_{cat} = 0.001\,01/2.22 \cdot 10^{-4} = 4.57\,\text{s}^{-1}$$

**Problem 9.4**  At a specific room temperature, the results for the enzymatic hydrolysis of sucrose (S) catalyzed by invertase (E $= 1 \cdot 10^{-2}$ mM) in a batch reactor can be summarized as follows:

| Cs (mmol/l) | 1 | 0.84 | 0.68 | 0.53 | 0.38 | 0.27 |
| --- | --- | --- | --- | --- | --- | --- |
| $t$ (h) | 0 | 1 | 2 | 3 | 4 | 5 |
| Cs (mmol/l) | 0.16 | 0.09 | 0.04 | 0.018 | 0.006 | 0.0025 |
| $t$ (h) | 6 | 7 | 8 | 9 | 10 | 11 |

Your task is to determine the values of the kinetic parameters $r_m$, $K_M$, and $k_{cat}$.

**Solution to Problem 9.4**

For details refer the Wiley website at http://www.wiley-vch.de/ISBN9783527354115

In Michaelis–Menten kinetics:

$$r = r_m \frac{S}{K_M + S}$$

Doing the inverse:

$$\frac{1}{r} = \frac{1}{r_m} + \frac{K_M}{r_m} \cdot \frac{1}{S}$$

Plotting $1/r$ versus $1/S$ we can do a linear fitting and obtain slope ($= K_M/r_m$) and intercept ($= 1/r_m$). For calculating "$r$" values, we can do:

$$r = \frac{\Delta S}{\Delta t} = \frac{S^{t+1} - S^t}{\Delta t}$$

Using the data:

| Cs (mmol/l) | 1 | 0.84 | 0.68 | 0.53 | ... | 0.006 | 0.0025 |
| --- | --- | --- | --- | --- | --- | --- | --- |
| $t$ (h) | 0 | 1 | 2 | 3 | ... | 10 | 11 |
| $r = \Delta S/\Delta t$ (mM/h) | 0.16 | 0.16 | 0.15 | 0.15 | ... | 0.0035 | 0.0025 |
| 1/S (1/mM) | 1.00 | 1.19 | 1.47 | 1.89 | ... | 166.67 | 400.00 |
| 1/r (h/mM) | 6.25 | 6.25 | 6.67 | 6.67 | ... | 285.71 | 400.00 |

The plot is:

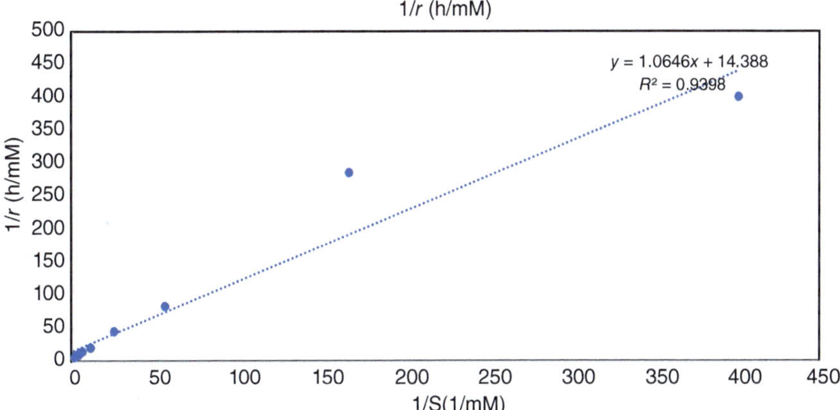


This indicates that:

$$1.0636 = K_M/r_m \quad \text{and} \quad 14.388 = 1/r_m$$

From this: $r_m = 0.068$ mM/h and $K_M = 0.074$ mM

Bearing in mind that $k_{cat} = r_m/E = 0.068\ (\text{mM/h})/10^{-2}\ \text{mM} = 6.8\ \text{h}^{-1}$.

**Problem 9.5** The hydrolysis rates of $p$-nitrophenyl-β-D-glucopyranoside by β-glucosidase, an irreversible unimolecular reaction, were measured at various substrate concentrations. The initial reaction rates obtained are presented in the following table. We need to determine the kinetic parameters of this enzyme reaction.

| Substrate concentration (mol/m³) | Reaction rate (mol/m³·s) |
| --- | --- |
| 5.00 | $4.84 \cdot 10^{-4}$ |
| 2.50 | $3.88 \cdot 10^{-4}$ |
| 1.67 | $3.18 \cdot 10^{-4}$ |
| 1.25 | $2.66 \cdot 10^{-4}$ |
| 1.00 | $2.14 \cdot 10^{-4}$ |

**Solution to Problem 9.5**

For details refer the Wiley website at http://www.wiley-vch.de/ISBN9783527354115

Assuming Michaelis–Menten kinetics, we have:

$$r = r_m \frac{S}{K_M + S}$$

The Lineweaver–Burk linearization of the previous equation is:

$$\frac{1}{r} = \frac{K_M}{r_m}\frac{1}{[S]}\frac{1}{r_m}$$

Using the data:

| $(1/S_0)$ (m³/mol$_S$) | $(1/r_0)$ (m³·s/mol$_P$) |
| --- | --- |
| 0.2000 | 2066 |
| 0.4000 | 2577 |
| 0.5988 | 3145 |
| 0.8000 | 3759 |
| 1.000 | 4673 |

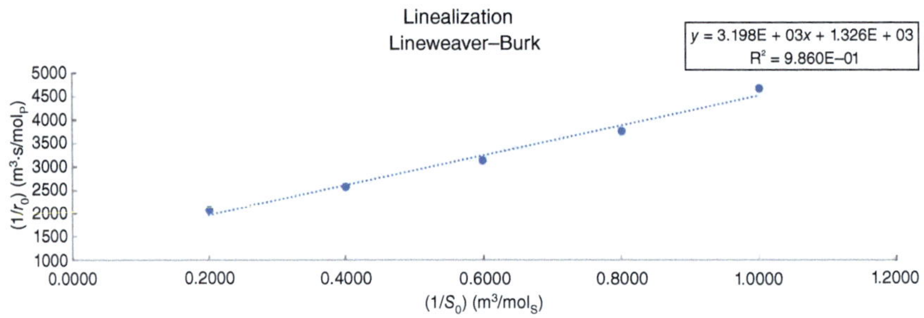


| Slope (s) | Intercept (m³·s/mol$_P$) |
| --- | --- |
| 3198 | 1326 |

| $(r_m)$ (mol$_P$/m³·s) | Michaelis–Menten constant $(K_M)$ (mol$_S$/m³) |
| --- | --- |
| $7.542 \cdot 10^{-4}$ | 2.412 |

And the Eadie–Hofstee linealization is:

$$r = -K_M \frac{r}{[S]} r_m$$

| $(r_0/S_0)$ (s$^{-1}$) | Initial rate $(r_0)$ (mol$_P$/m³·s) |
| --- | --- |
| $9.680 \cdot 10^{-5}$ | $4.840 \cdot 10^{-4}$ |
| $1.552 \cdot 10^{-4}$ | $3.880 \cdot 10^{-4}$ |
| $1.904 \cdot 10^{-4}$ | $3.180 \cdot 10^{-4}$ |
| $2.128 \cdot 10^{-4}$ | $2.660 \cdot 10^{-4}$ |
| $2.140 \cdot 10^{-4}$ | $2.140 \cdot 10^{-4}$ |

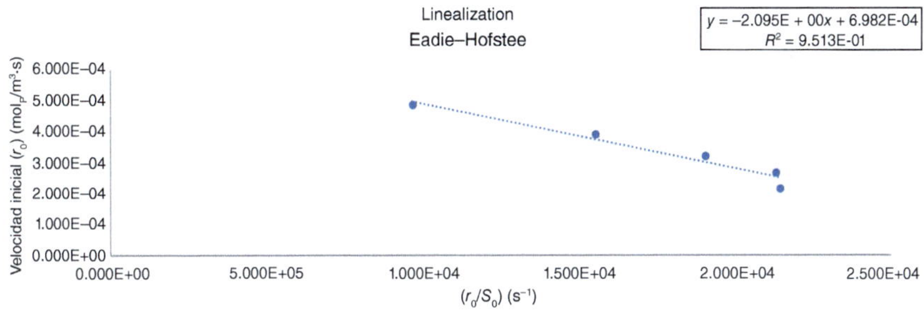


| Slope (mol$_S$/m³) | Intercept (mol$_p$/m³·s¹) |
|---|---|
| −2.095 | $6.982 \cdot 10^{-4}$ |

| $(r_m)$ (mol$_p$/m³·s) | Michaelis–Menten constant $(K_M)$ (mol$_S$/m³) |
|---|---|
| $6.982 \cdot 10^{-4}$ | 2.095 |

We can also try to calculate $r_m$ and $K_M$ by using an optimization of the values of the parameters and defining an objective function. In this way, the calculated value of the reaction rate $(r)_{cal}$ would depend on the supposed values of the parameters:

$$(r)_{cal} = (r_m)_{sup} \frac{S}{(K_M)_{sup} + S}$$

Defining an objective function by comparing the experimental and calculated values of the reaction rate:

$$OF = \sum_{all\ data} (r_{exp} - r_{cal})^2$$

and optimizing the parameters, we find:

| $(S_0)$ (mol$_S$/m³) | $(r_{0,experimental})$ (mol$_p$/m³·s) | $(r_{0,calculated})$ (mol$_p$/m³·s) | $(r_{exp} - r_{cal})^2$ ((mol$_p$/m³·s)²) |
|---|---|---|---|
| 5.000 | $4.840 \cdot 10^{-4}$ | $4.890 \cdot 10^{-4}$ | $2.517 \cdot 10^{-11}$ |
| 2.500 | $3.880 \cdot 10^{-4}$ | $3.803 \cdot 10^{-4}$ | $5.922 \cdot 10^{-11}$ |
| 1.670 | $3.180 \cdot 10^{-4}$ | $3.115 \cdot 10^{-4}$ | $4.257 \cdot 10^{-11}$ |
| 1.250 | $2.660 \cdot 10^{-4}$ | $2.633 \cdot 10^{-4}$ | $7.527 \cdot 10^{-12}$ |
| 1.000 | $2.140 \cdot 10^{-4}$ | $2.281 \cdot 10^{-4}$ | $2.002 \cdot 10^{-12}$ |
|  |  | Objective Function sum (OF) ((mol$_p$/m³·s)²) | $3.346 \cdot 10^{-10}$ |

And the optimized parameters are:

| $(r_m)$ (mol$_p$/m³·s) | Michaelis–Menten constant $(K_M)$ (mol$_S$/m³) |
|---|---|
| $6.848 \cdot 10^{-4}$ | 2.001 |

**Problem 9.6**  The hydrolysis of a substrate, ʟ-benzoyl arginine *p*-nitroanilide hydrochloride, by trypsin was studied using inhibitor concentrations of 0, 0.3, and 0.6 mmol/l. The corresponding hydrolysis rates (in µmol/l/s) are listed in the table below. Our goal is to determine the inhibition mechanism and the kinetic parameters ($K_M$, $V_{max}$, and $K_I$) of this enzyme reaction.

| Substrate concentration (mmol/l) | Inhibitor concentration (mmol/l) | | |
|---|---|---|---|
| | 0 | 0.3 | 0.6 |
| 0.1 | 0.79 | 0.57 | 0.45 |
| 0.15 | 1.11 | 0.84 | 0.66 |
| 0.2 | 1.45 | 1.06 | 0.86 |
| 0.3 | 2.00 | 1.52 | 1.22 |

**Solution to Problem 9.6**

For details refer the Wiley website at http://www.wiley-vch.de/ISBN9783527354115

Let us assume that:

$$r = \frac{r_m \cdot S}{K_M \left(1 + \frac{1}{k_{i1}}\right) + S}$$

That can be linearized by:

$$\frac{1}{r} = \frac{1}{r_m} + \frac{K_M}{r_m} \left(1 + \frac{1}{k_{i1}}\right) \frac{1}{S}$$

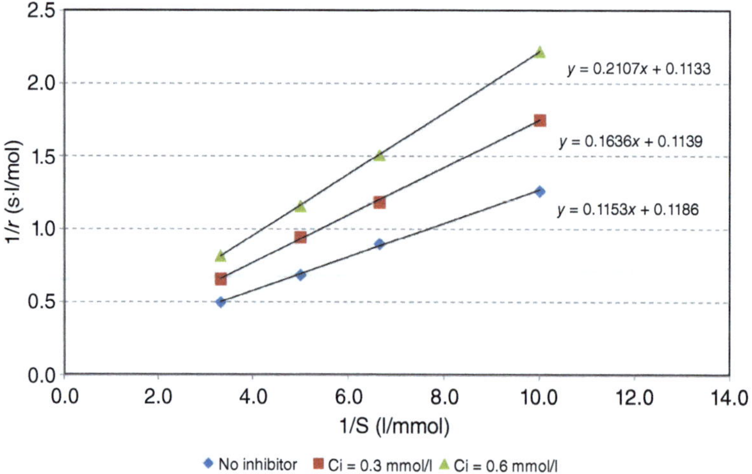


From the line, with no inhibition:

$$r_m = 8.4 \, \text{mmol/l} \cdot \text{s}$$

$$K_M = 1.0 \, \text{mmol/l}$$

From the line at 0.3 mmol/l concentration of inhibitor:

$$K_{i1} = 0.716 \text{ mmol/l}$$

From the line at 0.6 mmol/l concentration of inhibitor:

$$K_{i1} = 0.725 \text{ mmol/l}$$

**Problem 9.7**   The angiotensin-I converting enzyme (ACE) plays a crucial role in blood pressure regulation by catalyzing the hydrolysis of the amino acids His-Leu in angiotensin-I to produce the vasoconstrictor angiotensin-II. Additionally, ACE can hydrolyze a synthetic substrate, hippuryl-L-histidyl-L-leucine (HHL), to hippuric acid (HA). The initial rates of HA formation (in µmol/min) were measured at different concentrations of HHL solutions (pH 8.3) and are provided in the following table. The inhibitory effect of an inhibitory peptide (Ile-Lys-Tyr) at concentrations of 1.5 and 2.5 µmol/l on the formation of HA is also included in the table.

| HHL concentration (mol/m$^3$) | 20 | 8.0 | 4.0 | 2.0 |
| --- | --- | --- | --- | --- |
| Without peptide | 1.83 | 1.37 | 1.00 | 0.647 |
| 1.5 µmol/l peptide | 1.37 | 0.867 | 0.550 | 0.313 |
| 2.5 µmol/l peptide | 1.05 | 0.647 | 0.400 | 0.207 |

Determine the kinetic parameters of Michaelis–Menten reaction for the ACE reaction without the inhibitor. Determine the inhibition mechanism and the value of $K_I$.

**Solution to Problem 9.7**

For details refer the Wiley website at http://www.wiley-vch.de/ISBN9783527354115

In the previous problem:

$$\frac{1}{r} = \frac{1}{r_m} + \frac{K_M}{r_m}\left(1 + \frac{1}{k_{i1}}\right)\frac{1}{S}$$

| S (mol/m$^3$) | r (mmol/m$^3$ min) | 1/S (m$^3$/mol) | 1/r (min m$^3$/mol) |
| --- | --- | --- | --- |
| 20 | 1.830 | 0.050 | 546.4 |
| 8 | 1.370 | 0.125 | 729.9 |
| 4 | 1.000 | 0.250 | 1000.0 |
| 2 | 0.647 | 0.500 | 1545.6 |

From the data, by fitting a straight line:

| $r_m$ (1/min) | 0.0022 |
| --- | --- |
| $K_M$ (mol/m$^3$) | 4.962 |
| $K_{i1}$ | 0.069 |

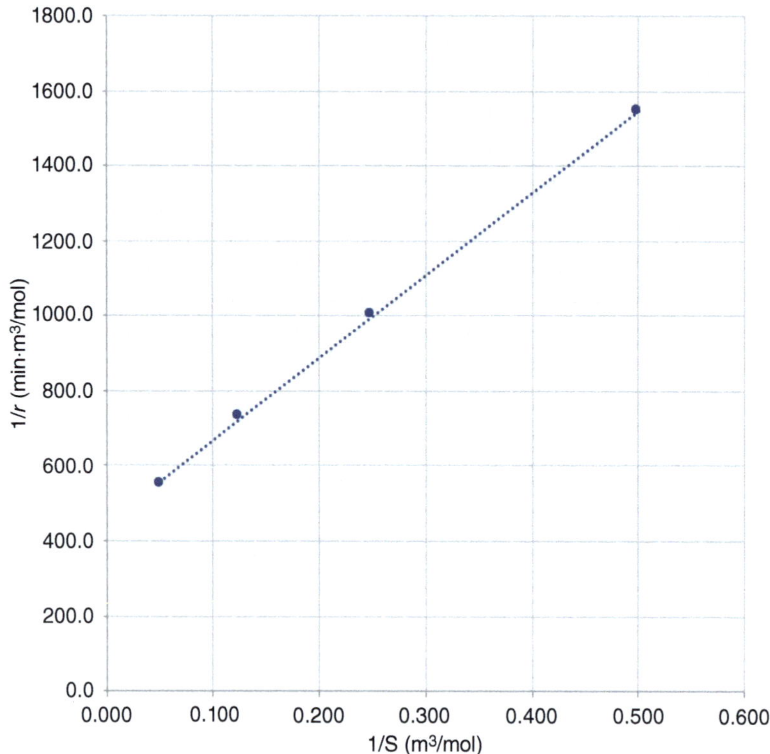


**Problem 9.8**   The enzyme invertase catalyzes the hydrolysis of sucrose into glucose and fructose. The rate of this enzymatic reaction decreases at higher substrate concentrations. Using the same amount of invertase, the initial rates at different sucrose concentrations are given in the table.

| Sucrose concentration (mol/m³) | Reaction rate (mol/m³·min) |
| --- | --- |
| 10 | 0.140 |
| 25 | 0.262 |
| 50 | 0.330 |
| 100 | 0.306 |
| 200 | 0.216 |

When the following reaction mechanism of the substrate inhibition is assumed, derive an equation for the rate of P species:

| | |
| --- | --- |
| $E + S \leftrightarrow ES$ | $k_{-1}/k_1 = K_m$ |
| $ES + S \leftrightarrow ES_2$ | $k_{-i}/k_i = K_I$ |
| $ES \rightarrow E + P$ | $k_p$ |

**Solution to Problem 9.8**

Using the definition of equilibrium constants, we have:

$$K_m = \frac{(ES)}{(E) \cdot (S)}$$

$$K_I = \frac{(ES_2)}{(ES) \cdot (S)}$$

So:

$$K_m \cdot (E) \cdot (S) = (ES)$$

$$K_I \cdot (ES) \cdot (S) = (ES_2) = K_I \cdot K_m \cdot (E) \cdot (S)^2$$

Bearing in mind that, from the reaction scheme:

$$\frac{d(ES)}{dt} = k_1 \cdot (E) \cdot (S) - k_{-1}(ES) + k_{-i}(ES_2) - k_i(ES) \cdot (S)$$

$$\frac{d(P)}{dt} = k_p \cdot (ES) = k_p \cdot K_m \cdot (E) \cdot (S)$$

Assuming QSSA for (ES) species:

$$k_1 \cdot (E) \cdot (S) + k_{-i}(ES_2) = k_{-1}(ES) + k_i(ES) \cdot (S)$$

And so:

$$k_1 \cdot (E) \cdot (S) + k_{-i} \cdot K_I \cdot K_m \cdot (E) \cdot (S)^2 = k_{-1} \cdot K_m \cdot (E) \cdot (S) + k_i \cdot K_m \cdot (E) \cdot (S)^2$$

We can get the substrate concentration in the equilibrium:

$$k_1 + k_{-i} \cdot K_I \cdot K_m \cdot (S) = k_{-1} \cdot K_m + k_i \cdot K_m \cdot (S)$$

$$(S) = \frac{k_1 - k_{-1} \cdot K_m}{k_{-i} \cdot K_I \cdot K_m + k_i \cdot K_m}$$

And finally:

$$\frac{d(P)}{dt} = k_p \cdot K_m \cdot (E) \cdot \frac{k_1 - k_{-1} \cdot K_m}{k_{-i} \cdot K_I \cdot K_m + k_i \cdot K_m}$$

**Problem 9.9**

(a) In a stirred-batch reactor, a substrate S with a concentration of $0.1\,\text{kmol/m}^3$ undergoes an irreversible unimolecular enzyme reaction. The reaction is characterized by a Michaelis constant $(K_M)$ of $0.010\,\text{kmol/m}^3$ and a maximum rate $(r_m)$ of $2.0 \cdot 10^{-5}\,\text{kmol/m}^3 \cdot \text{s}$. Determine the initial reaction rate and the conversion of the substrate after 10 minutes.

(b) In the same stirred-batch reactor, immobilized enzyme beads with a diameter of $10\,\text{mm}$, containing the same amount of the enzyme as in case (a), are utilized. Calculate the initial reaction rate for a substrate solution with a concentration of $0.1\,\text{kmol/m}^3$. Assume that the effective diffusion coefficient of the substrate within the catalyst beads is $1.0 \cdot 10^{-6}\,\text{cm}^2/\text{s}$.

(c) Under identical reaction conditions as in case (b), determine the minimum diameter required for the immobilized enzyme beads to achieve an effectiveness factor greater than 0.9.

**Solution to Problem 9.9**

(a) In Michaelis–Menten kinetics, we have:

$$r = r_m \cdot \frac{S}{K_M + S}$$

At the initial time:

$$r_0 = r_m \cdot \frac{S_0}{K_M + S_0} = 2 \cdot 10^{-5} \cdot \frac{0.1}{0.01 + 0.1} = 1.82 \cdot 10^{-5} \, \frac{kmol}{m^3 \cdot s}$$

For calculating the conversion, in a batch reactor:

$$r = \frac{dS}{dt} = r_m \cdot \frac{S}{K_M + S}$$

$$dS \cdot \frac{K_M + S}{S \cdot r_m} = dt$$

Integrating between "$t_0$" and "$t$" for the time variable and between "$S_0$" and "$S$" for the substrate, we obtain:

$$t - t_0 = \frac{K_M}{r_m} \cdot \ln\left(\frac{S}{S_0}\right) + \frac{1}{r_m} \cdot (S - S_0)$$

Using the known data:

$$600 - 0 = \frac{0.01}{2 \cdot 10^{-5}} \cdot \ln\left(\frac{S}{0.1}\right) + \frac{1}{2 \cdot 10^{-5}} \cdot (S - 0.1)$$

And we can check $S = 0.111 \, kmol/m^3$

The conversion is then $X_S = (0.111 - 0.1)/0.1 = 0.11$

(b) In the diffusional regime, an efficiency factor is needed for calculating the kinetics. In this situation:

$$r = \eta \cdot r_m \cdot \frac{S}{K_M + S}$$

The efficiency factor "$\eta$" can be estimated from the Thiele modulus, which for a system following Michaelis–Menten kinetics is:

$$mL = \left(\frac{R}{3}\right) \cdot \sqrt{\frac{r_m}{D_e \cdot K_M}} = \left(\frac{0.5}{3}\right) \cdot \sqrt{\frac{2 \cdot 10^{-5}}{1 \cdot 10^{-6} \cdot 0.01}} = 7.45$$

Using the approximation $\eta = \frac{\tanh(mL)}{mL} = 0.134$ and then:

$$r_0 = 0.134 \cdot 1.82 \cdot 10^{-5} = 2.44 \cdot 10^{-6} \, \frac{kmol}{m^3 \cdot s}$$

(c) For $\eta > 0.9$, we need a Thiele modulus higher than 2, and calculating the minimum radius, we get $R = 0.134 \, cm$, which is a diameter of $0.268 \, cm$.

**Problem 9.10**   In a batch reactor for the production of milk protein, it is synthesized from two lactine molecules, according to 2L $+H_3O^+ \rightarrow$ P, that is, in an acid medium, specifically at pH $= 6$. This synthesis takes place in pepsin enzymes, and follows Michaelis and Menten type kinetics, with $r_m = 10$ g lactine/l·min and an estimated value of $K_M$ of 5.2 g/l.

A lactine concentration of 200 g/l is initially provided in a 900 l reactor. Calculate the time required for a concentration of 1 g/l to be reached. If the yield of the reaction is 0.5 g protein/g lactine, how much milk protein will be synthesized?

**Solution to Problem 9.10**

In Michaelis–Menten kinetic equation, the reaction rate for the substrate consumption is:

$$r_S = r_m \frac{S}{K_M + S}$$

That, in a batch reactor, is used for:

$$\text{Input} + \text{Generation} = \text{Output} + \text{Accumulation}$$

$$0 + r_S \cdot V = 0 - \frac{dM_S}{dt} \left( \frac{g}{min} \right)$$

$$-r_S = \frac{dS}{dt}$$

being "S" the concentration $(M_S/V)$ of substrate.

$$\frac{dS}{dt} = -r_m \frac{S}{K_M + S}$$

$$\frac{K_M + S}{S} dS = -r_m dt$$

$$\int_{S_0}^{S_f} \frac{K_M + S}{S} dS = -r_m \int_0^t dt$$

$$K_M \cdot \ln \left( \frac{S_f}{S_0} \right) + (S_f - S_0) = -r_m t$$

We can calculate the time needed, as all other parameters are known. We obtain $t = 43.56$ min.

Bearing in mind the yield for protein production, we can do:

$$(P_f - P_0) = Y_{\frac{P}{S}}(S_0 - S_f) = 0.5\,(200 - 1) = 99.5\,g/l$$

As the initial amount of P is zero, this will be the total amount of product in the vessel. As we have 900 l of this culture, the total amount of protein is 89 550 g.

**Problem 9.11**   A 160-liter tubular bioreactor to produce ethanol from sugar enters a flow rate of $Q_0 = 10$ l/min and has a matrix with immobilized enzymes, with an average concentration of 5 g enzyme/l reactor. At the entrance, the sugar concentration is 100 g/l and the ethanol concentration is negligible, so the product leaves the reactor very diluted. For this reason, the recirculation of the product is proposed with a recirculation flow equivalent to three times the inlet flow. Under these conditions, the output of the system indicates that the concentration of ethanol is 40 g/l and that of sugar is 4 g/l. Since the enzymatic kinetics of this system follow Michaelis and Menten model, with $K_M = 2$ g/l, calculate the maximum conversion rate of sugar to ethanol ($r_m$) under the mentioned conditions.

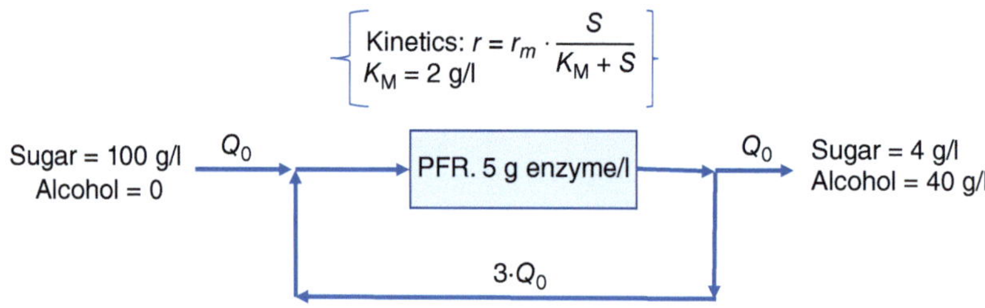



**Solution to Problem 9.11**

$$r = r_m \cdot \frac{S}{K_S + S}$$

At the input $S_0 = 100$ g/l. In the recirculation $S_f = 4$ g/l. At the junction of the two branches: $S' = (S_0 + 3S_f)/4 = 28$ g/l. Doing a differential balance in the reactor, where $E = 5$ g/l, we have:

$$n_A + r_A \cdot dV = n_A + dn_A$$

And:

$$dn_A = 4 \cdot Q_0 \cdot dC_A = r_A \cdot dV$$

This balance is referred to as the substrate (sugar), and so $C_A = S$. That is:

$$-r_m \cdot \frac{S}{K_S + S} \cdot dV = 4 \cdot Q_0 \cdot dS$$

In the whole reactor:

$$-\int_{S'}^{S_f} \frac{(K_M + S) \cdot dS}{r_m \cdot S} = \int_0^V \frac{dV}{4 \cdot Q_0}$$

And so:

$$\int_{S'}^{S_f} \frac{K_M \cdot dS}{S} + \int_{S'}^{S_f} \frac{S \cdot dS}{S} = -r_m \cdot \frac{V}{4 \cdot Q_0}$$

Hence:

$$K_M \cdot \ln\left(\frac{S'}{S_f}\right) + (S' - S_f) = r_m \cdot \frac{V}{4 \cdot Q_0}$$

Finally:

$$r_m = \frac{4 \cdot Q_0}{V} \cdot \left[ K_M \cdot \ln\left(\frac{S'}{S_f}\right) + (S' - S_f) \right]$$

Introducing the known parameters: $r_m = 6.97\,\text{g/l·min}$.

**Problem 9.12**  Wastewaters containing nitrates, such as those from septic tank sewage systems, agricultural runoff, and some food-related industries, pose significant pollution risks to marine environments and aquatic life. These wastewaters serve as a primary nutrient source for microorganisms that contribute to eutrophication of surface waters. To address this issue, a common approach involves a two-step process: aerobic nitrification followed by anaerobic denitrification. In the denitrification stage, facultative heterotrophic bacteria reduce nitrate ions to nitrogen gas.

The biological denitrification of wastewater streams has been studied in packed beds at various temperatures ranging from 5 to 20 °C. Under steady-state conditions with microbe populations and feed nitrate concentrations below 100 mg/l, the denitrification reaction follows a rate expression of the form:

$$r = \mu^* \cdot Y C_{NO_3}$$

Here, $\mu^*$ represents the rate constant for pseudo first-order nitrate removal, $Y$ is the biochemical yield coefficient, and $C_{NO_3}$ is the nitrate concentration.

Residence time distribution studies have indicated that a tube packed with 3-mm glass beads exhibits behavior consistent with a Plug Flow Reactor (PFR) model.

Now, let's consider a similar packed bed operating in a steady state at a temperature where the product of $\mu^*$ and $Y$ is 0.309 h$^{-1}$, and the inlet concentration of nitrate ions is 80 mg/l. We want to determine the residence time required to reduce this concentration to 20 mg /l.

**Solution to Problem 9.12**
A mass balance of nitrate ("A") in a differential volume:

$$n_A + r_A \cdot dV = n_A + dn_A$$

And:

$$dn_A = Q_0 \cdot dC_A = r_A \cdot dV$$

Substituting the kinetics:

$$-\mu^* Y C_A dV = Q_0 \cdot dC_A$$

In the whole reactor:

$$\int_0^V \frac{dV}{Q_0} = \frac{1}{-\mu^* Y} \int_{C_{A0}}^{C_A} \frac{dC_A}{C_A}$$

And so:

$$\frac{V}{Q_0} = \frac{1}{-\mu^* Y} \ln\left(\frac{C_{A0}}{C_A}\right)$$

Hence:

$$C_A = C_{A0} \exp\left(-\frac{\mu^* Y V}{Q_0}\right)$$

In the reactor being used:

$$20 = 80 \exp\left(-\frac{0.309\, V}{Q_0}\right)$$

So: $V/Q_0 = 4.48\,\text{h}.$

# 10

# Biochemical Reactor Design: Microbial Growth

## Summary of Kinetic Expressions and Mass Balances in Bioreactors

Microbial Growth

$$\frac{dX}{dt} = \mu \cdot X$$

Monod Kinetics

$$\mu = \mu_m \frac{S}{K_S + S}$$

$$\frac{1}{\mu} = \frac{1}{\mu_m} \frac{K_S + S}{S} = \frac{1}{\mu_m} + \frac{K_S}{\mu_m} \frac{1}{S}$$

| Symbol | Definition |
|---|---|
| $Y_{X/S}$ | Molar growth rate: gram of dry biomass per mole of substrate consumed |
| $Y_{X/O}$ | Gram of dry biomass per gram of oxygen consumed or per mole of oxygen consumed |
| $Y_{P/S}$ | Gram or mole of product per gram or mole of substrate consumed |
| $Y_{C/S}$ | Mole of $CO_2$ produced per mole of substrate consumed |

$$-Y_{\frac{X}{S}} = \frac{X_0 - X}{S_0 - S} = \frac{r_X}{r_S}$$

$$-Y_{\frac{P}{X}} = \frac{P_0 - P}{X_0 - X} = \frac{r_P}{r_X}$$

$$Y_{\frac{P}{S}} = \frac{P_0 - P}{S_0 - S} = \frac{r_P}{r_S}$$

*Problem Solving in Chemical Reactor Design*, First Edition. Juan A. Conesa.
© 2025 WILEY-VCH GmbH. Published 2025 by WILEY-VCH GmbH.

Bioreactors

Continuous Stirred Tank Bioreactor (CSTB) or Chemostat

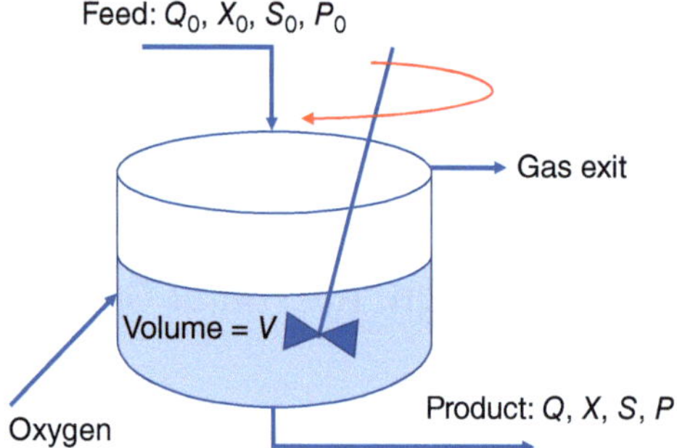


Mass balance of cells:

$$0 = Q_0\left(X_0 - X\right) + V\mu X$$

$$D = \frac{Q_0}{V} = \mu_m \frac{S}{K_S + S} = \mu = \frac{r_X}{X}$$

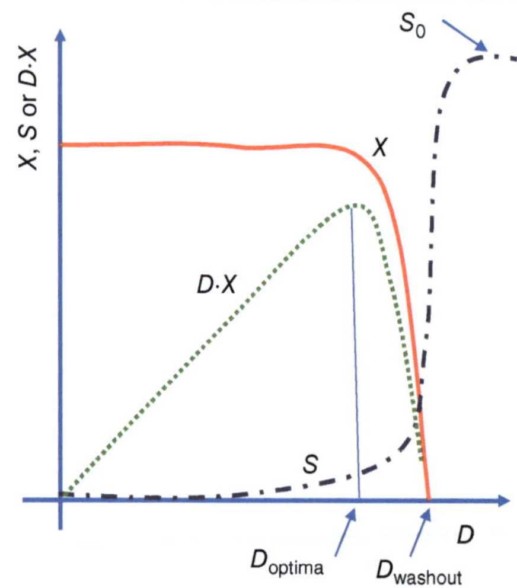


Variation of substrate, biomass, and cell production concentrations with the dilution rate.

$$S = D \cdot \frac{K_S}{\mu_m - D}$$

$$D_{\max} = D_{\text{washout}} = \mu_m \frac{S_0}{K_S + S_0}$$

$$D_{\text{opt}} = \mu_m \cdot \left(1 - \sqrt{\frac{K_S}{K_S + S_0}}\right)$$

Mass balance of substrate:

$$Q_0 \cdot S_0 + (-r_S) \cdot V = Q_0 \cdot S$$
$$Q_0 \cdot (S_0 - S) + (-r_X) \, Y_{S/X} \cdot V = 0$$
$$Q_0 \cdot (S_0 - S) + X \cdot \mu \cdot Y_{S/X} \cdot V = 0$$

**CSTB with Recirculation**

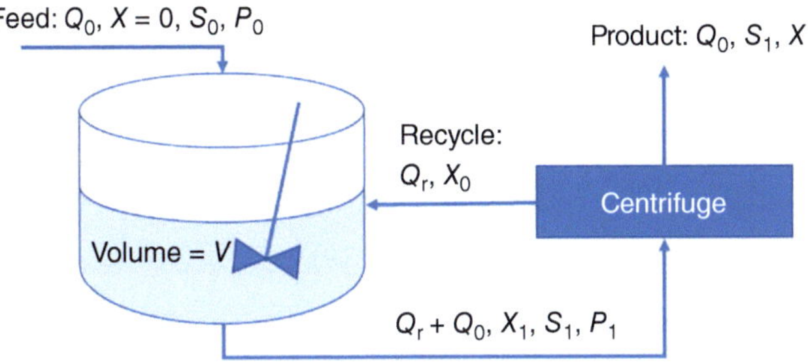


$$0 = Q_r X_0 - (Q_0 + Q_r) X_1 + r_X V = Q_r X_0 - (Q_0 + Q_r) X_1 + \mu X_1 V$$
$$\alpha = Q_r / Q_0$$
$$\beta = X_0 / X_1$$
$$D = \frac{\mu}{1 - \alpha (\beta - 1)}$$

**Tubular Fermenter**

$$Q \cdot dX = r_X \cdot dV$$
$$\frac{1}{D} = \frac{V}{Q} = \int_{X_0}^{X} \frac{dX}{r_X} = \int_{S_0}^{S} \frac{dS}{Y_{\frac{S}{X}} \cdot r_X}$$

**Fed-Batch Bioreactor**

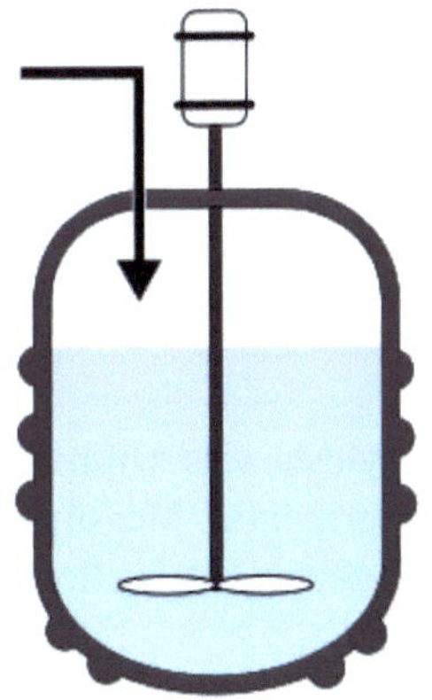

$$\frac{d\,(XV)}{dt} = X\frac{dV}{dt} + V\frac{dX}{dt} = V \cdot r_X \quad \text{(Cells)}$$

$$\frac{d\,(PV)}{dt} = P\frac{dV}{dt} + V\frac{dP}{dt} = V \cdot r_P \quad \text{(Product)}$$

$$\frac{d\,(SV)}{dt} = S\frac{dV}{dt} + V\frac{dS}{dt} = Q_0 S_0 - V \cdot r_S \quad \text{(Substrate)}$$

$$\frac{dX}{dt} = -\frac{Q_0 \cdot X}{V} + r_X$$

$$\frac{dP}{dt} = -\frac{Q_0 \cdot P}{V} + r_P$$

$$\frac{dS}{dt} = \frac{Q_0}{V}\left(S_0 - S\right) - \frac{r_X}{Y_{X/S}}$$

For more details, please consult Conesa (2019).

**Problem 10.1**

(a) Given that the maximum specific growth rate of an organism is $\mu_m = 1\,\text{h}^{-1}$ and its $K_S = 1.5\,\text{g/cm}^3$. If the input substrate concentration is $145\,\text{g/cm}^3$, determine the maximum output rate for this system in a continuous stirred tank reactor (CSTR).

(b) If the concentration of product at this maximum output rate is $10\,\text{g/cm}^3$, determine the maximum output rate of the product.

(c) If the concentration of biomass at this maximum output is $50\,\text{g/cm}^3$, determine the cell mass formation rate at the maximum output rate.

**Solution to Problem 10.1**

(a)

$$D_{\max} = \mu_m \frac{S_0}{K_S + S_0} = 1\frac{145}{1.5 + 145} = 0.989\,\text{h}^{-1}$$

(b) In the CSTB, a balance of product is:

$$Q_0 P_0 + \left(r_P\right)V = Q_0 P$$

Considering that $P_0 = 0$:

$$r_P = \frac{Q_0 P}{V} = D_{\max} \cdot P = 0.989 \cdot 10 = 9.89\,\frac{\text{g}}{\text{h} \cdot \text{cm}^3}$$

(c) A cell balance:

$$Q_0 X_0 + \left(r_X\right)V = Q_0 X$$

Considering that $X_0 = 0$:

$$r_P = D_{\max} \cdot X = 0.989 \cdot 50 = 49.45\,\frac{\text{g}}{\text{h} \cdot \text{cm}^3}$$

**Problem 10.2** The antibiotic Tylosin was synthesized in a CSTR using *Streptomyces fradiae* in a 5 l laboratory fermenter. Measurements of product and biomass concentrations were taken at different substrate flow rates:

| Flow rate (ml/h) | 50 | 100 | 150 | 200 | 350 |
|---|---|---|---|---|---|
| Biomass conc. (g DW/l) | 39 | 29 | 28 | 22 | 17 |
| Tylosin conc. (g/l) | 0.7 | 0.4 | 0.2 | 0.1 | 0.01 |

(DW = dry weight)

We need to determine the specific growth rate and the specific production rate using the provided experimental data.

**Solution to Problem 10.2**

For details refer the Wiley website at http://www.wiley-vch.de/ISBN9783527354115

In the reactor:

$$D = \frac{Q_0}{V} = \mu = \frac{r_X}{X}$$

---

**Tylosin production**

---

| CSTR fermenter volume (V) | | | 5 l | | | |
|---|---|---|---|---|---|---|

Experimental data

| Flow rate | $(Q_0)$ | (ml/h) | 50 | 100 | 150 | 200 | 350 |
|---|---|---|---|---|---|---|---|
| Biomass | $(X)$ | (g/l) | 39 | 29 | 28 | 22 | 17 |
| Tylosin | $(P)$ | (g/l) | 0.7 | 0.4 | 0.2 | 0.1 | 0.01 |
| | $\mu$ | $(h^{-1})$ | 0.01 | 0.02 | 0.03 | 0.04 | 0.07 |
| | $r_X$ | $(g/l \cdot h)$ | 0.39 | 0.58 | 0.84 | 0.88 | 1.19 |

---

**Problem 10.3** The growth of Methylomonas L3 on methanol was observed in a CSTR. The working volume of the tank was $V = 0.355 \, l$. The concentration of methanol in fresh medium was 1 g/l. The results of the experiment in a steady state are given in the following table:

| Flow rate (ml/h) | 67 | 70 | 120 | 155 | 165 |
|---|---|---|---|---|---|
| Biomass conc. (g DW/l) | 0.441 | 0.470 | 0.527 | 0.563 | 0.563 |
| Methanol conc. (mg/l) in output stream | 30 | 20 | 23 | 16 | 7 |

DW = dry weight.

Evaluate the $Y_{S/X}$ and the specific growth rate $\mu_m$.

**Solution to Problem 10.3**

For details refer the Wiley website at http://www.wiley-vch.de/ISBN9783527354115

The mass balance in the reactor is reduced to:

$$D = \frac{Q_0}{V} = \mu$$

so we can calculate the values of specific rate ($\mu$) at the different working conditions. Also, by definition, we know that:

$$-Y_{\frac{S}{X}} = \frac{S_0 - S}{X_0 - X} = \frac{\Delta S}{\Delta X}$$

The corresponding calculation is done:

| Q (ml/h) | Q (l/h) | X (g DW/l) | S (mg/l) | S (g/l) | $\mu$ (1/h) | $\Delta S$ (g/l) | $\Delta X$ (g/l) | $Y_{S/X}$ (g S/g X) |
| --- | --- | --- | --- | --- | --- | --- | --- | --- |
| 67 | 0.067 | 0.441 | 30 | 0.030 | 0.189 | −0.970 | 0.441 | 2.200 |
| 70 | 0.070 | 0.470 | 20 | 0.020 | 0.197 | −0.980 | 0.47 | 2.085 |
| 120 | 0.120 | 0.527 | 23 | 0.023 | 0.338 | −0.977 | 0.527 | 1.854 |
| 155 | 0.155 | 0.563 | 16 | 0.016 | 0.437 | −0.984 | 0.563 | 1.748 |
| 165 | 0.165 | 0.563 | 7 | 0.007 | 0.465 | −0.993 | 0.563 | 1.764 |

An average value of $Y_{S/X} = 1.930\,\mathrm{g\,S/g\,X}$ is found.

**Problem 10.4**  *Escherichia coli* grows with a doubling time of 0.5 h in the exponential growth phase.

(a) What is the value of the specific growth rate?
(b) How much time would be required to grow the cell culture from 0.1 kg-dry cell/m$^3$ to 10 kg-dry cell/m$^3$?

**Solution to Problem 10.4**

(a)

$$\frac{\mathrm{d}X}{\mathrm{d}t} = \mu X$$

$$\int_{X_0}^{X} \frac{\mathrm{d}X}{X} = \mu t$$

$$\ln\left(\frac{X}{X_0}\right) = \mu t$$

We know that $t = 0.5\,\mathrm{h}$ for $X = 2X_0$, so we have:

$$\ln\left(\frac{2X_0}{X_0}\right) = \mu \cdot 0.5$$

And finally, $\mu = 1.386\ \mathrm{h}^{-1}$

(b) Also, we can do:

$$\ln\left(\frac{10}{0.1}\right) = 1.386 \cdot t$$

And solving, $t = 3.32\,\mathrm{h}$.

**Problem 10.5**  *Pichia pastoris* (a yeast) was inoculated at $1.0\,\mathrm{kg\text{-}dry\ cell/m^3}$ and cultured in a medium containing 15 wt% glycerol (batch culture). The time-dependent concentrations of cells are shown in the table:

| Time (h) | 0 | 1 | 2 | 3 | 5 | 10 | 15 | 20 | 25 | 30 | 40 | 50 |
|---|---|---|---|---|---|---|---|---|---|---|---|---|
| Cell conc. (kg-dry cell/m³) | 1.0 | 1.0 | 1.0 | 1.1 | 1.7 | 4.1 | 8.3 | 18.2 | 36.2 | 64.3 | 86.1 | 98.4 |

(a) Draw a growth curve of the cells by plotting the logarithm of the cell concentrations against the cultivation time.
(b) Assign the accelerating, exponential growth, and decelerating phases of the growth curve in part (a).
(c) Evaluate the specific growth rate during the exponential growth phase.

**Solution to Problem 10.5**
For details refer the Wiley website at http://www.wiley-vch.de/ISBN9783527354115

(a) We should use a spreadsheet to do the calculation. The specific growth rate ($\mu$) is calculated as the incremental increase of the cell concentration ($r_X$) divided by the cell concentration.

$$\mu = \frac{X^{t+1} - X^t}{\Delta t \cdot X^t}$$

Upon constructing the semi-log plot, we found:

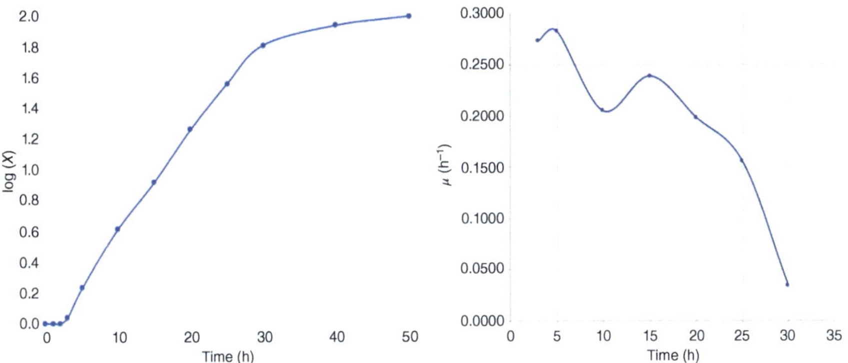

(b) As we can see, in the first three hours of culture time, the process is in the lag phase, where the cell concentration remains practically stable. From this moment on, the exponential growth phase can be observed for up to 30 hours. From this time, it returns to a deceleration phase that can be considered as a stationary phase. If more experimental data were available, the death phase could be observed later, when the cells begin to die and their concentration begins to decrease.
(c) To assess the specific growth rate during the exponential growth phase, we can calculate the derivative $dX/dt$ using the finite differences approximation. In the previous figure, it is plotted against time. It can be seen that the growth

rate is decreasing as time progresses. This means that at the beginning of the exponential phase, there will be more cell growth and less and less until you reach the stationary phase.

**Problem 10.6**

(a) In the fermentation process, yeast cells exhibited growth from 19 kg-dry cell/m$^3$ to 54 kg-dry cell/m$^3$ over a period of seven hours. During this time, 81 g of glycerol was consumed per liter of fermentation broth. Determine the average specific growth rate and the cell yield with respect to glycerol.

(b) In a continuous stirred tank fermenter operating as a chemostat, *E. coli* was cultured with a working volume of 1.0 l. The fermenter was supplied with a medium containing 4.0 g/l of glucose as the carbon source at a constant flow rate of 0.51 h$^{-1}$. The glucose concentration in the output stream was measured to be 0.20 g/l. The cell yield with respect to glucose ($Y_{x/s}$) was determined to be 0.42 g-dry cell/g-glucose. Determine the cell concentration in the output stream and the specific growth rate.

**Solution to Problem 10.6**

(a) As in previous problems, we have:

$$\ln\left(\frac{X}{X_0}\right) = \mu t$$

And, using the data:

$$\ln\left(\frac{54}{19}\right) = \mu \cdot 7$$

So $\mu = 0.1492\ \text{h}^{-1}$

For calculating the cell yield with respect to glycerol, we can do:

$$-Y_{\frac{x}{s}} = \frac{X_0 - X}{S_0 - S} = \frac{19 - 54}{81} = -0.43\ \frac{\text{kg dry cell}}{\text{kg substrate}}$$

bearing in mind that 81 g/l = 81 kg/m$^3$. So finally, $Y_{X/S} = 0.43$ kg/kg.

(b) In the CSTB, in steady state, we have that: $D = Q_0/V = \mu$

So in this case, $\mu = 0.51\ \text{h}^{-1}$

On the other hand:

$$-Y_{\frac{x}{s}} = \frac{X_0 - X}{S_0 - S}$$

$$-0.42 = \frac{-X}{4 - 0.2}$$

And we have: $X = 1.59$ g-dry cells/l.

**Problem 10.7**  Studies carried out in the laboratory showed that in continuous culture, *Azotobacter vinelandii* has a maximum specific growth rate of 0.45 h$^{-1}$ and a $K_S$ of 28 mg/l. If a 50 l bioreactor is operated, with a dilution rate of 0.32 h$^{-1}$,

$Y_{X/S} = 0.36\,\text{g}\,X/\text{g}\,S$, and a substrate concentration in the feed of $5\,\text{g/l}$, determine the steady-state cell concentration and the limiting substrate concentration in the effluent.

**Solution to Problem 10.7**

In this reactor, $D = \mu$, so we can do:

$$D = \mu = 0.32\,\text{h}^{-1} = \mu_m \cdot \frac{S}{K_S + S} = 0.45 \cdot \frac{S}{0.028 + S}$$

obtaining $S = 0.068\,\text{g/l}$. Doing a mass balance for the substrate:

$$Q_0 \cdot S_0 + \left(-r_S\right) \cdot V = Q_0 \cdot S$$

As we know $Q_0 = D \cdot V = 0.32 \cdot 50 = 16\,\text{l/h}$, so:

$$16 \cdot 5 + \left(-r_S\right) \cdot 50 = 16 \cdot 0.068$$

Solving $\left(-r_S\right) = 1.5779\,\text{g}\,S/\text{l} \cdot \text{h}$

Also, we have $Y_{\frac{X}{S}} = \frac{r_X}{r_S}$ and:

$$0.36 = r_X/1.5779$$

So $r_X = 0.5681\,\text{g}\,X/\text{l} \cdot \text{h}$

Finally, we can do $X = r_X/\mu = 0.5681/0.32 = 1.775\,\text{g/l}$.

**Problem 10.8** A new yeast strain is being considered for biomass production. To characterize the yeast strain, experiments were carried out in a one-stage bioreactor. In the experimentation, a concentration of substrate at the bioreactor inlet of $800\,\text{mg/l}$ was used, operating with excess oxygen, pH of 8.5, and a temperature of $35\,^\circ\text{C}$. Using the data in the following table, estimate:

| $D$ (h$^{-1}$) | $S$ (mg/l) | $X$ (mg/l) |
| --- | --- | --- |
| 0.1 | 16.7 | 366 |
| 0.2 | 33.5 | 407 |
| 0.3 | 59.4 | 408 |
| 0.4 | 101 | 404 |
| 0.5 | 169 | 371 |
| 0.6 | 298 | 299 |
| 0.7 | 702 | 59 |

(a) Values of $\mu_m$ and $K_S$.
(b) What is the $D$ that maximizes cell productivity? Compare theoretical and graphic values.
(c) Also, determine the volumetric productivity of biomass under optimal conditions and the $D_{\text{critical}}$ for the growth of the strain.

**Solution to Problem 10.8**

(a) In the reactor, $D = \mu$, and then $D = \mu_m \cdot S/(K_S + S)$. Rearranging this expression:

$$\frac{1}{D} = \frac{K_S}{\mu_m} \cdot \frac{1}{S} + \frac{1}{\mu_m}$$

We can plot $1/D$ versus $1/S$ and calculate the slope and ordinate.

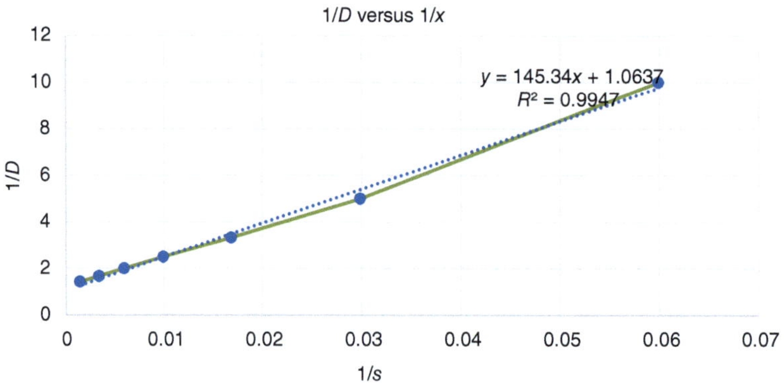



From that plot:

$$\frac{K_S}{\mu_m} \cdot = 145.34$$

$$\frac{1}{\mu_m} = 1.0637$$

Obtaining: $\mu_m = 0.94\,\text{h}^{-1}$ and $K_S = 136.63\,\text{mg/l}$

(b) For estimating the value of "$D$" that maximizes the cell growth, a graphical plot of "$D \cdot X$" versus $D$ is done, locating its maximum that corresponds to:

$$D_{\text{opt}} = \mu_m \cdot \left(1 - \sqrt{\frac{K_S}{K_S + S_0}}\right) = 0.58\,\text{h}^{-1}$$

This would be the theoretical value. In the table with the experimental results, we see that the value of $D$ for which the $X$ is higher corresponds to $0.5\,\text{h}^{-1}$.

(c)

$$S = D \cdot \frac{K_S}{\mu_m - D}$$

Using the value of $D = 0.5\,\text{h}^{-1}$, we get $S = 155.22\,\text{mg/l}$
Also, $S = S_0 - X$ for what $X = 644.7\,\text{mg/l}$ for the critical conditions.
Finally:

$$D_{\text{washout}} = \mu_m \frac{S_0}{K_S + S_0} = 0.94 \frac{800}{800 + 136.6} = 0.81\,\text{h}^{-1}$$

**Problem 10.9**   The following data were obtained on the oxidation of pesticides present in wastewater. Data were obtained using a mixture of microorganisms in a continuous operation in a chemostat-type aeration lagoon where the pesticide is the limiting reagent.

| D (h$^{-1}$) | S (mq/l) |
|---|---|
| 0.08 | 15 |
| 0.11 | 25 |
| 0.24 | 50 |
| 0.39 | 100 |
| 0.52 | 140 |
| 0.70 | 180 |
| 0.82 | 240 |

If you have a liquid residue to treat with a flow rate of $0.5\,\mathrm{m^3/s}$ and a pesticide concentration of 500 mg/l, and you need to reduce its concentration by 95%,

(a) What should be the dilution rate and the volume of the aeration pond?
(b) What is the microbial mass in tons produced per day of operation? Take $Y_{x/s}$ to be $0.6\,\mathrm{g\,X/g\,S}$.

**Solution To Problem 10.9**

(a) As in the previous problem:

$$\frac{1}{D} = \frac{K_S}{\mu_m} \cdot \frac{1}{S} + \frac{1}{\mu_m}$$

And we can use a spreadsheet to calculate $K_S = 755.5\,\mathrm{mg/l}$ and $\mu_m = 3.434\,\mathrm{h^{-1}}$
If we need to convert 95% of the substrate: $S_0 = 500\,\mathrm{mg/l}$ and $S = 25\,\mathrm{mg/l}$

$$D = \mu = \mu_m \cdot \frac{S}{K_S + S} = 3.434 \cdot \frac{25}{755.5 + 25} = 0.11\,\mathrm{h^{-1}}$$

From that, the volume would be:

$$V = \frac{Q}{D} = \frac{500\,\frac{l}{s} \cdot 3600\,\frac{s}{h}}{0.11\,\mathrm{h^{-1}}} = 16\,363\,636\,l = 16\,363\,\mathrm{m^3}$$

(b) For the substrate: $Q_0 \cdot S_0 + (-r_S) \cdot V = Q_0 \cdot S$
Using the definition: $Y_{\frac{x}{s}} = \frac{r_X}{r_S} = \frac{X \cdot \mu}{r_S}$
Then:

$$Q_0 \cdot S_0 + \frac{X \cdot \mu}{Y_{\frac{x}{s}}} \cdot V = Q_0 \cdot S$$

$$Q_0 \cdot S_0 + \frac{X \cdot \mu_m}{Y_{\frac{x}{s}}} \cdot \frac{S}{K_S + S} \cdot V = Q_0 \cdot S$$

$$500 \cdot 3600 \cdot (500 - 25) + \frac{X \cdot 3.434}{0.6} \cdot \frac{25}{755.5 + 25} \cdot 16\,363\,636 = 0$$

Obtaining $X = 285.25\,\mathrm{mg/l}$

The production of cells would be: $285.25 \, \text{mg/l} \cdot 500 \, \text{l/s} \cdot 3600 \, \text{s/h} \cdot 24 \, \text{h/day} \cdot 1/10^9$ ton/mg $= 12.31$ ton/day.

**Problem 10.10**   The bacterium *Pseudomonas* sp. has a maximum specific growth rate of $0.4 \, \text{h}^{-1}$ when it is cultivated in acetate. The saturation constant using this substrate is $1.3 \, \text{g/l}$, and the coefficient of yield of cells in acetate is $0.46 \, \text{g cel/g acetate}$. If this system is operated as a chemostat, with an $S_0$ of $38 \, \text{g/l}$, perform the following analysis:

(a) What is the critical dilution rate?
(b) What is the concentration of cells when the dilution rate is half critical?
(c) What is the substrate concentration when D is 80% critical?
(d) What is the cell productivity at $D$ value calculated in (c)?
(e) Find $D_{\text{opt}}$ for cell productivity; what percentage of $D_{\text{critical}}$ is this?

**Solution to Problem 10.10**

(a) A balance of cells, assuming $X_0 = 0$, in the reactor is: $\mu = D$, and then:

$$D = \mu_m \cdot \frac{S}{K_S + S}$$

Assuming that $D_{\text{critical}}$ is obtained when the substrate concentration at the exit equals S0:

$$D_{\text{critical}} = 0.4 \cdot \frac{38}{1.3 + 38} = 0.3867 \, \text{h}^{-1}$$

(b) Value of "$X$" can be estimated as:

$$D = D_{\text{critical}}/2 = 0.1935 \, \text{h}^{-1}$$

As we have:

$$S = D \cdot \frac{K_S}{\mu_m - D}$$

$$S = 0.1935 \cdot \frac{1.3}{0.4 - 0.1935} = 1.218 \, \frac{\text{g}}{\text{l}}$$

Using the $Y_{X/S}$ value:

$$-Y_{\frac{X}{S}} = \frac{X_0 - X}{S_0 - S}$$

$$-0.46 = \frac{0 - X}{38 - 1.218}$$

And $X = 16.92 \, \text{g/l}$.

(c) and (d) For $D = 0.8 \cdot 0.3867 = 0.3096 \, \text{h}^{-1}$
Using the same expressions, $S = 4.45 \, \text{g/l}$
The cell concentration is:

$$X = 15.43 \, \text{g/l}$$

and:

$$\text{Cell production} = 15.43 \text{ g/l} \cdot 0.3096 \text{ h}^{-1} = 4.778 \text{ g/l} \cdot \text{h}$$

(e) We have:

$$D_{opt} = \mu_m \cdot \left( 1 - \sqrt{\frac{K_S}{K_S + S_0}} \right) = 0.327 \text{ h}^{-1}$$

$$D_{opt}/D_{critical} \cdot 100 = 84.5\%$$

**Problem 10.11**  A continuous bioreactor is operated with a working volume of 120 l and a feed flow of 20 l/h. The bacterial population presents a minimum doubling time of 3.15 h and a $K_S$ of 1 g/l. The yield expressed in gram of cells per gram of substrate has been estimated at 0.28. Preliminary tests make it advisable to work at a dilution rate equal to 82% of the critical value.

(a) What is the maximum specific velocity, $\mu_m$?
(b) What is the substrate concentration in the feed?
(c) What is the substrate concentration in the discharge?
(d) What is the cell concentration at steady state?

**Solution to Problem 10.11**

(a) During the exponential growth:

$$\ln\left(\frac{X}{X_0}\right) = \mu t$$

As we know, $t = 3.15$ h for $X = 2X_0$, so: $\mu = 0.22 \text{ h}^{-1}$ that would actually represent $\mu_m$ as the time is the minimum, as indicated in the statement.

(b), (c), and (d) When $D = \mu_m$, the state of the reactor is critical, as it operates at the maximum rate. In the present case, $D = 0.82 \cdot D_{critical} = 0.82 \cdot 0.22 = 0.1804 \text{ h}^{-1}$

From previous problems, we know that:

$$S = D \cdot \frac{K_S}{\mu_m - D} = 0.1804 \cdot \left( \frac{1}{0.22 - 0.1804} \right) = 4.55 \frac{\text{g}}{\text{l}}$$

And:

$$Q_0 \cdot S_0 + \frac{X \cdot \mu_m}{Y_{\frac{X}{S}}} \cdot \frac{S}{K_S + S} \cdot V = Q_0 \cdot S$$

$$20 \cdot \left( S_0 - 4.55 \right) + \frac{X \cdot 0.22}{0.28} \cdot \frac{4.55}{1 + 4.55} \cdot 120 = 0$$

From this equation: $S_0 = 4.55 - 3.865 \cdot X$
We also know that $X = Y_{\frac{X}{S}} \cdot \left( S_0 - S \right)$, so:

$$X = 0.28 \cdot \left( S_0 - 4.55 \right)$$

$$S_0 = 4.55 - 3.865 \cdot 0.28 \cdot \left( S_0 - 4.55 \right)$$

Solving:

$$S_0 = 10.63 \, \frac{g}{l}$$

$$X = 1.71 \, \frac{g}{l}$$

**Problem 10.12**  *Pseudomonas* sp. has a mass doubling time of 2.4 h when grown on acetate. The saturation constant using this substrate is 1.3 g/l, and the cell yield in acetate is 0.6 g cells/g acetate. If a chemostat is operated with a feed stream containing 38 g/l acetate, find:

(a) The cell concentration when the dilution rate is one-half of the maximum.
(b) The concentration of the substrate when the dilution rate is $0.8\,D_{\max}$.
(c) The maximum rate of dilution.
(d) Cell productivity at $0.8\,D_{\max}$.

**Solution to Problem 10.12**

(a)

$$S = D \cdot \frac{K_S}{\mu_m - D}$$

In the present case, $D = \mu_m/2$, so:

$$S = \frac{\mu_m}{2} \cdot \frac{K_S}{\mu_m - \frac{\mu_m}{2}} = K_S = 1.3 \, \frac{g}{l}$$

(b) $D = 0.8 \cdot D_{\max}$

$$D_{\max} = \mu_m \frac{S_0}{K_S + S_0} = 38 \frac{\mu_m}{38 + 1.3} = 0.967 \mu_m$$

$$D = 0.7735 \, \mu m$$

$$S = 0.7735 \mu_m \cdot \frac{K_S}{\mu_m - 0.7735\mu_m} = 4.44 \, \frac{g}{l}$$

(c)

$$\ln\left(\frac{2X_0}{X_0}\right) = \mu_m \cdot 2.4$$

$$\mu_m = 0.288 \, h^{-1}$$

$$D_{\max} = 0.967 \mu_m = 0.2785 \, h^{-1}$$

(d)

$$X = Y_{\frac{x}{s}} \cdot (S_0 - S) = 0.6 \cdot (38 - 4.44) = 20.136 \, \frac{g}{l}$$

$$DX = 0.8 \, D_{\max} \cdot 20.136 = 4.486 \, \frac{g}{l \cdot h}$$

**Problem 10.13**  A 5 m$^3$ stirred tank fermenter operates continuously with a fed substrate concentration of 20 kg/m$^3$. The microorganism cultivated in the reactor presents the following characteristics: $\mu_m = 0.45\,h^{-1}$, $K_S = 0.8\,kg/m^3$, $Y_{x/s} = 0.55$ kg/kg. The Monod equation is assumed to hold. Calculate:

(a) What feed rate is needed to achieve 90% substrate conversion?
(b) What is the biomass production corresponding to a substrate conversion of 90% compared to the maximum possible?

**Solution to Problem 10.13**

(a) For a conversion of 90%: $S = (1 - 0.9) \cdot S_0 = 2 \, \text{kg/m}^3$
   In that situation

$$D = \mu = \mu_m \frac{S}{K_S + S} = 0.45 \left( \frac{2}{0.8 + 2} \right) = 0.321 \, \frac{\text{kg}}{\text{m}^3}$$

$$Q_0 = D \cdot V = 0.321 \cdot 5 = 1.6 \, \frac{\text{m}^3}{\text{h}}$$

(b) Biomass production:

$$X = Y_{X/S} \cdot (S_0 - S) = 0.55 \cdot (20 - 2) = 9.9 \, \frac{\text{kg}}{\text{m}^3}$$

$$D \cdot X = 0.321 \cdot 9.9 = 3.18 \frac{\text{kg}}{\text{m}^3 \cdot \text{h}}$$

If the conversion were 100%: $S = 0$, $X = Y_{X/S} \cdot S_0 = 11 \, \text{kg/m}^3$

$$D = \mu_m \frac{0}{K_S + 0} = \frac{\mu_m}{K_S} = 0.5625 \, \text{h}^{-1}$$

$$D \cdot X = 6.18 \, \frac{\text{kg}}{\text{m}^3 \cdot \text{h}}$$

This would be the maximum production of biomass in the bioreactor.

**Problem 10.14**   A simple chemostat with a useful volume of 5 l presents some design problems that cause imperfect agitation. The equipment is going to be used to grow yeast with $\mu_m = 0.48 \, \text{h}^{-1}$ and $K_S = 0.072 \, \text{g/l}$. The feed flow will be 1.1 l/h, with a limiting nutrient concentration of 6 g/l and $Y_{X/S}$ of 0.43 g/g. It is assumed that 35% of the feed stream bypasses the fermenter but joins the output stream, and the remainder enters the chemostat. Determine $(D_{\text{opt}}, X_{\text{opt}})$, $\mu$, and $S$ at the exit of the fermenter.

**Solution to Problem 10.14**

In the chemostat, we can write:

$$D_{\text{opt}} = \mu_m \cdot \left( 1 - \sqrt{\frac{K_S}{K_S + S_0}} \right) = 0.48 \cdot \left( 1 - \sqrt{\frac{0.072}{0.072 + 6}} \right) = 0.4227 \, \text{s}^{-1}$$

The rate of microbial growth at this dilution rate is:

$$r_{\text{opt}} = \mu_m \frac{S_{\text{opt}}}{K_S + S_{\text{opt}}} X_{\text{opt}}$$

$$S_{\text{opt}} = D_{\text{opt}} \cdot \frac{K_S}{\mu_m - D_{\text{opt}}} = \frac{0.4227 \cdot 0.072}{0.48 - 0.4227} = 0.589 \, \frac{\text{g}}{\text{l}}$$

$$X_{\text{opt}} = \left(S_0 - S_{\text{opt}}\right) \cdot Y_{X/S} = (6 - 0.589) \cdot 0.43 = 2.32 \, \frac{g}{l}$$

$$r_{\text{opt}} = 0.48 \frac{0.589}{0.072 + 0.589} 2.32 = 0.9951 \, \frac{g}{h \cdot l}$$

In the fermenter, we have:

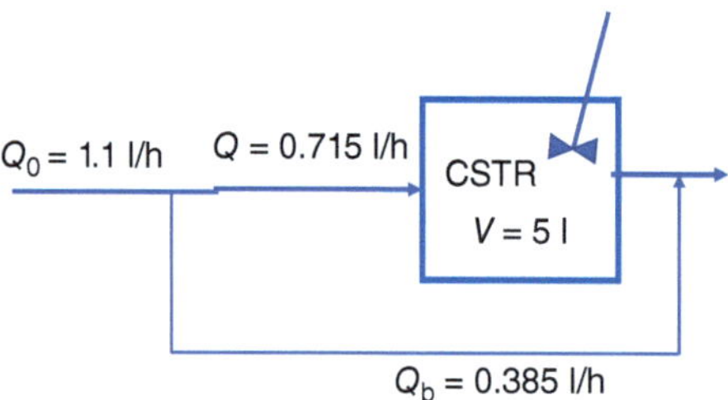


So actually, $D = Q/V = 0.143 \, \text{h}^{-1} = \mu$

In that case:

$$S = D \cdot \frac{K_S}{\mu_m - D} = 0.031 \, \frac{g}{l}$$

$$X = \left(S_0 - S\right) \cdot Y_{X/S} = 2.566 \, \frac{g}{l}$$

$$r = \mu_m \frac{S}{K_S + S} X = 0.371 \, \frac{g}{h \cdot l}$$

The concentration at the exit of the fermenter must account for the bypass current, so:

$$S_{\text{exit}} = \frac{0.031 \cdot 0.715 + 6 \cdot 0.385}{1.1} = 0.230 \, \frac{g}{l}$$

**Problem 10.15**  The growth of *E. coli* in glucose can be described by the Monod kinetics model, with a substrate saturation constant ($K_S$) of 4 g/m$^3$ and a maximum specific growth rate ($\mu_m$) of 1.33 h$^{-1}$. The yield coefficient ($Y_{X/S}$) is 0.1 g cells/g glucose. Calculate the glucose feed rate (with a concentration of 60 g/m$^3$) necessary to achieve the maximum rate of glucose consumption and cell production in a 1 m$^3$ continuous stirred tank fermenter. Determine these values.

**Solution to Problem 10.15**

$$D_{\text{max}} = \mu_m \frac{S_0}{K_S + S_0} = 1.33 \frac{60}{4 + 60} = 1.2468 \, \text{h}^{-1} = \frac{Q_{\text{max}}}{V}$$

With $V = 1$ m$^3$, the flow rate to the reactor is 1.2468 m$^3$/h. The concentrations at the exit are:

$$S = D \cdot \frac{K_S}{\mu_m - D} = 1.2468 \frac{4}{1.33 - 1.2468} = 59.94 \, \frac{g}{l}$$

$$X = \left(S_0 - S\right) \cdot Y_{X/S} = (60 - 59.94) \cdot 0.1 = 0.006 \, \frac{g}{l}$$

On the contrary, at the optimum values:

$$D_{\mathrm{opt}} = \mu_m \cdot \left(1 - \sqrt{\frac{K_S}{K_S + S_0}}\right) = 1.33\left(1 - \sqrt{\frac{4}{4+60}}\right) = 0.9975\,\mathrm{h}^{-1}$$

The concentrations at the exit are:

$$S = D \cdot \frac{K_S}{\mu_m - D} = 0.9975\frac{4}{1.33 - 0.9975} = 12\,\frac{\mathrm{g}}{\mathrm{l}}$$

$$X = \left(S_0 - S\right) \cdot Y_{X/S} = (60 - 12) \cdot 0.1 = 4.8\,\frac{\mathrm{g}}{\mathrm{l}}$$

**Problem 10.16**  Microbial growth of *E. coli* on glucose follows Monod kinetics. In the laboratory, it is observed that for a glucose concentration of $1\,\mathrm{g/m^3}$, the growth rate of the microorganism is $0.26\,\mathrm{g\,cells/m^3 \cdot h}$, while if the concentration is $4\,\mathrm{g/m^3}$, the rate is $0.65\,\mathrm{g\,cells/m^3 \cdot h}$.

(a) Determine the values of $K_S$ and maximum rate of the kinetics of Monod.
(b) Study the behavior of a continuous stirred tank fermenter of $1\,\mathrm{m^3}$ in which this process takes place by changing the glucose feed flow (concentration $S_0 = 60\,\mathrm{g/m^3}$).

*Data:* The value of $Y_{X/S}$ is 0.1 g/g.

**Solution to Problem 10.16**

(a)

$$\mu = \mu_m \frac{S}{K_S + S}$$

$$r_X \cdot X = r_m \cdot X \frac{S}{K_S + S}$$

$$\frac{1}{r_X} = \frac{1}{r_m} + \frac{K_S}{\mu_m}\frac{1}{S}$$

With the data:

$$\frac{1}{0.26} = \frac{1}{r_m} + \frac{K_S}{r_m}\frac{1}{1}$$

$$\frac{1}{0.65} = \frac{1}{r_m} + \frac{K_S}{r_m}\frac{1}{4}$$

We obtain that: $1/r_m = 0.796$ and $K_S/r_m = 3.08$ so $r_m = 1.3\,\mathrm{g\,cells/m^3 \cdot h}$ and $K_S = 4\,\mathrm{g\,cells/m^3}$.

(b) Study the behavior of a continuous stirred tank fermenter of $1\,\mathrm{m^3}$ in which this process takes place by changing the glucose feed flow (concentration $S_0 = 60\,\mathrm{g/m^3}$). For this, we will use a spreadsheet, and the equations are:

$$D = \frac{Q_0}{V}$$

$$S = D \cdot \frac{K_S}{\mu_m - D}$$

$$X = \left(S_0 - S\right) \cdot Y_{X/S}$$

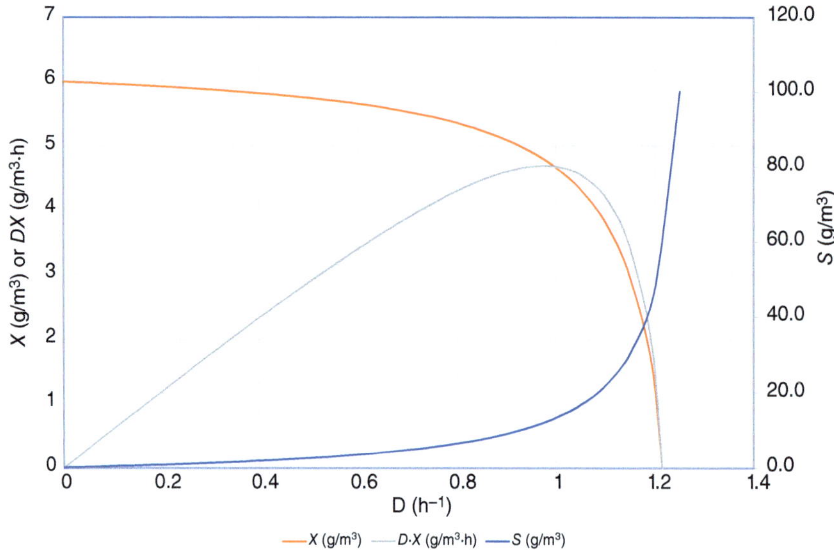



**Problem 10.17**  In a tubular biochemical reactor, biomass is being produced according to the process:

$$\text{Substrate (S)} + \text{Cells (X)} \rightarrow \text{More cells (X)} + \text{Product (P)}$$

In the system, there is a recirculation of cells according to the scheme:

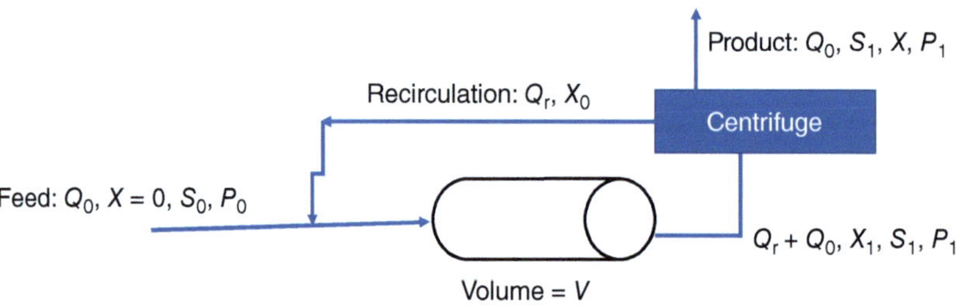



(a) Do a material balance for the X cells in the reactor.
(b) If the kinetics is of the Monod type with excess substrate, indicate how this balance can be simplified.

**Solution to Problem 10.17**

(a) In the centrifuge, the balance is:

$$\left(Q_r + Q_0\right) \cdot X_1 = Q_0 \cdot X + Q_r \cdot X_0$$

Before the reactor, the concentration of cells entering, $X'$, is:

$$\left(Q_r + Q_0\right) \cdot X' = Q_0 \cdot 0 + Q_r \cdot X_0$$

So:

$$X' = \frac{Q_r}{Q_r + Q_0} \cdot X_0$$

In the PFR, the balance in a differential element is:

$$\text{Input} = \left(Q_0 + Q_r\right) \cdot X$$

$$\text{Output} = \left(Q_0 + Q_r\right) \cdot (X + dX)$$

$$\text{Generation} = r_X \cdot dV = \mu \cdot X \cdot dV$$

Doing the mass balance:

$$\left(Q_0 + Q_r\right) \cdot dX = \mu \cdot X \cdot dV = \frac{\mu_m \cdot S}{\left(K_S + S\right)} \cdot X \cdot dV$$

Separating:

$$\left(Q_0 + Q_r\right) \cdot \frac{dX}{X} = \frac{\mu_m \cdot S}{\left(K_S + S\right)} \cdot dV$$

(b) If $S \gg K_S$, we have that $(K_S + S)$ is approximately equal to $S$, and:

$$\left(Q_0 + Q_r\right) \cdot \frac{dX}{X} = \mu_m \cdot dV$$

Integrating into the whole system:

$$\left(Q_0 + Q_r\right) \cdot \int_{X'}^{X_1} \frac{dX}{X} = \mu_m \cdot V$$

$$\left(Q_0 + Q_r\right) \cdot \left( \ln \left( \frac{X'}{X_1} \right) \right) = \mu_m \cdot V$$

Substituting:

$$\left(Q_0 + Q_r\right) \cdot \left( \ln \left( \frac{Q_r}{Q_r + Q_0} \cdot \frac{X_0}{X_1} \right) \right) = \mu_m \cdot V$$

**Problem 10.18**

| $t$ (d) | 0 | 0.5 | 1 |
|---|---|---|---|
| $S$ (g/l) | 40 | 36 | 28 |
| $X$ (g/l) | 2 | 5 | 10 |

A series of preliminary experiments have been carried out in a batch fermentation reactor, measuring the concentration of microorganisms ($X$) and substrate ($S$) throughout the operation time, with the data that appear in the table. Since the inoculum had previously been acclimatized to the substrate used, the induction time can be considered to be negligible.

(a) Determine the parameters of the Monod equation.

(b) Calculate the initial speed of a process in which a substrate with an initial concentration of 35 g/l and an initial inoculum concentration of 2.5 g/l is going to be used.

**Solution to Problem 10.18**
For details refer the Wiley website at http://www.wiley-vch.de/ISBN9783527354115

(a) We can calculate, by increments, the reaction rate:

| $t$ (d) | 0 | 0.5 | 1 |
| --- | --- | --- | --- |
| $S$ (g/l) | 40 | 36 | 28 |
| $X$ (g/l) | 2 | 5 | 10 |
| $\frac{\Delta X}{\Delta t}$ (g/l·d) | 6 | 10 | — |
| $\mu = \frac{1}{X}\frac{\Delta X}{\Delta t}$ (d$^{-1}$) | 3 | 2 | — |

So, we can write:

$$\mu = \mu_m \frac{S}{K_S + S}$$

$$3 = \mu_m \frac{40}{K_S + 40}$$

$$2 = \mu_m \frac{36}{K_S + 36}$$

Obtaining: $\mu_m = 3.08 \frac{g}{l \cdot d}$ and $K_S = 1.07 \frac{g}{l}$

(b) With $S_0 = 35$ g/l and $X_0 = 2.5$ g/l:

$$r_X = \mu \cdot X = \mu_m \frac{S \cdot X}{K_S + S} = 3.08 \frac{35 \cdot 2.5}{1.07 + 35} = 7.47 \frac{\text{mol}}{l \cdot d}$$

**Problem 10.19**   In a system of two chemostats in series, the volumes of the first and second reactors are $V_1 = 500$ l and $V_2 = 300$ l, respectively. The first reactor is used for biomass production, and the second is used for the formation of a secondary metabolite. The feed stream of the first reactor is $Q_0 = 100$ l/h, and the glucose concentration in the feed is $S_0 = 5.0$ g/l. Use the following constants for the cells: $\mu_m = 0.3$ h$^{-1}$, $K_S = 0.1$ g/l, $Y_{X/S} = 0.4$ g cells/g glucose. Determine:

(a) The concentration of biomass and glucose in the effluent of the first stage.
(b) Assuming that growth is negligible in the second stage and that the specific rate of product formation is $Y_{P/X} = 0.02$ g P/g cells and that $Y_{P/S} = 0.6$ g P/g S, determine the concentration of the product and the substrate in the effluent from the second reactor.

**Solution to Problem 10.19**
For details refer the Wiley website at http://www.wiley-vch.de/ISBN9783527354115

(a) Let us draw a scheme:

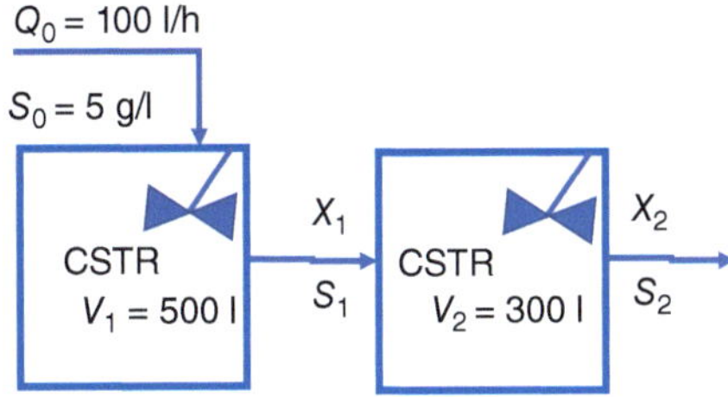


In the first tank:

$$S_1 = D_1 \cdot \frac{K_S}{\mu_m - D_1}$$

With $D_1 = Q_0/V_1 = 0.2\,\mathrm{h}^{-1}$, we obtain $S_1 = 0.2\,\mathrm{g/l}$, and:

$$X_1 = \left(S_0 - S_1\right) \cdot Y_{X/S} = (5 - 0.2) \cdot 0.4 = 1.92\,\frac{\mathrm{g}}{\mathrm{l}}$$

(b) Determine the concentration of the product and the substrate in the effluent from the second reactor. In that reactor, at the input, we have:

$$\text{Substrate} = S_1 Q_0 = 0.2 \cdot 100 = 20\,\frac{\mathrm{g}}{\mathrm{h}}$$

$$\text{Cells} = X_1 Q_0 = 1.92 \cdot 100 = 192\,\frac{\mathrm{g}}{\mathrm{h}}$$

With $Y_{P/X} = 0.02\,\mathrm{g\ P/g\ cells}$ and $Y_{P/S} = 0.6\,\mathrm{g\ P/g\ S}$, we obtain:

$$-Y_{\frac{P}{X}} = \frac{P_0 - P}{X_0 - X} = -0.02 = \frac{0 - P}{1.92 - 0}$$

And so: $P = 0.0384\,\mathrm{g/l}$, which means a production of $3.84\,\mathrm{g/h}$. In a similar way:

$$Y_{\frac{P}{S}} = \frac{P_0 - P}{S_0 - S} = 0.6 = \frac{0 - 0.0384}{0.2 - S}$$

Obtaining $S = 0.136\,\mathrm{g/l}$, which means that the effluent contains $13.6\,\mathrm{g/h}$ of substrate.

**Problem 10.20**  Two stirred tank fermenters are connected in series. The first has an operating volume of 100 l and the second of 50 l. The feeding of the first fermenter is sterile, contains 5 g/l of substrate, and is fed to the fermenter with a flow rate of 18 l/h. Microbial growth follows Monod kinetics, with the values of $K_S$ and $\mu_m$ being 120 mg/l and 0.25 h$^{-1}$, respectively. Calculate the concentration of substrate in the second fermenter under steady-state conditions and the yield of the process. $Y_{X/S} = 0.7\,\mathrm{g/g}$.

**Solution to Problem 10.20**

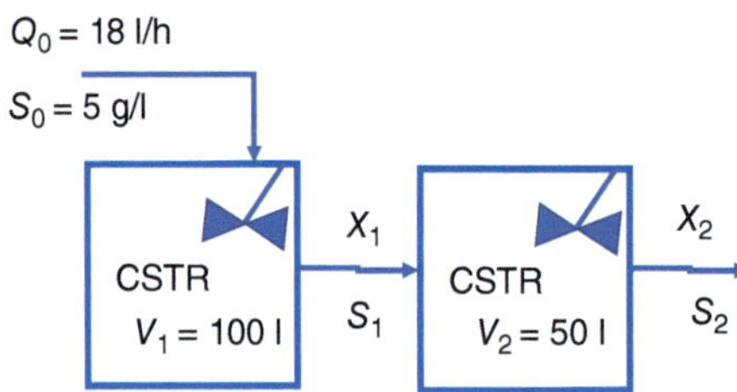


In the first tank:

$$S_1 = D_1 \cdot \frac{K_S}{\mu_m - D_1}$$

With $D_1 = Q_0/V_1 = 18/100 = 0.18\,\text{h}^{-1}$, we obtain, with $K_S = 0.12\,\text{g/l}$, $S_1 = 0.31\,\text{g/l}$, and:

$$X_1 = (S_0 - S_1) \cdot Y_{X/S} = (5 - 0.31) \cdot 0.7 = 3.283\,\frac{\text{g}}{\text{l}}$$

In the second tank:

$$0 = Q_0 (X_1 - X_2) + V_2 \mu X_2$$
$$(X_2 - X_1) = (S_1 - S_2) \cdot Y_{X/S}$$

And then:

$$0 = Q_0 (X_1 - X_2) + V_2 \left( \mu_m \frac{S_2}{K_S + S_2} \right) X_2$$

With the data:

$$0 = 18 (3.283 - X_2) + 50 \left( 0.25 \frac{S_2}{0.12 + S_2} \right) X_2$$

$$(X_2 - 3.283) = (0.31 - S_2) \cdot 0.7$$

Obtaining: $X_2 = 3.28\,\text{g/l}$, $S_2 = 0$.

**Problem 10.21**   Let us derive the equations for the concentrations of biomass and the limiting substrate in the effluent from the second reactor in the cascade of two chemostats shown in the figure, assuming that the limiting substrate is the same in both chemostats.

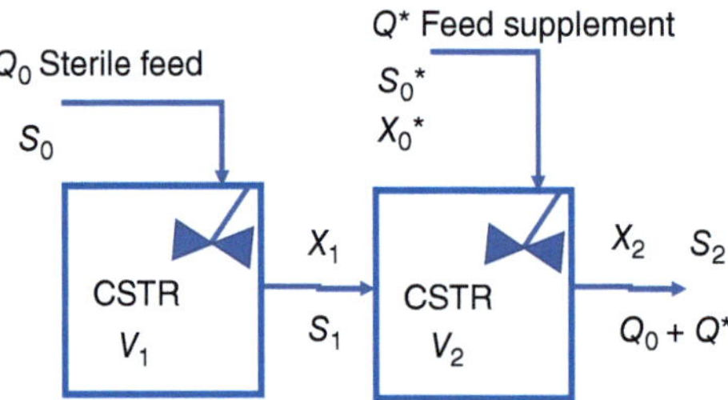

**Solution to Problem 10.21**

In the first reactor, a balance of cells gives:

$$0 = Q_0 \left( X_0 - X_1 \right) + V_1 \mu_1 X_1$$

$$\frac{Q_0}{V_1} \left( 0 - X_1 \right) = -\mu_1 X_1$$

$$D_1 = \frac{Q_0}{V_1} = \mu_1$$

$$\mu_m \frac{S_1}{K_S + S_1} = \mu_1$$

In the second one:

$$Q_0 X_1 + Q^* X_0^* = - \left( Q_0 + Q^* \right) X_2 + V_2 \mu_2 X_2$$

$$\frac{Q_0 X_1 + Q^* X_0^* + \left( Q_0 + Q^* \right) X_2}{V_2} = \mu_2 X_2$$

$$\mu_2 = \frac{Q_0 X_1 + Q^* X_0^*}{V_2 X_2} + \frac{\left( Q_0 + Q^* \right)}{V_2}$$

$$D_2 = \frac{Q_0 + Q^*}{V_2}$$

$$\mu_2 - \frac{Q_0 X_1 + Q^* X_0^*}{V_2 X_2} = D_2$$

$$\mu_2 = \mu_m \frac{S_2}{K_S + S_2}$$

For the substrate in the second reactor:

$$Q_0 S_1 + Q^* S_0^* = - \left( Q_0 + Q^* \right) S_2 + \frac{V_2 \mu_2 X_2}{Y_{\frac{X}{S}}}$$

Then:

$$S_2 = \frac{\frac{V_2 \mu_2 X_2}{Y_{\frac{X}{S}}} - Q_0 S_1 - Q^* S_0^*}{\left( Q_0 + Q^* \right)}$$

**Problem 10.22**  A biochemical reaction is being performed for an autocatalytic biochemical reaction whose rate law is of the logistic form:

$$r_X = \mu_m \left( 1 - \frac{X}{X_\infty} \right) X$$

with $\mu_m = 1.2 \, \text{g/l} \cdot \text{d}$ and $X_\infty = 140 \, \text{g/l}$. Consider carrying out this reaction in a cascade of three CSTRs that differ from each other in size. Use a graphical approach to ascertain the dilution rates and working volumes for each of these reactors if the volumetric flow rate to the first CSTR is 4000 l/d. You may assume that (i) the cascade operates at steady state; (ii) the feed to the first CSTR is sterile; (iii) the working volume of the second reactor is 50% larger than that of the first; (iv) the working volume of the third reactor is 50% larger than that of the second; and (v) the first CSTR is operated at conditions corresponding to the maximum in the rate curve.

**Solution to Problem 10.22**
For details refer the Wiley website at http://www.wiley-vch.de/ISBN9783527354115

In the system, we have:

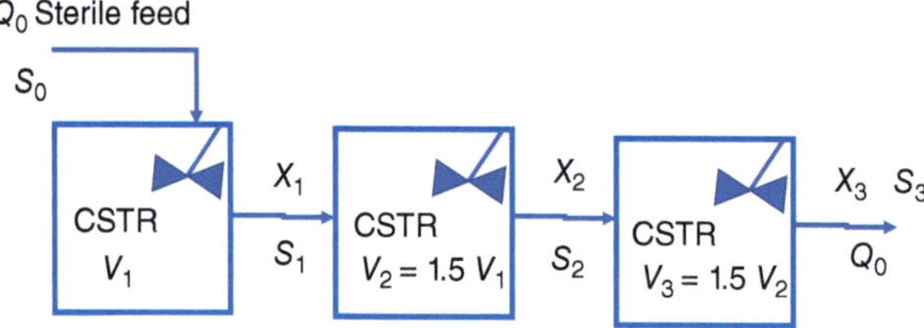


The maximum reaction rate corresponds to a condition for which the derivative of the rate law with respect to the concentration of biomass is zero:

$$\frac{\mathrm{d}r_X}{\mathrm{d}X} = \mu_m \frac{\mathrm{d}\left(\left(1 - \frac{X}{X_\infty}\right)X\right)}{\mathrm{d}X} = \mu_m\left(\left(1 - \frac{X}{X_\infty}\right) - \frac{X}{X_\infty}\right) = 0$$

Thus, at the maximum rate:

$$X = \frac{X_\infty}{2} = 70\ \frac{\mathrm{g}}{\mathrm{l}}$$

The corresponding rate is:

$$r_X = \mu_m\left(1 - \frac{\frac{X_\infty}{2}}{X_\infty}\right)\frac{X_\infty}{2} = \frac{\mu_m X_\infty}{4} = 42\ \frac{\mathrm{g}}{\mathrm{l}\cdot\mathrm{d}}$$

A steady-state material balance on the biomass around the first reactor in the cascade yields:

$$\text{input} = \text{output} + \text{disappearance by reaction}$$

$$0 = X_1 Q_0 - r_{X1}\cdot V_1$$

where we have designated the concentration of biomass, leaving the first CSTR as $X_1$. The corresponding rate of production of biomass per unit volume of this reactor is $r_{X1}$. Rearrangement of equation and introduction of the dilution rate ($D_1 = Q_0/V_1$) give:

$$r_{X1} = D_1 X_1$$

As we know that $X_1 = 70\,\mathrm{g/l}$ and $r_{X1} = 42\,\mathrm{g/l}\cdot\mathrm{d}$, we should operate at $D_1 = 42/70 = 0.6\,\mathrm{d}^{-1}$. Bearing in mind the relation with the volumes:

$$V_1 = \frac{Q_0}{D_1} = \frac{4000}{0.6} = 6667\,\mathrm{l}$$

So, $V_2 = 10\,000$ l and $V_3 = 15\,000$ l. The corresponding dilution rates are:

$$D_2 = \frac{Q_0}{V_2} = 0.4\,\text{d}^{-1}$$

$$D_3 = \frac{Q_0}{V_3} = 0.2667\,\text{d}^{-1}$$

Material balances around the second and third CSTRs lead to the following equations:

$$r_{X2} = D_2\left(X_2 - X_1\right) = \mu_m\left(1 - \frac{X_2}{X_\infty}\right)X_2$$

$$r_{X3} = D_3\left(X_3 - X_2\right) = \mu_m\left(1 - \frac{X_3}{X_\infty}\right)X_3$$

With all the parameters known, we finally get:

$$
\begin{aligned}
X_2 &= 120.45\,\text{g/l} & r_{X2} &= 20.18\,\text{g/l}\cdot\text{d}\\
X_3 &= 136.37\,\text{g/l} & r_{X3} &= 4.24\,\text{g/l}\cdot\text{d}
\end{aligned}
$$

We can plot the kinetics and the points where the reactors are being used.

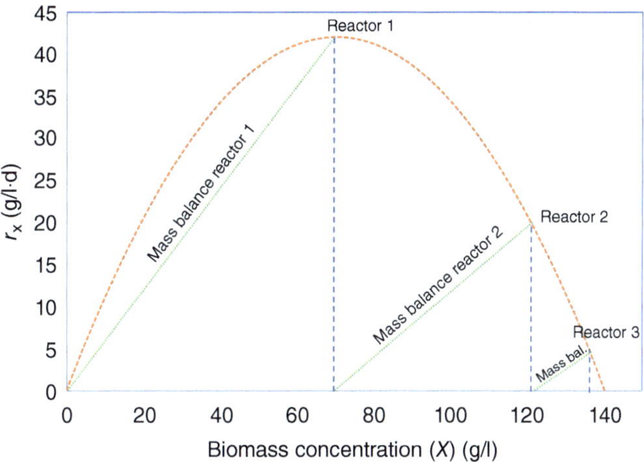



**Problem 10.23** A fed-batch reactor is being set up in order to produce a product "*P*" in a cell culture. The following parameters were chosen: input cell concentration $= 0.05$ g/l, input substrate concentration $= 10$ g/l, initial volume $= 1$ l, and flow rate input $= 1$ l/h. The feed has no concentration of the desired product. Other data are available: $\mu_m = 0.20\,\text{h}^{-1}$; $K_S = 1.00$ g/l; $Y_{X/S} = 0.5$ g/g; $Y_{P/X} = 0.2$. Simulate the behavior of this reactor by calculating the evolution of all concentrations and the volume with time.

**Solution to Problem 10.23**

For details refer the Wiley website at http://www.wiley-vch.de/ISBN9783527354115

In the present design, the input flow has a known concentration of cells $X_0 = 0.05$ g/l. So, in this case, the balance of cells is:

$$\frac{d(XV)}{dt} = QX_0 + V \cdot r_X$$

Using the chain rule for derivation and bearing in mind that $dV/dt = Q$, we can see that:

$$V\frac{dX}{dt} = Q(X_0 - X) + V \cdot r_X$$

And then, using the previous equation, we can write:

$$\frac{X^{t+1} - X^t}{\Delta t} = \frac{Q_0 \cdot (X_0 - X)}{V} + r_X$$

$$X^{t+1} = X^t + \Delta t \left( \frac{Q_0 \cdot (X_0 - X^t)}{V} + r_X \right)$$

In the same way:

$$P^{t+1} = P^t + \Delta t \left( -\frac{Q_0 \cdot P}{V} + Y_{P/X} r_X \right)$$

$$S^{t+1} = S^t + \Delta t \left( \frac{Q_0}{V}(S_0 - S) - \frac{r_X}{Y_{X/S}} \right)$$

Using Matlab for doing the calculation (it can also be done in a spreadsheet), the results are shown in the figure.

```
clear all
close all
        mumax = 0.20; % h-1 Maximum Growth Rate
        Ks    = 1.00; % g/L Monod Constant
        Yxs   = 0.5; % g/g Cell yield
        Ypx   = 0.2; % g/g Product yield
        S0    = 10.0; % g/L Feed Substrate concentration
    Q=1; % L/h Input Flow rate

V0=1; % L Initial Volume

    V(1)=V0;
    X(1) = 0; % g/L Initial concent. of cells in the tank
    X0=0.05; % g/L concentration of cells in the input current
    S(1) = 0; % g/L
    P(1) =0; % g/L
    t=0; % h
    tfinal=50; % h
    inct=0.1; % h

    i=1;
    tv=0:inct:tfinal; % Values of time used in the calculation

    while t<tfinal
```

```
    i=i+1;
    t=t+inct;
    rx=mumax*S(i-1)*X(i-1)/(Ks+S(i-1));
    rp=rx*Ypx;
    rs=rx/Yxs;
    X(i)=X(i-1)+inct*(Q*(X0-X(i-1))/V(i-1)+rx);
    P(i)=P(i-1)+inct*(-Q*P(i-1)/V(i-1)+rp);
    S(i)=S(i-1)+inct*(Q*(S0-S(i-1))/V(i-1)-rs);
    V(i)=V(i-1)+Q*inct;
end
plot(tv,X,tv,S,'*',tv,P)
legend('Cells','Substrate','Product')
xlabel('Time (h)')
ylabel('Concentration (g/L)')
```

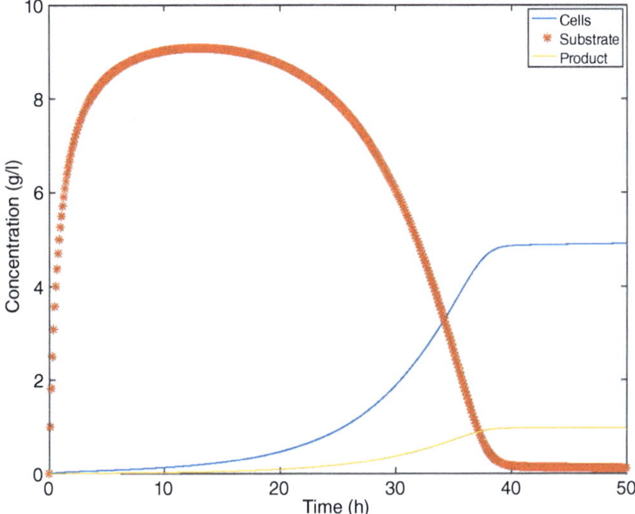


As we can see in the figure, the consumption of substrate is too high when time approaches 40 hours, so the best operation time is less than 40 hours when the product concentration does not increase anymore and the concentration of cells behaves in a similar way. In that case, the volume of the reactor needed will be approximately 40 l.

# Bibliography


Conesa, J.A. (2019). *Chemical Reactor Design: Mathematical Modeling and Applications.* Wiley. https://doi.org/10.1002/9783527823376.

Conesa, J.A. and Font Montesinos, R. (2002). *Reactores heterogéneos.* Universidad de Alicante. http://rua.ua.es/dspace/handle/10045/15296.

Fogler, H.S. (1980). *Elements of Chemical Reaction Engineering.* Prentice-Hall International. ISBN 978-0-130-47394-3.

Fogler, H.S. (2017). *Essentials of Chemical Reaction Engineering* (International Series in the Physical and Chemical Engineering Sciences), Pearson. ISBN 978-0-134-66389-0.

Froment, G.F., de Wilde, J., and Bischoff, K.B. (2011). *Chemical Reactor Analysis and Design.* Wiley. ISBN 978-0-470-56541-4.

Izquierdo, F. (2021). *Problemas resueltos de cinética de las reacciones químicas.* Ediciones Libreria Universitaria. ISBN 978-84-17452-51-3.

Katoh, S., Horiuchi, J., and Yoshida, F. (2015). *Biochemical Engineering: A Textbook for Engineers, Chemists and Biologists.* Blackwell Verlag GmbH. ISBN 978-3527338047.

Kayode Coker, A. (2001). *Modeling of Chemical Kinetics and Reactor Design.* Gulf Professional Publishing. ISBN 9780884154815. https://doi.org/10.1016/B978-088415481-5/50000-0.

Levenspiel, O. (1979). *The Chemical Reactor Omnibook.* OSU Book Stores.

Levenspiel, O. (1999). *Chemical Reaction Engineering*, 3e. New York: Wiley, 54. http://dx.doi.org/10.1021/ie990488g.

Levenspiel, O. and Bischoff, K.B. (1959). Backmixing in the design of chemical reactors. *Industrial and Engineering Chemistry* 51 (12): 1431–1434. https://doi.org/10.1021/ie50600a023.

Li, S. (2017). Residence time distribution and flow models for reactors. In: *Reaction Engineering* (ed. S. Li). Butterworth-Heinemann. https://doi.org/10.1016/B978-0-12-410416-7.00005-7.

Richardson, J.F. and Peacock, D.G. (2017). Coulson and Richardson's chemical engineering. In: *Chemical and Biochemical Reactors and Process Control*, 4e, vol. 3. Butterworth-Heinemann.

Tarhan, M.O. (1983). *Catalytic Reactor Design*. McGraw Hill. ISBN 978-0070628717.

Tiscareño Lechuga, F. (2012). *ABC para comprender Reactores Químicos con Multireacción* (ed. Reverte). https://www.reverte.com/libro/abc-para-comprender-reactores-quimicos-con-multireaccion_146664/.

Trambouze, P. and Euzen, J.P. (2004). *Chemical Reactors: From Design to Operation*. Editions Technip. ISBN 9782710808459.

# Index




*Problem Solving in Chemical Reactor Design*, First Edition. Juan A. Conesa.
© 2025 WILEY-VCH GmbH. Published 2025 by WILEY-VCH GmbH.